MULTIPLE CRITERIA DECISION ANALYSIS FOR

INDUSTRIAL ENGINEERING

METHODOLOGY AND APPLICATIONS

The Operations Research Series

Series Editor: A. Ravi Ravindran

Professor, Department of Industrial and Manufacturing Engineering

The Pennsylvania State University – University Park, PA

Published Titles:

Multiple Criteria Decision Analysis for Industrial Engineering: Methodology and Applications

Gerald W. Evans

Multiple Criteria Decision Making in Supply Chain Management

A. Ravi Ravindran

Operations Planning: Mixed Integer Optimization Models

Joseph Geunes

Introduction to Linear Optimization and Extensions with MATLAB®

Roy H. Kwon

Supply Chain Engineering: Models and Applications

A. Ravi Ravindran & Donald Paul Warsing

Analysis of Queues: Methods and Applications

Natarajan Gautam

Integer Programming: Theory and Practice

John K. Karlof

Operations Research and Management Science Handbook

A. Ravi Ravindran

Operations Research Applications

A. Ravi Ravindran

Operations Research: A Practical Introduction

Michael W. Carter & Camille C. Price

Operations Research Calculations Handbook, Second Edition

Dennis Blumenfeld

Operations Research Methodologies

A. Ravi Ravindran

Probability Models in Operations Research

C. Richard Cassady & Joel A. Nachlas

MULTIPLE CRITERIA DECISION ANALYSIS FOR INDUSTRIAL ENGINEERING

METHODOLOGY AND APPLICATIONS

Gerald W. Evans

CRC Press is an imprint of the
Taylor & Francis Group, an **informa** business

CRC Press
Taylor & Francis Group
6000 Broken Sound Parkway NW, Suite 300
Boca Raton, FL 33487-2742

Printed on acid-free paper
Version Date: 20160509

International Standard Book Number-13: 978-1-4987-3982-5 (Pack - Book and Ebook)

Library of Congress Cataloging-in-Publication Data

Names: Evans, Gerald W., author.
Title: Multiple criteria decision analysis for industrial engineering : methodology and applications / : Gerald William Evans.
Description: Boca Raton : Taylor & Francis, a CRC title, part of the Taylor & Francis imprint, a member of the Taylor & Francis Group, the academic division of T&F Informa, plc, [2016] | Series: Operations research series ; 12 | Includes bibliographical references and index.
Identifiers: LCCN 2016000238 | ISBN 9781498739825 (alk. paper)
Subjects: LCSH: Industrial engineering. | Multiple criteria decision making.
Classification: LCC T57 .E924 2016 | DDC 620.0068/4--dc23
LC record available at http://lccn.loc.gov/2016000238

Visit the Taylor & Francis Web site at
http://www.taylorandfrancis.com

and the CRC Press Web site at
http://www.crcpress.com

To Linda, Matt, and Brian

Contents

List of Examples

Preface

Industrial engineers deal with a variety of problems which require making decisions involving multiple criteria, uncertainty, and risk. Hence, the subject of this book, multiple criteria decision analysis, is essential for many areas of industrial engineering. In fact, a primary activity of industrial engineers is making decisions with respect to a system's design and operation.

This book is meant to be used as a text for a one-semester course in decision analysis for advanced undergraduate or graduate-level students. It is especially designed for students who are studying industrial engineering, but it can also be used in a program designed for business and management students.

The book arises from the author's experience of more than 30 years in teaching, research, and consulting in decision analysis, simulation modeling and analysis, and industrial engineering in general. Two unique features of the book are, first, its comprehensive coverage of multiple criteria decision analysis and related areas, including simulation optimization; and, second, its emphasis on applications involving industrial engineering.

Following an introductory chapter, Chapter 2 focuses on problem structuring, including the generation of alternatives, objectives, and attributes (performance measures) for a decision situation; the problem structuring activity is an essential first step in a decision analysis. Chapter 3 covers multiobjective methods for decision making in deterministic situations with only a few alternatives to consider. These methods include the assessment and use of multiattribute value functions, Technique for Order Preference by Similarity to Ideal Solution (TOPSIS), the analytic hierarchy process (AHP), and outranking methods.

Chapter 4 addresses deterministic decision situations with a large number of alternatives—that is, multiple objective mathematical programming, with a focus on goal programming. In this case, the alternatives are defined in terms of combinations of values assigned to decision variables; hence, because of the large number of "alternatives," not all of them can be explicitly evaluated.

Chapter 5 provides a brief review of probability. This chapter may be skipped, or just covered in a cursory fashion for students with a good background in probability and statistics. Chapter 6 covers the modeling of a decision-maker's preferences in probabilistic situations involving (1) a single attribute (single attribute utility functions) and (2) multiple attributes (multiattribute utility functions). Chapter 7 presents methodologies which allow for the "probabilistic mapping" of alternatives into a single attribute or multiple attributes, and includes decision trees and influence diagrams.

Chapter 8 discusses the determination of probabilistic inputs for simulation models (including models of decision trees and influence diagrams) of decision situations. These probabilistic inputs are represented as probability distributions. This chapter is divided into two main parts, depending upon whether or not data are available for developing these probabilistic inputs.

Finally, Chapter 9 presents the use of simulation as a tool to execute the modeling methodologies (including decision trees and influence diagrams) presented previously. Included in this chapter is a discussion on simulation optimization.

Each methodology and technique (e.g., assessment of a multiattribute value or utility function, evaluation of a decision tree to choose an initial best alternative, simulation optimization via ranking and selection) is described with an easy to follow "step-by-step" process. In addition, through the use of numerous cited references, information on the theoretical backgrounds, as well as criticisms and comparisons to other techniques are provided.

Numerous examples of the various methodologies are presented throughout the book in order to illustrate the applications in various areas of industrial engineering. In addition, each chapter contains both Material Review Questions and Exercises that students can use to test their knowledge of the material.

Various software packages were employed to develop the examples found in the book and to solve the exercises at the end of the chapters. These packages include *Expert Choice*™ for executing the analytic hierarchy process, *Arena*™ for developing and executing simulation models, *OptQuest*™ for solving optimization models interfaced with simulation, *LINGO*™ for solving optimization models, and *Precision Tree*™ for developing and solving decision trees.

Materials for the instructor are also provided. These materials include files containing PowerPoint slides that correspond to each respective chapter, and solutions to the exercises at the end of each chapter.

Acknowledgments

The knowledge that I have gained from several different books has served as a basis for this book. Particularly helpful have been Keeney's *Value Focused Thinking* (Harvard University Press, 1992); Keeney and Raiffa's *Decisions with Multiple Objectives*, 2nd Edition (Cambridge University Press, 1993); Law's *Simulation Modeling and Analysis*, 4th Edition (McGraw-Hill, 2007); Banks, Carson, Nelson, and Nicol's *Discrete Event System Simulation*, 4th Edition (Prentice Hall, 2005); Kelton, Sadowski, and Zupick's *Simulation with Arena*, 6th Edition (McGraw-Hill, 2015); and Clemen and Reilly's *Making Hard Decisions with Decision Tools* (South-Western Cengage Learning, 2001). (Also, see Clemen and Reilly's *Making Hard Decisions with Decision Tools*, 3rd Edition, South-Western Cengage Learning, 2013). The book by Jones and Tamiz: *Practical Goal Programming* (Springer, 2010) was particularly helpful for Chapter 4 of the book.

I have also benefited from comments and suggestions provided by several colleagues including Ki-Hwan Bae, Lihui Bai, Aman Gupta, Tim Hardin, Sunderesh Heragu, Ehsan Khodabandeh, and Mike Lankert.

Prajwal Khadgi, Sathya Sundaresan, and Han Wu, graduate students in the Department of Industrial Engineering at the University of Louisville, were also very helpful with their suggestions and especially through their efforts in developing the PowerPoint presentation files for the book.

I welcome comments, suggestions, and especially the discovery of any errors found from readers. Please send these to me at gerald.evans@louisville.edu.

Author

Gerald W. Evans, PhD, is Professor Emeritus in the Department of Industrial Engineering at the University of Louisville (UL). His research and teaching interests lie in the areas of multicriteria decision analysis, simulation modeling and analysis, optimization, logistics, and project management.

His previous positions include working as an industrial engineer for the Department of the Army, and senior research engineer for General Motors Research Laboratories. He has also served as an American Society for Engineering Education (ASEE) Faculty Fellow at the National Aeronautics and Space Administration's (NASA) Langley Research Center and Kennedy Space Center.

Dr. Evans received a BS in mathematics in 1972, an MS in industrial engineering in 1974, and a PhD in industrial engineering in 1979, all from Purdue University (West Lafayette, Indiana).

He has published approximately 100 papers in various journals and conference proceedings. His paper, "An Overview of Techniques for Solving Multiobjective Mathematical Programs," published in 1984 was listed as one of the most cited publications in *Management Science* over the last 50 years (from the 2004 Commemorative CD: *Celebrating 50 Years of Management Science*, INFORMS Publication).

Dr. Evans has served as coeditor of *Applications of Fuzzy Set Methodologies in Industrial Engineering* (Elsevier Science, 1989), the *Proceedings of the Winter Simulation Conference* in 1993 and 1999, and for a special issue of *Computers and Industrial Engineering* entitled "Multi-Criteria Decision Making in Industrial Engineering" (November 1999). He has also served as a referee for approximately 25 different journals, as associate editor for the *Institute of Industrial Engineers (IIE) Transactions*, as director of the Operations Research Division of IIE, and as a vice president of IIE.

Dr. Evans has received the Fellow Award and the Operations Research Division Award from IIE and the Moving Spirit Award from INFORMS for his work with the UL INFORMS Student Chapter.

1

Introduction

1.1 Decision Making Is Important

Decision making is an important aspect of any human endeavor, whether that endeavor is accomplished by an individual or as part of an organization. Analyzing a decision situation and then making a "good decision" is a prelude to accomplishing important endeavors. As such, both individuals and organizations make decisions in order to solve problems and to take advantage of opportunities. Some examples of personal decisions made by individuals include the following:

1. A high school student decides where to attend college and what to study.
2. A consumer decides which type of automobile to purchase.
3. A newly married couple decides whether to honeymoon in Florida or Hawaii.
4. A patient, in conjunction with his or her doctor, decides whether to use chemotherapy or radiation to treat his or her cancer.

In fact, personal decisions, especially those having to do with our health and safety, can even result in death (Keeney, 2008).

Here are some examples of decision problems faced by private industry:

1. General Electric decides whether or not to develop a new MRI machine with a larger opening for the patient (Welch, 2005, pp. 74, 75).
2. An electric utility cooperative decides whether or not to add an additional transmission line to link with an electric utility (Borison, 1995).
3. A corporation decides where to locate a new production facility.
4. An automobile manufacturer decides whether or not to produce a new line of cars.

Examples of decision problems faced by individuals (within professional settings) include the following:

1. A machine operator decides which part to process next.
2. An industrial engineer designs a layout for a new production facility.
3. A plant manager selects a new quality engineer from among several applicants.
4. An information systems manager selects an accounting system for his or her organization.

Finally, some examples of decisions made by public institutions such as local or federal governments include the following:

1. A state government (in conjunction with the local metropolitan government and the federal government) chooses a location for a new bridge in the metropolitan area.
2. The federal government decides whether or not commercial airplanes should have special protection against surface-to-air missile attacks by terrorists (von Winterfeldt and O'Sullivan, 2006).
3. The Department of Energy decides how to allocate a limited budget among various R&D projects (Parnell, 2001).

In its most elemental form, a *decision problem* involves a situation in which one alternative must be selected from among several feasible alternatives. In fact, the word "decide" is derived from the Latin *decidere*, which means "to cut off." When we decide on something, we "cut off" the other alternatives from further consideration by selecting *one* alternative. For example, the corporation in the aforementioned example might be considering three alternatives for the location of a new production facility: Evansville, Indiana; Lexington, Kentucky; and Nashville, Tennessee. When they choose one of the locations (i.e., they make a decision), they *cut off* the other two from further consideration.

This book is concerned with problems and situations involving decision making and optimization. Of particular interest are situations in which multiple objectives and uncertainty (and resulting risk) must be considered. When the number of alternative decisions in a particular situation is so large that evaluation of all alternatives is not possible, then some type of optimization technique is desirable as part of the solution process. Examples of situations involving such a large number of alternatives would be one involving continuous "decision variables," such as the amount of product produced by a company in the next month, or a combination of discrete/integer decision variables, such as the number of machines of various types to include in a production facility.

1.2 Characteristics of Decision Situations

1.2.1 Decisions Are Made to Solve Problems or to Take Advantage of Opportunities

As mentioned in the previous section, a decision is made to solve a problem or, alternatively, to take advantage of an opportunity. A problem can be defined as a gap between a current state of affairs and some desired state of affairs. For example, in the situation described earlier involving the protection of commercial airplanes, the "problem" might be defined as "there is a gap between the perceived current level of safety and the desired level of safety because of a possible terrorist attack on a commercial airplane." As another example, a personal problem might be defined as "there is a gap between my current salary and the salary I would like to receive."

Note that in many cases, the "state of affairs" is defined in a fairly nebulous fashion so that in order to proceed in the evaluation of various alternatives, one must clearly define the performance measures that determine the state of affairs. The concept of a "problem" will be discussed in much more detail in Chapter 2.

1.2.2 A Good Decision May Not Always Result in a Good Outcome (and a Bad Decision May Not Always Result in a Bad Outcome)

The timed sequence of processes/events that occur in a decision situation might be defined as follows:

1. Decision analysis
2. Selection of a decision to implement
3. Implementation of a decision
4. Occurrence of the results of the decision

The results of a decision are typically uncertain in nature, often depending on many uncertain parameters.

As a simple example, consider a friend of yours named Harry who is down to his last $2000. Suppose that Harry is considering two alternatives for the use of this money:

1. Invest the money in an educational course that will give him a 9 out of 10 chance of procuring a better job with an increase in annual salary of $10,000 over his current salary.
2. Purchase lottery tickets that will give him a probability of 1 chance in 10 million of winning $1 million.

Most people would consider the first alternative the best one. But if Harry does *not* procure a better job after the course (a 10% chance), then the result (or outcome) would be bad. One of the purposes of a decision analysis process would be to maximize the chances of getting a good outcome. Although it may not always turn out this way, over the long run the results of your decisions should be better outcomes through the use of a good decision analysis process.

1.2.3 Decisions Can Often Be Categorized as Being Strategic, Tactical, or Operational in Nature

Decisions must be made in a variety of situations. For example, one categorization of decision situations is strategic (e.g., decisions that influence outcomes more than a period of 1–10 years or longer), tactical (affecting outcomes more than a period of 1–12 months), and operational (affecting outcomes more than a period of one to a few days). Typically, decisions made at the strategic level place constraints on decisions made at the tactical and operational levels, and decisions made at the tactical level constrain decisions made at the operational level.

As an example, an organization's strategic decision of where to locate a new manufacturing facility will probably affect tactical decisions regarding transportation policy. Hence, one wants to at least implicitly consider the tactical and operational decisions resulting from strategic decisions in a strategic decision analysis. By the same token, one wants to implicitly consider the relevant operational decisions in an analysis for a tactical decision.

1.3 Steps in the Process of Decision Making

The major steps in the process of decision making are as follows:

1. Generate alternative solutions (or just alternatives) for the problem.
2. Determine the performance measures for the problem situation.
3. Rank the alternatives in terms of the performance measures.
4. Implement the first-ranked alternative.

Of course, each of these major steps involves several related activities. For example, steps 1 and 2 involve the consideration of the values and objectives associated with the person/organization making the decision as well as the decision situation. In order to generate the values and objectives of an organization or a decision situation, one must consider the decision makers and stakeholders involved.

For example, the first thing that the couple who were about to marry had to do was to generate potential locations for their honeymoon. This could have been accomplished through the gathering of information (e.g., through the Internet, by talking with friends and travel agents, and by visiting the library) about different possibilities. The next step, as described earlier, involves the determination of performance measures. The performance measures in this case could have involved such measures as the projected cost of the trip and the amount of "fun" that the couple could expect to have. Note that the first performance measure is quantitative (or objective) in nature, while the second one is qualitative. In addition, there may be some uncertainty about the outcome associated with a performance measure for a particular alternative. For example, the amount of "fun" that the couple has on a trip to Florida may be dependent on the weather in Florida while they are there.

The third step in the process involves ranking the alternatives. This could, and probably would, involve making trade-offs between pairs of performance measures. For example, the honeymooning couple would probably have to trade off between "fun/enjoyment" and "cost."

Finally, the decision must be implemented. In the case of the honeymooning couple, this would involve the purchase of tickets, making reservations, and so on.

Of course, most decisions are made without the *explicit* consideration of these formal steps. The contention associated with decision analysis is that better decisions will be made through the explicit use of the methodologies and processes such as those described in this book.

1.4 Elements in a Decision Analysis Process

The steps of the decision-making process described earlier involve several elements/agents. These elements include the following:

1. The people involved: decision maker(s), stakeholders, and analysts
2. Alternative solutions to the problem
3. Ways to measure the performance of an alternative: criteria, performance measures, and attributes
4. Constraints on the alternatives
5. A forecast associated with various states of nature
6. Models/techniques for the evaluation of alternatives
7. Models/techniques for the ranking of feasible alternatives and/or the selection of a best alternative

8. An optimization technique for use in situations where it is impossible to explicitly enumerate all of the feasible alternatives
9. A procedure for implementing the chosen solution

A *decision maker* is a person responsible for making the decision about a best course of action (i.e., alternative) for solving a problem. Typically, a problem will have several decision makers. For example, a group of corporate executives may be decision makers with respect to determining the projects on which the company should bid. As such, decision makers should be involved in all elements of the decision-making process. In particular, decision makers should be heavily involved in the development of the criterion for ranking alternatives since a criterion must account for the decision makers' trade-offs among multiple performance measures.

A *stakeholder* is someone who is affected by the decision problem and its ultimate solution but has no *direct* say in how the problem should be solved. For example, in a decision problem involving the bidding problem discussed in the aforementioned paragraph, the workers in the company involved would be stakeholders, since they would be performing the work associated with the projects bid on and received. As such, the decision makers should certainly consider the preferences of the stakeholders as a part of their decision analysis process.

An *analyst* is someone who aids the decision makers in the decision-making process. This can include the use of methods for the development and definition of alternatives, defining the performance measures, evaluating the alternatives in terms of the performance measures, quantifying trade-offs that the decision makers are willing to make between the various pairs of performance measures, and finally choosing the best alternative or ranking the alternatives. As such, the analyst should be educated in the methodology of decision analysis as well as related methodologies, such as probability and statistics, optimization, and simulation. The analyst does not necessarily have to be trained in the application area of the decision situation, since the decision makers themselves add this expertise.

Alternative solutions are the various possible (mutually exclusive) courses of action for solving a problem. Depending upon how well defined the problem is, these alternative solutions can be very general in nature or very specific. For example, a general alternative to a transportation problem for a metropolitan area might be "construct another bridge across the river that divides the area," while a more specific alternative might be "construct a bridge of a particular type at a particular location on the river." Sometimes, alternative solutions are defined in terms of a set of values given to a particular set of *decision variables*; such is the case with mathematical programming (optimization) problems.

Performance measures or *attributes* represent ways of measuring how well an alternative does with respect to important problem objectives. For example, if an important problem objective is to maximize the production rate of a system, an attribute corresponding to this objective could be the average number of parts produced in a typical day.

Constraints are used to restrict the set of alternative solutions. For example, the decision makers for a particular problem may determine that no solution with an initial cost of more than $100,000 should be considered. As such, constraints can be used as inputs to a process for generating alternative solutions or as inputs to an evaluation model for the automatic restriction of the solution space. In the second case, constraints are an important part of a mathematical programming model. As noted by Miser and Quade (1985), one should be very careful in the use of constraints, especially in the early stages of problem solving, so that potentially optimal solutions are not cut off from further consideration.

A *state of nature* is represented as a set of values assigned to important problem parameters. These problem parameters are variables that affect the outcome (in terms of performance measure values) associated with an alternative, but the decision maker(s) have no control over these variables. An example would be a situation in which a company must decide whether or not to expand its main factory. The future demand for the items produced by the factory will affect the amount of profit associated with either alternative. The decision maker has no control over this future demand; hence, this would be one of the variables that would have to be considered as part of the state of nature. For example, one might have

Probability of high demand	0.2
Probability of medium demand	0.6
Probability of low demand	0.2

An *evaluation model* allows a decision maker to "map" the various alternatives into their associated *outcomes* (as measured by the attributes) given a particular state of nature. Many types of evaluation models exist, and the type to use for a particular situation depends upon many factors including how well defined the alternatives and attributes are, the amount of time available for making the decision, the abilities of the decision makers to understand quantitative methodologies, the availability of data for building the model, the number of alternative solutions and performance measures to consider, and the importance of the decision to be made, among other factors. For example, one would probably not want to build an elaborate simulation model for a decision that has to be made by the end of the day.

Examples of types of evaluation models include

- Simulation models
- Analytic queuing network models
- Decision trees

- Influence diagrams
- Analytic models (e.g., consisting of a set of equations)
- Judgmental models

A judgmental model is one that employs the expertise of one or more experts in order to predict the outcome associated with the various feasible alternatives. As such, a judgmental model is not explicit in nature and therefore cannot be explicitly analyzed in the same ways as the other models.

Once the outcomes, in terms of multiple attributes, have been forecast for each of the alternatives, a *criterion* or *ranking/optimization model* can be used for selection of the best alternative. As such, a criterion should consider all of the attributes simultaneously through the use of objectives, goals, constraints, and so on.

As an example of a ranking/optimization model, consider a decision-making situation where the two relevant attributes are the following:

1. Initial system cost, measured in thousands of dollars
2. System reliability, measured in terms of probability of no failure within the first 6 months of system operation

A criterion for this situation could be expressed by the following model:

Minimize initial system cost
Subject to
System Reliability $\geq .95$

Note that given a set of performance measures/attributes, objectives, goals, and so on, for a particular situation, there are typically a large number of ways to combine these into a ranking/optimization model. For example, one could employ a format where one attribute is optimized, subject to constraints on the other attributes (as in the aforementioned example), or one could employ a utility/value function approach that combines all of the attributes into a single performance measure function. A third approach would involve a goal programming model.

In solving a ranking/optimization model representing a criterion (i.e., determining the best alternative for the given criterion), one sometimes has to use a sophisticated *optimization technique*. For example, when the number of feasible alternatives as defined by the constraints is very large, it may not be practical to evaluate every alternative solution since this evaluation process might require more time than is available. In these cases, there are many optimization techniques available for use depending upon the type of criterion model one is considering. For example, if one has a linear program as an optimization model, the revised simplex method can be used. If the

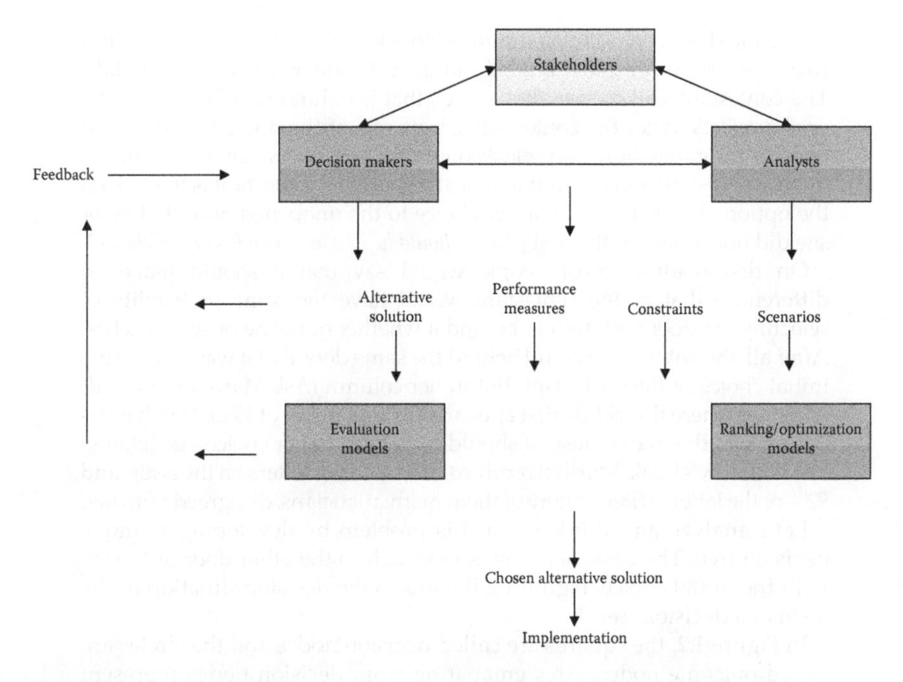

FIGURE 1.1
Elements in the process of decision making.

optimization model is an integer program, the branch-and-bound optimization method can be used.

Once a particular alternative has been chosen (i.e., a decision has been made) through the use of the ranking/optimization model, that alternative must be implemented. Implementation of an alternative may involve a whole new decision situation involving the question of

What is the best way to implement the decision?

Implementing the decision may involve the use of project management techniques as described in Meredith and Mantel (2012).

Figure 1.1 illustrates some of the relationships among the various elements discussed in this section.

Example 1.1: A Game Show Example

The purpose of this example is to illustrate how just the development of a model is helpful in thinking about a decision situation. The example involves the construction of a simple *decision tree* in order to answer a question concerning the best of two alternatives. The example is derived from Leonard Mlodinow's excellent book, *The Drunkard's Walk: How Randomness Rules Our Lives* (2008, 43–45).

A game show contestant is supposed to select one of three doors. Behind two of the doors are goats, but behind the third door is a new automobile. The contestant will receive the "prize" that is behind whichever door he or she selects. After the contestant selects one of the doors (but does not open it), the game show host selects one of the two doors that wasn't chosen by the contestant and opens it to reveal a goat. The contestant is then given the option of switching his or her choice to the unopened door that he or she did not choose in the first place. *Should he or she switch his or her choice?*

On first reading, most people would say that it should make no difference—that is, the contestant would have the same probability of selecting the door with the car behind it whether or not he or she switches. After all, the automobile is still behind the same door that it was prior to the initial choice of the contestant. But in her column (Ask Marilyn) in *Parade Magazine* where this riddle first appeared in September of 1990, Marilyn vos Savant said that the contestant should switch his or her choice. As detailed in Mlodinow's book, Marilyn received thousands of letters on the issue, and 92% of the letter writers, many of them mathematicians, disagreed with her.

Let's analyze and think about this problem by developing a simple decision tree. The basic decision is to switch to the other door or to stay with the initial choice. Figure 1.2 illustrates the decision situation in the form of a decision tree.

In Figure 1.2, the squares are called decision nodes and the circles are called outcome nodes. Arcs emanating from decision nodes represent mutually exclusive, alternative decisions, while the set of arcs emanating from outcome nodes represent a partition of the sample space—that is, a set of mutually exclusive outcomes that represent all of the possible outcomes. The numbers associated with the arcs emanating from the outcome nodes are probabilities. The decision tree indicates that there is a 2/3 (or about 67%) chance of getting the automobile if the contestant switches his or her choice of doors, but only a 1/3 chance if he or she doesn't. The question is "why these probabilities?"

First, note that Figure 1.2 does not exactly represent a decision tree (which will be discussed later in Chapter 7) since it does not represent time moving forward as one moves from left to right in the figure.

Also, without loss of generality, we are assuming that the contestant initially chooses door 1 in the diagram. If the contestant chose door 2 or door 3 initially, we could draw equivalent respective diagrams that would lead to the same conclusions regarding the probabilities associated with the outcomes of car or goat as a function of the choice made with respect to switching doors or not.

Now, thinking about Figure 1.2, the contestant has a choice of switching doors or not switching doors (following the opening of a door by the host); these switching or not switching alternatives are represented by the arcs emanating from node 1. For either alternative, there is a 1/3 chance of the car being behind each of the three respective doors, as shown by the sets of arcs emanating from nodes 2 and 3, respectively.

Now, consider the arcs emanating from node 4, which assumes that the car is behind door 1, according to the figure. The host has a choice of opening either door 2 or door 3, for which we can assume that there is a 50% (1/2) chance of each. If the host opens door 2, then the contestant,

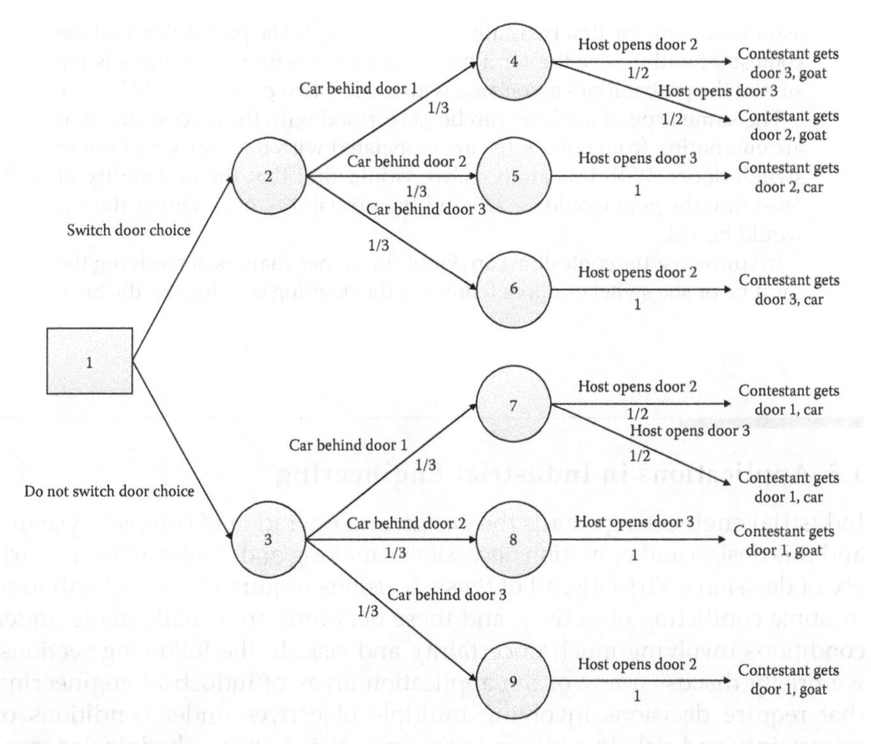

FIGURE 1.2
A "decision tree representation" of the door switch decision (assuming the contestant's initial choice is door 1).

who initially chose door 1, will switch and open door 3 and will therefore receive the goat. If the host opens door 3, then the contestant will switch to door 2 and still receive a goat.

Similar reasoning can be done for the rest of the figure.

The probability of any particular outcome can be computed by multiplying the probabilities along the arcs of a path, since the events associated with these arcs are independent. So, if the contestant decides to switch doors, there are four different paths that can be followed:

1. 2–4—The host opens door 2 and the contestant opens door 3 and receives a goat: Probability of (1/3) * (1/2) = 1/6.
2. 2–4—The host opens door 3 and the contestant opens door 2 and receives a goat: Probability of (1/3) * (1/2) = 1/6.
3. 2–5—The host opens door 3 and the contestant opens door 2 and receives a car: Probability of (1/3) * (1) = 1/3.
4. 2–6—The host opens door 2 and the contestant opens door 3 and receives a car: Probability of (1/3) * (1) = 1/3.

So, the probability that the contestant will receive the goat if he or she switches his or her choice from door 1 is the sum of the probabilities associated with the first two paths: 1/6 + 1/6 = 1/3. The probability that the contestant will receive the car if he or she switches his or her choice is the sum of the probabilities associated with the last two paths: 1/3 + 1/3 = 2/3.

The same type of analysis can be performed with the second decision arc emanating from node 1, the arc associated with the decision of not to switch doors. With this analysis, we would find that the probability of receiving the goat would be 2/3 and the probability of receiving the car would be 1/3.

In summary, the contestant can double his or her chances of receiving the car if he or she switches doors following the opening of a door by the host.

1.5 Applications in Industrial Engineering

Industrial engineering entails the design and operation of complex systems, and this design and operation necessitates making and implementing a variety of decisions. Virtually, all of these decisions require the consideration of multiple conflicting objectives, and these decisions are usually made under conditions involving much uncertainty and risk. In the following sections, we briefly discuss a few of the application areas of industrial engineering that require decisions involving multiple objectives under conditions of uncertainty and risk. In addition to the areas listed, the methodologies associated with multiple criteria decision analysis are also important for areas such as routing and distribution, production and inventory control, production scheduling, and facility location and layout.

1.5.1 Project Management and Control

A project is a temporary endeavor involving the development of a product or service. Decisions made in project management include (1) determining which projects to undertake (project selection); (2) deciding when to schedule the various activities of a project; (3) selecting vendors, suppliers, and subcontractors; and (4) allocating resources to project activities, among others. Uncertainties/risks in making these decisions arise from uncertainties associated with activity durations, finish times of preceding activities, vendor/supplier delivery times, strike probabilities, weather considerations, costs and benefits associated with projects, and so on. Objectives/performance measures associated with the various decisions relate to project duration (typically set so that a specific duration is achieved), achievement of a bonus, for example, as associated with completing a project early, avoidance of a penalty (e.g., related to not exceeding a maximum project duration), minimization of direct/indirect costs, and leveling of resource usage over time.

1.5.2 Quality Control: Acceptance Sampling and Process Control Charts

Acceptance sampling and process control charts are important subject areas within quality control. There are several types of acceptance sampling schemes, including single sample acceptance sampling and multiple sample acceptance sampling.

Single sample acceptance sampling involves (randomly) selecting a few items from a lot of items (say as received from a vendor/supplier), inspecting those items from the randomly selected sample, and then "accepting" the entire lot if the number of defective items is less than or equal to a so-called acceptance number. Even if the lot is accepted, the defective items in the sample are reworked, discarded, or returned to the supplier. If the lot is not accepted, then either (1) the entire lot is inspected (and the defective items are reworked, discarded, or returned to the supplier) or (2) the entire lot is just returned to the supplier.

The decisions to be made in single sample acceptance sampling include the values set for the sample size (typically denoted as n) and the acceptance number (typically denoted as a). Uncertainties in the outcome of the process result from the laws of probability; for example, the number of defective items in a randomly selected sample of 20 items selected from a lot of 1000 items that contain, say, 50 defective items will be a random variable with a hypergeometric distribution. Objectives/performance measures associated with this decision situation include the sampling, inspection, and rework costs, along with the quality of the outgoing lot (e.g., as measured by the number of defective items in this lot). See Evans and Alexander (1987) for a more detailed discussion of multiobjective decision analysis applied to single sample acceptance sampling.

There are also several different types of process control charts, including $\bar{X}$ control charts, R control charts, and p control charts, among others. Just as the name indicates, the purpose of these process control charts is to control production processes. The basic decisions associated with these charts have to do with the setting of control parameters. For example, in $\bar{X}$ control charts, the values for the upper control limit and lower control limit must be set. As with acceptance sampling plans, the uncertainties associated with the outcome (expressed in terms of the costs of implementing the control chart and the quality of the process output) associated with this process can be derived from the basic concepts of probability and statistics.

1.5.3 Design and Operation of Supply Chains

There are many complex and difficult decisions associated with the design and operation of a supply chain. These various decisions can be categorized as strategic, tactical, or operational, as referred to in the earlier discussion in this chapter. Also, as noted in Section 1.2.3, strategic decisions constrain tactical decisions, which in turn constrain the operational decisions.

See Shapiro (2007) or Bramel and Simchi-Levi (1997) for a discussion of the various decisions associated with a supply chain.

Examples of some of the many decisions associated with the design and operation of supply chains include those related to the locations and capacities of facilities (including production plants, warehouses, and distribution centers); selection of vendors/suppliers; production schedules; inventory policies, for example, as related to reorder points and reorder levels; routings; and shipping schedules.

Uncertainties associated with the outcomes of these decisions are a result of uncertainties in customer demands and in the reliability of production/other resources, among other entities. Outcomes are measured through the use of many different performance measures, including lateness/tardiness related to deliveries/production, inventory costs (holding, ordering, back orders, and lost sales), and customer satisfaction, among others.

Uncertainties/risks result from the demands of different customers, reliability of production and other resources, task durations (e.g., related to production and shipping), travel times, and so on.

1.5.4 Workforce Scheduling

Workforce scheduling involves determining work shifts and assigning workers of various skills and skill levels to these shifts. This application area is especially important in the operation of service-oriented systems such as call centers, hospitals, and restaurants. Important performance measures could be categorized according to the stakeholder/decision maker: system manager/owner, worker, and customer. Uncertainty/risk arises from uncertainty associated with demands (e.g., related to customer arrival rate) over time.

1.6 Taxonomy and a Look Ahead

The various modeling and solution methodologies for multiple criteria decision analysis can be classified according to the following problem or representation categories:

1. Number of objectives/performance measures
    a. One objective/performance measure
    b. Multiple objectives/performance measures
2. Mapping from the alternative space to the outcome space
    a. Deterministic, but not explicit
    b. Deterministic, but represented as a closed-form function from a decision variable space to an outcome space

   c. Probabilistic, but modeled analytically
   d. Probabilistic, but modeled using simulation
3. Number of alternatives
   a. A finite, but "small" number
   b. A very large, perhaps infinite, number

Choosing one descriptor from each of the three categories allows us to define the type of modeling/solution methodology to be used.

In this book, we will not address the modeling/solution methodologies that involve descriptor a (one objective/performance measure) from category 1 and either descriptor a or b from category 2—that is, anytime, we are considering only one objective/performance measure, then we will also be considering uncertainty/risk in the process.

In addition, in our terminology, when considering the area of multiattribute value (MAV) functions, which allows the combination of several performance measures into a "value," we will consider this to be within the domain of descriptor b from category 1.

The simpler probabilistic situations of category 2 will allow the use of analytic models, while the more complex circumstances require the use of simulation—hence the differentiation associated with descriptors c and d in category 2.

With respect to category 3, some situations allow for an "evaluation" of every alternative (the case of descriptor a, involving a relative small number of alternatives), while other situations (category b involving a very large or even an infinite number of alternatives) do not allow for the explicit evaluation. Hence, the case of category b requires the use of some type of optimization scheme that allows for an "implicit evaluation."

Each chapter of this book, except for Chapters 2, 5, and 8, can be identified with a combination from the taxonomy shown earlier. Chapter 2 addresses the area of problem structuring, which discusses how to better define or gain a perspective on a problem; more specifically, this chapter presents methodologies for generating alternative solutions to a problem and for determining good performance measures to use in solving a problem. Chapter 5 provides a brief review of probability theory. Chapter 8 discusses the determination of probabilistic inputs for the models required for situations involving categories 2c and 2d.

Chapter 3 addresses the area given by a combination of 1b, 2a, and 3a. Of particular interest in this chapter is the formation/assessment of an MAV function, which represents a mapping from the outcome (or attribute value) space to a value space. In this chapter, we assume that each alternative has an associated (deterministic) outcome, and the number of alternatives/outcomes is relatively small. Hence, each of these alternatives/outcomes can be explicitly evaluated (i.e., without an optimization scheme).

Chapter 4 covers the areas given by a combination of 1b, 2b, and 3b. The title used for this method of study is multiple objective optimization or multiple objective mathematical programming. The particular emphasis in this chapter is goal programming. As the categorization indicates, the number of alternatives allowed is large or even infinite; these alternatives are represented as respective combinations of values assigned to decision variables. The mapping from the alternative space to the outcome space is deterministic in nature.

Chapter 6 addresses the area of utility functions—both single attribute and multiattribute utility functions are presented in this chapter. The focus here is on representing a decision maker's preferences over probabilistic outcomes. As such, the categorization addressed is 1a (for single attribute utility functions) or 1b (for multiattribute utility functions), while for the second category, the categorization scheme would be either 2c or 2d, depending on whether one wants to use an analytic model or a simulation model to represent the mapping from the alternative space to the outcome space.

The main topics for Chapter 7 are decision trees and influence diagrams. Both of these areas present models that allow for a probabilistic mapping from the alternative space to the outcome space. In many cases, these mappings can be evaluated analytically (i.e., without the use of simulation as a tool). As such, the categorization scheme for this chapter is 1a/1b, 2c, and 3a.

Chapter 9 discusses the use of simulation, in particular Monte Carlo simulation, as a tool for multiple criteria decision analysis. The emphasis in this chapter will be on the analysis of the output from these simulation models. Simulation modeling is a particularly useful tool for a decision tree or an influence diagram, as discussed in Chapter 7, which has continuous parameter values and/or continuous decision variables. As such, this chapter addresses the areas categorized as 1a/1b, 2d, and 3a/3b.

Material Review Questions

1.1 What Latin word is the word "decide" derived from?

1.2 With respect to a time frame, what are the three categorizations of a decision?

1.3 What is the difference between a stakeholder and a decision maker?

1.4 What is a state of nature for a decision situation?

1.5 What is the difference between an "evaluation model" and a "ranking/optimization model?"

1.6 What are the two types of nodes in a decision tree?

1.7 List several areas of industrial engineering for which multiple criteria decision analysis can be used. For what types of decisions in these areas can decision analysis be applied?

Exercises

1.1 Identify a past situation in which you made a decision that turned out to have a good outcome. Were you just lucky, or was there some process you went through that helped you have a good result?

1.2 Consider an important decision recently made (or to be made soon) by your organization or by you personally. (An organizational decision is preferred.) The decision might have to do with where to locate a new facility, a major expansion, investment in a new technology/product, where to live, which job offer to accept, and so on. Write *brief* descriptions of the following important elements associated with this decision (there is no correct or incorrect answer associated with any of these—there are some you may not be able to answer—don't worry about that).

a. Most decisions address a particular problem or problem area (e.g., losing market share). How did the problem addressed by this decision arise? How is this problem related to other problems?
b. Who were the decision makers for the situation? Who were the stakeholders (people affected by the decision)?
c. What alternative solutions were analyzed (e.g., alternate plant locations)? How were they generated?
d. What are the attributes (performance measures) that were considered? How were they determined?
e. What types of evaluation models were used, if any (i.e., how were the performance measure values determined as a function of the various alternative solutions)?
f. What are the sources of uncertainty/risk in the situation? How are these modeled?
g. What criterion model was used? (i.e., how were trade-offs that existed between the various performances considered?).
h. How could the decision-making process (which was used) be improved?

2

Problem Structuring

2.1 Introduction

As initially encountered, most problems do not lend themselves to the classical techniques of decision analysis. More specifically, when first encountering most decision situations, an analyst will not know the identities of all of the decision makers (DMs) and stakeholders, many of the good alternative solutions, sources of uncertainty and resulting risk, or the relevant performance measures, among other important elements. The process of deriving these elements is often called *problem structuring*.

Problem structuring has been applied in a large number of areas, including policy analysis for the Home Office Prison Department of the United Kingdom (Eden and Ackermann, 2004), reduction of electromagnetic field exposure from electric power lines (von Winterfeldt and Fasolo, 2009), organizational restructuring of the manufacturing function at Shell (Checkland and Scholes, 1990), quantification of delay and disruption of a litigation process (Ackermann et al., 1997), and management of a housing co-op (Thunhurst and Ritchie, 1992). See Table 2.1, recreated from Mingers and Rosenhead (2004) for a listing of applications. In general, the area of problem structuring has received much more attention in Europe than in the United States.

In this chapter, we present, first, some general concepts associated with problems and problem structuring, including the important concept that all problems are contained within a network of related problems. Following this initial discussion, specific problem structuring techniques/methodologies will be presented. These techniques/methodologies can be divided into two categories. The first category of methods is helpful in terms of providing perspective for a problem; that is, they can be used to better define an ill-structured problem. In particular, we focus on two approaches: the *Why-What's Stopping* (WWS) technique and *Breakthrough Thinking*. To a lesser extent, we discuss other methods, such as the Kepner and Tregoe approach and cognitive mapping, among others. These specific techniques are discussed in Sections 2.3 through 2.5.

TABLE 2.1
Applications of Problem Structuring Methods

Application Area	Methods/Techniques Used	References
General Organizational		
Mining performance evaluation	Soft systems methodology (SSM) cognitive maps, queueing theory	Pauley and Ormerod (1998)
Evaluating organizational performance	SSM critical systems	Gregory and Jackson (1992)
Career management	SSM	Bolton and Gold (1994)
Developing competence profiles	SSM	Brocklesby (1995)
Industrial psychology	SSM	Kennedy (1996)
TQM	SSM system dynamics	Bennett and Kerr (1996)
Developing R&D strategies	SSM	Nakano et al. (1997)
Organizational planning	SSM	O'Connor (1992)
Designing a parliamentary briefing system	Cognitive maps	Bennett (1994)
System for organizational learning	Cognitive maps	Lee et al. (1992)
Assisting community groups	Interactive planning Systems dynamics (SD)	Magidson (1992)
Teaching entrepreneurship	Interactive planning	Robbins (1994)
Modeling the San Francisco Zoo	Viable systems model (VSM)	Dickover (1994)
Organizational change	VSM	Brocklesby and Cummings (1996)
Modeling a municipal organization	VSM	Rasegard (1991)
Performance improvement in a multibusiness	VSM	Haynes et al. (1997)
Analysis of the drug trade	SD SSM	Coyle and Alexander (1997)
Organizational restructuring	VSM	Walker (1990)
Litigation/project management	Cognitive mapSD	Ackermann et al. (1997)
Facilities relocation	System dynamics soft systems	Vos and Akkermans (1996)
Developing a business strategy	System dynamics soft systems	Winch (1993)
Information Systems		
Accounting information system	SSM	Ledington (1992)
Analysis of a CD-ROM network	SSM	Knowles (1993)
Information systems strategy	VSM	Schuhman (1990)
Capturing process knowledge	SSM process models	Boardman and Cole (1996)
Building process models	SSM grounded theory	Platt (1996)
Developing an information systems strategy	Interactive planning SSMVSM strategic choice	Ormerod (1996a,b, 1998)

(*Continued*)

TABLE 2.1 (*Continued*)
Applications of Problem Structuring Methods

Application Area	Methods/Techniques Used	References
Technology, Resources, Planning		
New technology and culture conflict	SSM	Kartowisastro and Kijima (1994)
Planning livestock management in Nepal	SSM	Macadam et al. (1995)
Transport planning	SSM	Khisty (1995)
Agrotechnology transfer in Hawaii	SSM	Millspacko et al. (1991)
Natural resource management	SSM nonequilibrium ecology	Brown and Macleod (1996)
Lake management	SSM DSS	Gough and Ward (1996)
Energy rationalization	SSM QQT	Fielden and Jacques (1998)
Integration in transport planning	Cognitive maps	Ulengin and Topcu (1997)
Regional planning in S. Africa	Interactive planning	Strumpfer (1997)
Health Services		
Outpatient—clinics	Systems thinking data analysis, queueing, simulation	Bennett and Worthington (1998)
Problems of disabled users	Systems thinking	Thoren (1996)
Modeling outpatient services	SSM simulation	Lehaney and Paul (1994, 1996)
Nurse management	SSM	Wells (1995)
Contract management in the NHS	SSM	Hindle et al. (1995)
Health care information system	SSM	Maciaschapula (1995)
Resource planning and allocation	SSM simulation	Lehaney and Hlupic (1995)
Employment for those with mental health	Critical systems	Midgley and Milne (1995)
Planning hospital organization	Interactive planning	Lartindrake and Curran (1996)
General Research		
Qualitative survey research	Cognitive maps	Brown (1992)
CEO's cognitive capacity	Cognitive maps	Calori et al. (1994)
Eliciting knowledge about pesticides	Cognitive maps	Popper et al. (1996)
Automated knowledge discovery	Cognitive maps	Billman and Courtney (1993)

Source: Recreated from Mingers, J. and Rosenhead, J., *Eur. J. Oper. Res.*, 152, 530, 2004.

The results associated with applying the methods in the first category can be useful as input to the second category of methods, which are used to generate the objectives and attributes (usually in the form of a hierarchy) for the decision situation. The objectives and attributes (performance measures) are used to evaluate the overall quality of a decision. These techniques are discussed in Section 2.6.

Finally, we present important attributes for specific categories of decision situations in Section 2.7.

2.2 Problems and Problem Structuring Methods: General Concepts

The concept of a *problem* can be defined in any of several different ways. For example, Evans (1991, p. 11) notes various definitions from the literature, including (1) a gap to be circumvented, (2) a felt difficulty, (3) a dissatisfaction with the current state of affairs, (4) a perception of a variance between the present and some desired state of affairs, and (5) an undesirable situation. Each of these definitions connotes the idea of a *gap* between the current state of affairs and a desired state of affairs. The phrase "state of affairs" implies a set of performance measures used to define the current state of affairs.

This gap may be positive, negative, or unknown (Basadur et al., 1994). A *negative* gap occurs from a drop in performance (e.g., a machine has all of a sudden started producing many defective parts). A *positive* gap exists when an opportunity is perceived. For example, Land (Callahan, 1972) attributed his Polaroid camera invention to his ability to discover and define a problem in terms of an opportunity. *Unknown* gaps often result from a significant change in the base state of affairs, that is, when a policy, technology, or some other change causes the previous baseline to become irrelevant.

Simon (1960) identified three types of problems: well structured, semistructured, and ill structured. Well-structured problems come with complete information and are typically repetitive or routine. In a well-structured problem, the objectives are clear, and the workable alternative solutions are often obvious. As examples, most of the problems given in college-level engineering/business courses are well structured in nature. (This, at least, makes it easy for the instructor to grade the problems.)

As noted by Ackoff (1979), however, most problems encountered in real life are ill-structured problems. These ill-structured problems have little or no data and unclear performance measures, with no readily apparent specific alternatives. In addition, the appropriate DMs may not be apparent. These problems tend to be complex, nonroutine, and difficult to define. Potential alternative solutions, objective(s) associated with solving these problems, and the outcomes associated even with those alternatives that have been defined are not known, and the relevant DMs and stakeholders are often not obvious. The data required to model the problem are usually not readily available.

These ill-structured problems have also been termed as *messes* (Ackoff, 1979), *wicked* (as opposed to tame) problems (Rittel and Webber, 1973), or *swamps* or swampy situations (Schon, 1987). In more technical terminology, an ill-structured problem is "a dynamic situation consisting of complex systems of changing problems that interact with each other" (Ellspermann et al., 2007). Rosenhead (2006, p. 759) notes that the types of problems addressed by problem structuring methods include those that involve "multiple actors, differing perspectives, partially conflicting interests, and perplexing uncertainties."

As an example of an ill-structured problem, suppose you are a newly hired industrial engineer for the XYZ Manufacturing Corporation. On your first day at work, your supervisor tells you that there have been too many "safety incidents" in the plant and asks you to "solve the problem." Such a problem might be called an ill-structured problem because you probably do not know the identities of the DMs and stakeholders; the relevant performance measures/attributes, objectives, goals, and constraints; or good alternative solutions for the problem. In addition, you probably do not have a good feel for how this problem is connected to other problems. For example, this "number of safety incidents" problem may well be connected to the problem of "poor training." You need to define, formulate, or structure the problem first in order to determine (1) how this problem is related to other problems, (2) the relevant DMs and stakeholders, (3) the performance measures, (4) some good alternative solutions, and so on.

Case studies have documented situations where a project involving some application of decision analysis has failed, not because of a lack of expertise or effort in solving the problem as defined, but because the problem was not defined correctly in the first place. As noted by Churchman et al. (1957):

> There is an old saying that a problem well put is half-solved. This much is obvious. What is not so obvious is how to put a problem well.

In addition, Larson has stated that (Horner, 2004):

> 70% of the value added of operations research is the correct framing and formulation of the problem.

Finally, Keeney (1992) notes that although "alternative-focused thinking" is more prevalent and less difficult than "value-focused thinking" (in which more thought is given to discovering the underlying values of the stakeholders and DMs prior to generating alternatives), value-focused thinking will usually lead to better results.

A classic example of not defining a problem correctly in the first place is given in Hesse and Woolsey (1980, p. 3). This problem involved "slow elevators" in a tall building. The problem was not that the elevators were moving too slowly but that the riders *perceived* the elevators to be moving too slowly. The difference in problem definitions is subtle but significant in terms of generating good alternative solutions. That is, instead of concentrating on making the elevators move faster, alternative solutions should concentrate on making the passenger wait more pleasant. The solution then turned out to be placing mirrors next to the elevators; attractive people would look at

TABLE 2.2

Examples of Well-Structured Problems

We must choose one of two possible vendors to supply a particular purchased part. We have all of the relevant data on both vendors (cost, quality, schedule performances, etc.). Which vendor should be chosen?
Our firm needs to locate a new manufacturing facility in one of four possible locations. All relevant data (e.g., transportation costs, taxes, incentives, potential pool of personnel) have been collected, and we know our criterion. Which location should be chosen?

TABLE 2.3

Examples of Ill-Structured Problems

Our employees seem dissatisfied. What, if anything, should we do about it?
Our company is losing market shares. How can we change this trend?
The country's social security system will go broke within 25 years if nothing is done. Should we do something about this, and, if so, what?

themselves in the mirrors and the less attractive people would look at the more attractive people, and hence, the perceived wait for each was less than it would have been without the mirrors. (Of course, attractiveness is in the eye of the beholder!) Examples of well-structured problems are shown in Table 2.2, while examples of ill-structured problems are shown in Table 2.3.

Problem structuring is a critical step in the problem-solving process, especially for ill-structured problems. For example, Pidd (1989) noted that problem structuring should be preliminary to the other aspects of the problem-solving process, such as data collection, interviews, modeling, and so on.

The term *problem structuring* has been defined in a number of ways. For example, Pitz et al. (1980) have defined problem structuring as "the activity of identifying relevant variables in the problem situation, as well as establishing relationships between the variables." Shwenk and Thomas (1983) employed an operational definition for problem structuring: "the problem of formulating the present set of conditions, symptoms, and causes and triggering events into a problem or set of problems sufficiently well specified so that the risk of using analytic procedures to solve the wrong problem has been minimized." Note how Shwenk and Thomas implicitly stress the importance of problem structuring early in the problem-solving process. Finally, Pidd (1989) gives a somewhat cryptic definition for problem structuring by defining it as a process "by which some initially presented conditions and requests become a set of issues for further research."

One important point to note with respect to problem structuring is that every problem exists within a system of problems (sometimes called a *problematique*). That is, every problem is somehow connected to other problems. Realizing these connections can be helpful in generating alternative solutions and in evaluating the system-wide effect of a solution. These problem connections can be illustrated through a "network of problems." In fact, the concepts associated with hierarchies, networks, and matrices, which allow one to represent the relationships among problem elements, are important for many techniques associated with problem structuring, as we shall see later in this chapter.

Another important aspect of problem structuring is the idea of assuming an optimistic attitude in addressing the problem. For example, this idea is stressed in many problem structuring methods, including Breakthrough Thinking (Nadler and Hibino, 1990) and the Why–What's Stopping method (Basadur et al., 1994). Keeney (1992) in fact refers to the fact that prior to choosing an alternative in solving a "problem," there is always an *opportunity* to create alternatives. Thinking of problems as opportunities helps to place an optimistic light on the problem structuring process.

Finally, the importance of "divergent thinking" as opposed to "convergent thinking" is stressed in most problem structuring methods. Keeney (1992) refers to this as "constraint-free thinking." Most science and engineering-based courses rely on convergent thinking, in which a problem statement is given in terms of various data, and a student derives a single solution. Many students have difficulty with this process since many engineering courses focus on convergent thinking. In divergent thinking, for example, one needs to generate a whole set of problem statements from a single statement and/or a whole set of objectives and attributes from a single, overall, objective.

Most of the techniques for problem structuring address one or more of the following characterizing features of ill-structured problems:

1. The existence of several DMs and stakeholders, each with their own subjective perspective of the problem.
2. Closely related to the first feature, the existence of multiple performance measures, several of which may be conflicting in nature.
3. Large amounts of uncertainty (and related risk) associated with parameters and relationships of the problem.
4. The existence of an entire network of problems to which the original ill-structured problem is related.
5. The fact that good alternative solutions to the problem are not readily apparent.

Since there are typically several DMs and stakeholders for ill-structured problems (the first characteristic noted earlier), each with their own respective viewpoints of the problem situation, problem structuring methods should be "participative and interactive" in nature (Rosenhead, 1996). This often implies the use of a "facilitator" working with a group of DMs and stakeholders. The facilitator will moderate a problem structuring meeting to make sure that everyone is allowed to express their viewpoints and that a DM/stakeholder with a forceful personality will not have disproportionate effect on the outcome of the process. The facilitator should be well versed in the problem structuring method being used but should have no personal stake in the problem.

In the next two respective sections of this chapter, we will discuss two specific problem structuring methods: Breakthrough Thinking (Nadler and Hibino, 1990) and Why–What's Stopping analysis (Basadur et al., 1994). This will be followed by a section that will briefly detail other problem structuring methods including the Kepner and Tregoe Method (Kepner and Tregoe, 1981) and cognitive mapping (Eden and Ackermann, 2004).

In addition, see Miser and Quade (1985, Chapter 5), Checkland (2001), Joldersma and Roelofs (2004), Mingers and White (2010), Nutt (2001), Shaw et al. (2004), and White (2009) for a more detailed discussion of some of these and other problem structuring techniques.

2.3 Breakthrough Thinking

Breakthrough Thinking (Nadler and Hibino, 1990) is a technique not only for problem structuring but also for problem solving in general. It does however place a great emphasis on problem structuring. Its roots lie in the *Purpose Design Approach* pioneered by Nadler and Hibino.

The application of Breakthrough Thinking relies on seven core principles:

1. The Uniqueness Principle
2. The Purposes Principle
3. The Solution-after-Next (SAN) Principle
4. The Systems Principle
5. The Limited Information Collection Principle
6. The People Design Principle
7. The Betterment Timeline Principle

The *uniqueness principle* basically states that every problem is unique, which runs counter to the old adage "do not reinvent the wheel." Often, problems that may appear at the outset to be the same are actually unique, due to the fact that they occur at different times in different environments. This results in having different sets of stakeholders, DMs, available technologies for solution, and so on. For example, the "problems/opportunities" of building a simulation model of a distribution system for two different organizations may be quite different due to the facts that (1) the DMs for the organizations have quite different technical backgrounds, (2) data availabilities for the two organizations are different, and (3) simulation software packages owned by the two organizations are quite different. Certainly, there may be much value in having already solved one problem when it comes to solving a similar problem, but "turnkey solutions" normally require substantial modification in order to work well in a new environment.

The *purposes principle* refers to the fact that prior to doing anything (solving a problem, collecting data, etc.), one should always have a purpose. As noted by Nadler and Hibino, there are four elements associated with a purpose: function, values/goals, performance measures, and objectives. In thinking about the purpose of doing something, such as solving a problem, one can develop a "purpose hierarchy," which allows for varying perspectives on a problem. For example, let's suppose that a manufacturing organization is having difficulty with achieving on-time deliveries from one of its suppliers. In this case, a purpose hierarchy might be constructed as shown in Figure 2.1.

Note that a purpose hierarchy might allow for the generation of alternative solutions not readily apparent from an isolated consideration of the initial problem only.

The *SAN principle* refers to the fact that often there is another problem that will occur after a solution to the initial problem is implemented and that this problem should be foreseen and planned for. An example of this situation would be the installation of a new computerized production control system, installed to address inefficiencies in a production system. The use of the new system might require that new procedures be employed by the production system's personnel, which might best be addressed through a training program. Of course, this leads to the problems associated with designing and scheduling the training program.

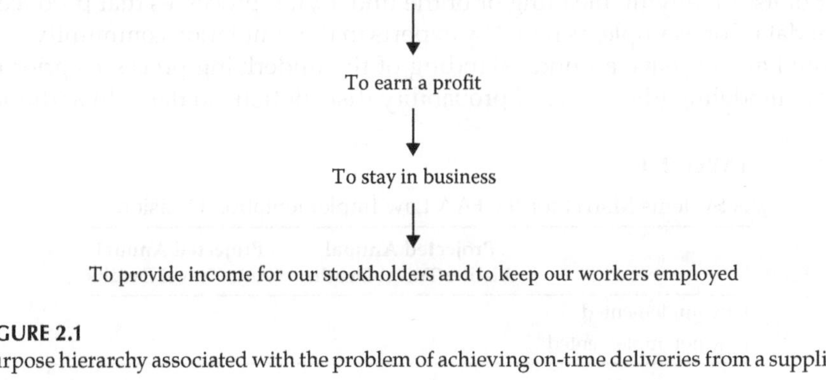

What is the purpose of achieving on-time delivery from the supplier in question?

↓

To meet our production schedule for products that contain the parts delivered by the supplier

↓

To satisfy our customers

↓

To earn a profit

↓

To stay in business

↓

To provide income for our stockholders and to keep our workers employed

FIGURE 2.1
Purpose hierarchy associated with the problem of achieving on-time deliveries from a supplier.

The *systems principle* addresses the unintended consequences that often occur when a solution to a problem is implemented. For example, most people would think that a Federal Aviation Administration (FAA) law requiring that parents be required to purchase a separate ticket for their children under the age of two for airline flights (instead of having the child held on the parent's lap) would be a good thing in terms of the infant's safety. However, an analysis (Machol, 1996) showed that since many more parents would drive rather than fly due to the cost, and since driving is much less safe than flying, many more infants would be injured or killed. One's initial thought process in addressing this situation may not have considered the entire transportation system, nor the interface between the passenger air transportation system and the private auto transportation system, and thereby not realize this unintended consequence.

An aid in the consideration of the systems principle is the *systems matrix*, which allows one to show the relationships between two different categories of elements associated with a system. Any specific problem/situation might have several different systems matrices developed for it. For example, the methodology associated with quality function deployment (Bossert, 1991) employs a series of "houses of quality" (essentially systems matrices) that allow one to relate product characteristics to engineering characteristics, engineering characteristics to part characteristics, part characteristics to process plans, and so on. There are many different types of systems matrices that one can use to implement the systems principle, but often a systems matrix will relate alternative solutions to performance measure values. For example, in the case of the FAA law referred to earlier, a systems matrix might appear as in Table 2.4. The rows represent alternative decisions, while the columns represent performance measures.

The *limited information collection principle* suggests that one should always have some purpose for collecting data, prior to the collection/gathering process. Too much data often obscure the problem structuring and problem-solving processes, and in this day and age, data are typically collected in an automatic fashion and are therefore relatively easy to obtain. One of the difficulties that often occurs when much data are available is the extensive analysis of that data; this analysis often occurs without a clear understanding of its underlying meaning or of the underlying processes that produced that data. For example, as noted by experts in the simulation community, one should always have an understanding of the underlying processes prior to input modeling (the fitting of probability distributions to data). In addition,

TABLE 2.4

A Systems Matrix for the FAA Law Implementation Decision

	Projected Annual Airline Travel Deaths	**Projected Annual Auto Travel Deaths**
Law implemented		
Law not implemented		

using large amounts of data in the problem structuring/solving process may give one a false sense of security concerning the quality of that process.

The *people design principle* suggests that the different perspectives of different people are usually extremely useful in the problem structuring process. Allowing for various perspectives also aids the divergent thinking processes that are important in problem structuring. In addition, many solution implementations for organizations require buy-in from a large number of people within the organization. When these people have been involved in the problem structuring process, they are much more likely to give their buy-in.

Finally, the *betterment timeline principle* is closely aligned with the SAN principle, in that this principle implies that one should develop a schedule for improving on a solution once it is initially implemented. In addition, this principle nicely complements the concept of "continuous improvement," employed by many organizations. The benefits realized from the application of this principle result from the facts that solutions are implemented over time, and these solutions result in effects over time within changing systems. Such effects should be monitored and control systems put in place to allow for system modifications.

As can be realized from the earlier discussion, Breakthrough Thinking is not only an approach for problem structuring but also for problem solution and solution implementation.

2.4 Why–What's Stopping Technique

As noted earlier, every problem exists within a system or network of problems (a problematique, or a mess) (Ackoff, 1981). In order to be sure that the "right" problem is being addressed, one should be aware of this network of problems. In addition, being aware of the network of problems is helpful in generating alternatives and performance measures and in evaluating the effect of various alternatives on a system-wide basis.

The Why–What's Stopping technique (Basadur et al., 1994) is one approach for generating this network of problems. The technique typically involves a facilitator and a group of DMs and stakeholders meeting for a "brainstorming session." The output of the session is a network of related problems from which one may obtain a set of performance measures, DMs, stakeholders, preliminary alternatives, constraints, and so on. By a network, we mean a set of "nodes" (each node being a problem statement) connected with arcs. An arc connecting two nodes implies a direct connection between these two problem statements. A particular node (problem statement) may be connected to several other nodes.

The session commences, following an explanation of the technique by the facilitator, with a brief statement of what is thought to be the problem by one of the participants (i.e., stakeholder or DM). The facilitator will write this statement on a blackboard (or some other device). For example, suppose that the initial problem statement is given by:

> There are too many defects in our main product.

This problem statement is transformed into a "challenge" by attaching the phrase "how might we …":

> How might we reduce the number of defects in our main product?

The purpose of attaching the phrase "how might we …" is to allow a viewpoint in which the problem is perceived as an opportunity, not a difficulty. In addition, rephrasing the problem statement in this fashion allows an optimistic tone to the problem structuring process, an important aspect in problem structuring as noted earlier.

In order to expand on this initial problem statement and therefore start the development of the problem network, two types of questions are asked, corresponding to a *"why"* question and a *"what's stopping us"* question. For example, for the problem statement earlier, a "why" question would be:

> Why do we want to reduce the number of defects in our main product?

and a *"what's stopping us"* question would be:

> What's stopping us from reducing the number of defects in our main product?

Note that upon receiving an initial answer to each of the earlier questions, the facilitator would most likely solicit alternative answers to these questions by asking the session participants why else they would want to reduce defects and what else is stopping them from reducing defects.

Answers to the "why" question might be:

> So that our main product has a better reputation.
> So that we can spend less money in reworking operations.

Answers to the "what's stopping us" question might be:

> Our workers are not well trained.
> Our standards are too high.
> We have no incentive program to reward our workers for quality work.

Note that each of these answers would be translated into problem statements such as the following, respectively, using the "how might we" phrase to lend an optimistic tone to the problem statement:

> How might we develop a better reputation for our main product?
> How might we spend less money on reworking?
> How might we train our workers better?
> How might we correctly set our standards?
> How might we develop an incentive program?

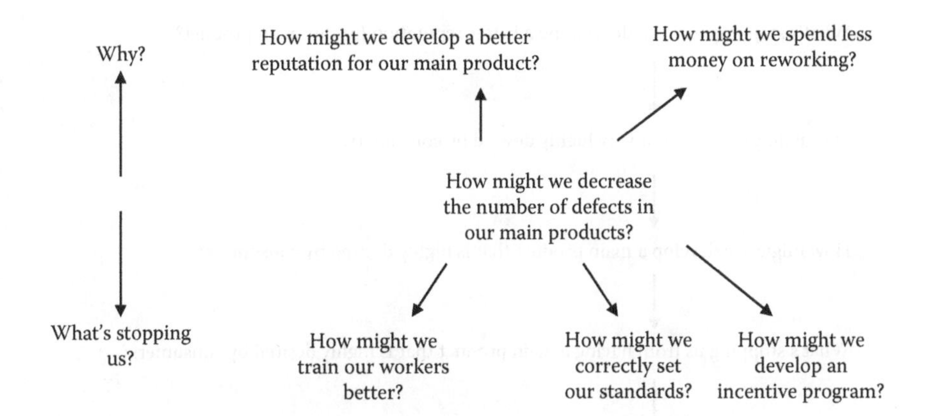

FIGURE 2.2
Part of a problem network resulting from applying the Why–What's Stopping technique.

These "how might we" problem statements would be arranged in the form of a problem network with the problem statements represented as the nodes and arcs pointing from one node to another node if the second node corresponded to an answer to the question from the first node problem statement. The answers corresponding to the "why" questions generated from a particular problem statement would be placed *above* that problem statement. The answers corresponding to "what's stopping us" questions generated from a particular problem statement would be placed *below* that problem statement. Hence, one would obtain a problem network such as shown in Figure 2.2.

Note that the reverse process should also work for the question–answer sequence. That is, if the answer to a "why" question originating from problem statement A resulted in problem statement B, then a reasonable answer to a "what's stopping" question originating from problem statement B should result in problem statement A. In viewing the network in Figure 2.3, this is clearly the case; for example, a reasonable answer to the question:

> What's stopping us from developing a better reputation for our main product?

would be:

> We have too many defects in our main product.

This problem statement could be restated using the "how might we" phrase as:

> How might we decrease the number of defects in our main product?

Throughout the process of generating the problem network, the "reverse" question should be asked as a check on the answers given.

Note that the answers given to the "why" questions result in problem statements corresponding to a higher-level perspective than the problem

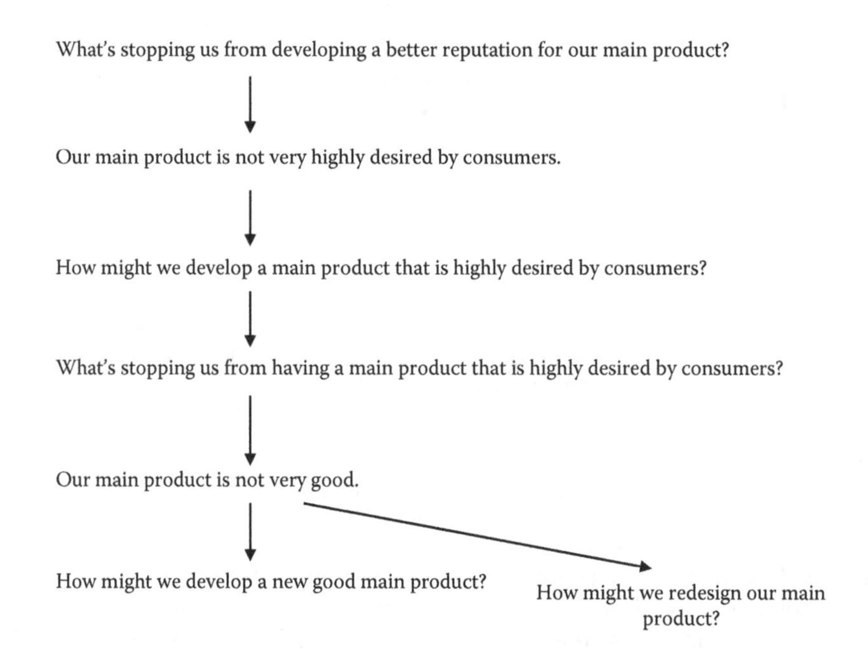

FIGURE 2.3
A partial problem network.

statement used to generate the "why" question. Conversely, problem statements resulting from the "what's stopping us" questions result in problem statements corresponding to a lower-level perspective than the problem statement used to generate the "what's stopping us" question.

Note also that one could extend either upward (by asking the "why" question) or downward (by asking the "what's stopping" question) from any problem statement in the network. Exactly which question is asked at which point in time during the process typically requires an experienced facilitator as well as engaged participants. Focusing on a specific part of the problem network rather than some other part can very easily lead to solutions that would not have been thought of otherwise. For example, upon attaining the problem statement "How might we develop a better reputation for our main product?," one might focus on extending downward from this problem statement, thus leading to a partial problem network as shown in Figure 2.3. Also, as noted by Keeney (1992, p. 27), conversion of a decision *problem* into a decision *opportunity* (which allows for the creation of additional alternatives) can be accomplished by considering the problem from a broader perspective. See Exercise 2.4 for another example of a situation in which viewing a problem from a higher-level perspective leads to a different set of alternative solutions.

Note that Figure 2.3 allows the focus to shift from addressing defects in the main product to redesigning the product, or developing a completely new product.

Note also that some of the answers, in addition to providing alternative problem statements, also provide solution alternatives. For example, a solution to decrease the number of defects in the main product is to develop a new training program for workers. Of course, this "solution" leads to a whole host of additional problems involving, for example,

1. What type of training should be done?
2. How should the training be scheduled?
3. Should we do the training in-house or contract an outside vendor?

This corresponds to the "SAN" principle of Nadler and Hibino.

Finally, one needs to be concerned with how far to extend the problem map. In particular, the map stops at the "top" when we reach the problem statement of:

> How might we reach happiness and bliss?

This problem statement represents the theoretically broadest challenge and in a business setting might equate to:

> How might we increase the profitability of our company?

We reach the lowest point (i.e., the most narrowly defined problem statement) when we define some very specific course of action—that is, an idea that can be easily executed. For the example problem being addressed, such a lowest-level problem statement might be:

> How might we contact three vendors to bid on a specific training program for our employer?

Note that it's not absolutely necessary to extend the problem map to its extremes in order for the map to be useful. Keep in mind that the basic purpose of the technique is to allow a "better" problem formulation—that is, to give a better understanding of the system of problems to the DMs, to allow a new "angle" on the problem, to help generate a good set of performance measures, and to generate some initial solution alternatives. Figure 2.4 depicts an overview of the Why–What's Stopping technique.

As another example of a Why–What's Stopping analysis, consider the problem of childhood obesity in Louisville, Kentucky. A multiresource medical clinic in Louisville addresses this unstructured problem. The staff of the clinic includes physicians, nurse practitioners, a nutritionist, an exercise

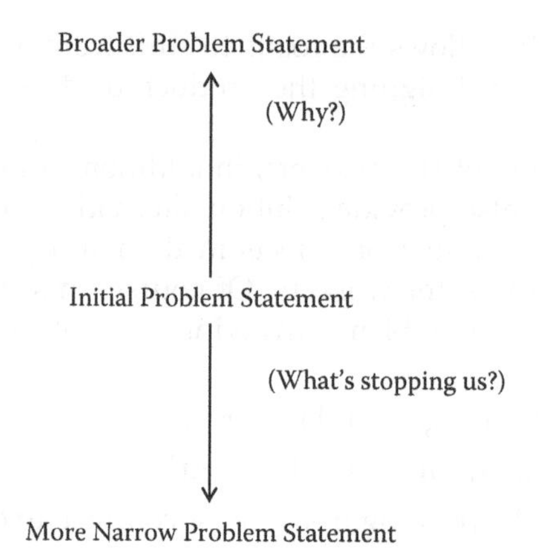

FIGURE 2.4
Network of problems resulting from the Why–What's Stopping technique.

physiologist, and a psychiatrist. Two Why–What's Stopping sessions were conducted: first, with the staff of the clinic and, second, with parents of some of the children who were patients at the clinic. The networks that arose from the sessions are shown in Figure 2.5 for the clinic's staff and in Figure 2.6 for the parents' of children who are the clinic's patients, respectively.

The ideal situation would have allowed for both groups (clinic staff and parents) to meet together in order to generate a single network. However, one can see the different perspectives in each of the two respective networks. For example, the clinic staff network (Figure 2.5) places greater emphasis on activities directly controlled at the clinic. On the other hand, the parents' network (Figure 2.6) emphasizes family and other social relationships. Each network does include some of each perspective however. Of course, both perspectives are important in order to address the problem from a systems viewpoint.

In addition, each of the networks would allow for the generation of a wide variety of alternative solutions to the problem. These alternative solutions would basically be generated from the lower-level nodes of the network. For example, consider the lower-level node in Figure 2.5:

> How might we improve motivational coaching with primary care providers?

A solution associated with this node might involve training programs in motivational coaching for primary care providers.

Consider a lower-level problem statement from Figure 2.6:

> How might we recruit volunteers for after-school activities?

A solution associated with this problem statement would be: "Develop a marketing campaign to recruit and train volunteers for after-school activities."

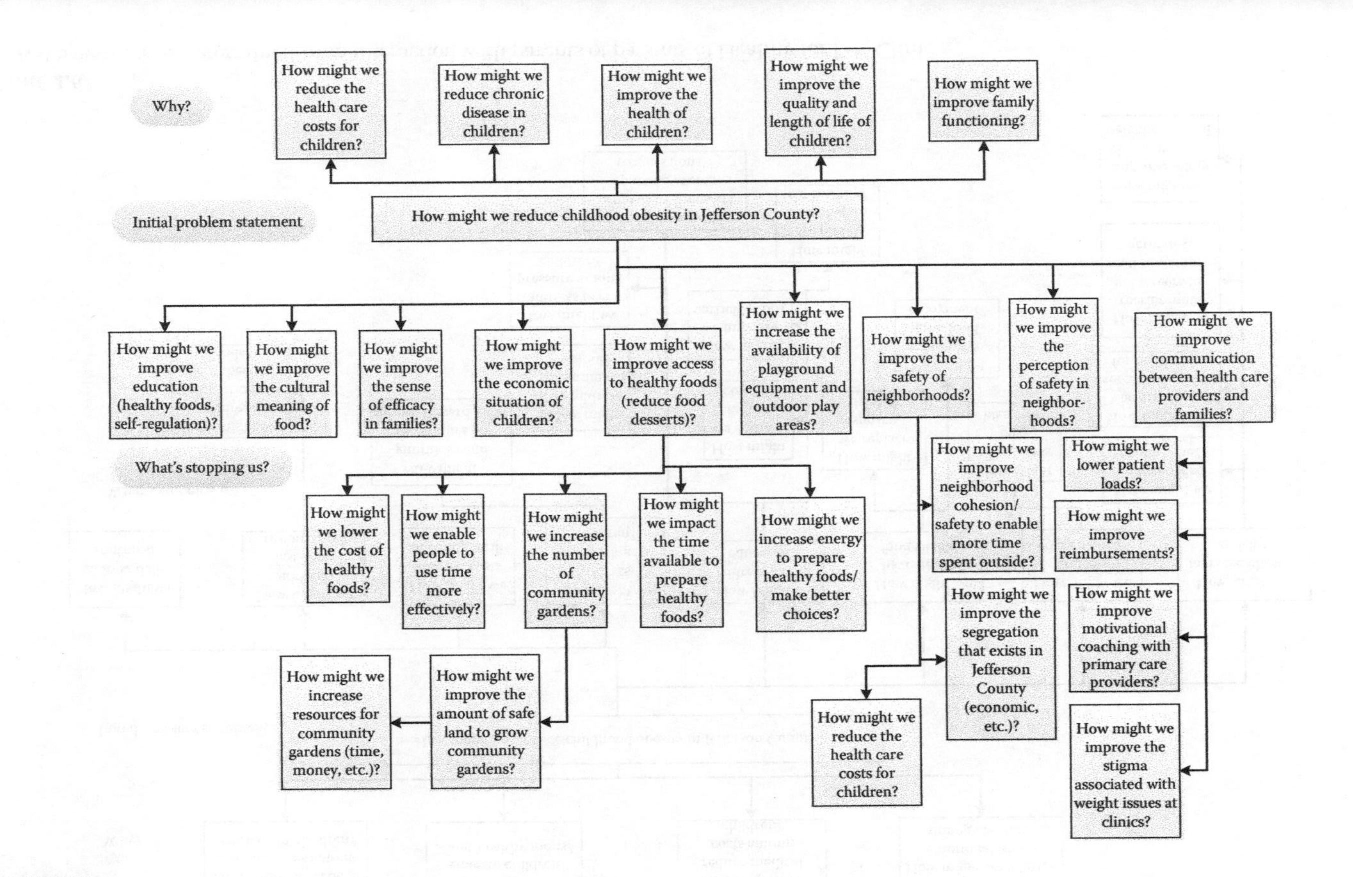

FIGURE 2.5
The WWS network as determined from interaction with staff of Healthy for Life Clinic.

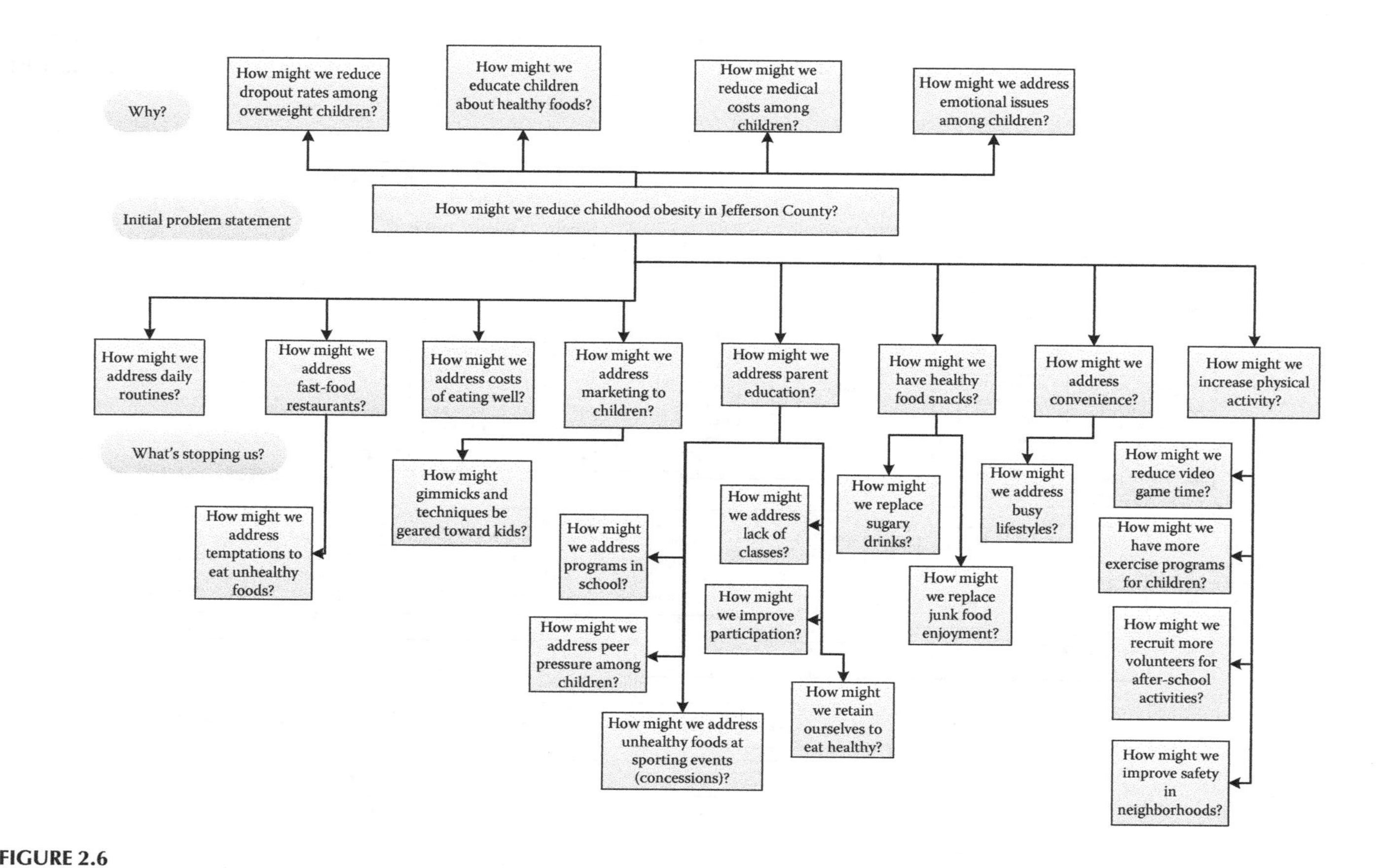

FIGURE 2.6
The WWS network as determined from interaction with parents of patients of Healthy for Life Clinic.

2.5 Additional Methods for Problem Structuring

Some additional problem structuring methods are discussed in this section, but in less detail than Breakthrough Thinking and the Why–What's Stopping technique just presented. The reader is referred to the references cited in the discussion for additional details. In addition, Ackermann (2012) provides a relatively recent review of the field. We will briefly discuss two of the more popular methods in this section: the Kepner and Tregoe method and cognitive mapping.

The *Kepner and Tregoe method* originated from social science research performed by Charles H. Kepner and Benjamin B. Tregoe at the RAND Corporation. It can be thought of as not only an approach to problem structuring but also broadly as an approach to problem solving. The overall approach is described in the book *The New Rational Manager* (Kepner and Tregoe, 1981). The book contains numerous examples of actual problems and solution processes that are used to illustrate various aspects of the method.

The method involves the application of four "patterns of thinking" that are related to four questions that should be asked by managers every day. These four patterns, along with the four questions in parentheses, are as follows:

1. Assessing and clarifying (What's going on?)
2. Cause and effect (Why did this happen?)
3. Making choices (Which course of action should we take?)
4. Anticipating the future (What lies ahead?)

The first two patterns listed might be thought of as being related to the area of problem structuring, as described in this chapter.

Cognitive maps are "generally records of verbal accounts of issues given in interviews by members of the management team" (Eden and Ackermann, 2004). It consists of nodes and arcs (i.e., a cognitive map is a directed graph), where the nodes represent statements concerning a problem and the arcs represent causality. For this reason, cognitive maps are sometimes called causal maps. The basis for cognitive maps is derived from the personal construct theory of Kelly (1955). A cognitive map often leads to the later development of a related influence diagram, which will be discussed in Chapter 7.

The concept of cognitive mapping has been formalized through the development of Strategic Options Development and Analysis (*SODA*) (Eden and Ackermann, 2001) and *JO*intly *U*nderstanding *R*eflecting and *NE*gotiating Strateg*Y* (*JOURNEY*) Making (Eden and Ackermann, 2004). A computer software package (decision explorer) has also been developed as an aid for cognitive mapping.

Some other problem structuring techniques/approaches that have been developed include mind-mapping (Buzan, 1991), cause-and-effect diagrams (sometimes called fishbone diagrams or Ishikawa diagrams) (Ishikawa, 1990), stairstepping (Huge, 1990), Brightman's alternative worldview method (Brightman, 1988), the five Ws and H technique (VanGundy, 1988, Chapter 3), dimensional analysis (VanGundy, 1988, Chapter 3), strengths, weaknesses, opportunities, and threats (SWOT) analysis (Fine, 2009), and the Smith framework (Smith, 1988).

2.6 Generating Objectives and Attributes for a Decision Situation

The methodologies discussed in Section 2.3 through Section 2.5 can be thought of as a prelude to the activity discussed in this section: generation of objectives and attributes for a decision situation. In order to rank a group of feasible alternatives, one must evaluate these alternatives in some way. Evaluation implies the use of performance measures or *attributes*, corresponding in some sense to the objectives of the system under consideration. In this section, we present methods by which to obtain the objectives and attributes for a decision situation.

Various studies have indicated that, without methods that make DMs focus on their objectives and attributes, these DMs omit many relevant objectives and attributes for a decision problem (Bond et al., 2008). Without a complete set of objectives and attributes, relevant alternatives might not be generated for consideration, and alternatives that are evaluated may not be ranked appropriately.

2.6.1 Terminology: Criteria, Values, Mission Statements, Vision Statements, Objectives, Goals, Constraints, and Attributes

Some of the terminology discussed in this section is often used in an unclear way. A reason for this is that terms such as criteria, values, objectives, goals, constraints, and attributes are, for the most part, basic terms to people in all walks of life. In order to have a single frame of reference throughout this section, we provide definitions for these terms.

> A *criterion* is "a standard on which a judgment or decision may be based." (See: http://www.merriam-webster.com/dictionary/criteria.)

The word is often used interchangeably with criteria since typically several things go into making up a criterion. Since many different things go into making up a criterion (e.g., objectives, goals, constraints), criterion is a very general term.

As defined by Keeney (1992, p. 6), *values* are "principles used for evaluation." One of the tenets of Keeney's value-focused thinking is that by focusing on one's values, more and better alternatives can be generated for a decision problem. In addition, focusing on values is helpful in the generation of a good set of attributes for a decision problem.

The values of organizations are sometimes indicated by their mission/vision statements but are more likely indicated by their actions; in particular, an organization's actions indicate their preferences toward risk and their value trade-offs, for example, between the environment and profits. Some examples of organizational mission/vision statements are given in the following:

General Motors Mission Statement

> GM is a multinational corporation engaged in socially responsible operations, worldwide. It is dedicated to provide products and services of such quality that our customers will receive superior value while our employees and business partners will share in our success and our stock holders will receive a sustained superior return on their investment
>
> **(See: http://wiki.answers.com/Q/What_is_General_Motors_mission_statement)**

General Motors Vision Statement

> Over the past 100 years, GM has been a leader in the global automotive industry. And the next 100 years will be no different. GM is committed to leading the industry in alternative fuel propulsion.
>
> GM's vision is to be the world leader in transportation products and related services. We will earn our customers' enthusiasm through continuous improvement driven by the integrity, teamwork, and innovation of GM people.
>
> Over the past 100 years, GM has been a leader in the global automotive industry. And the next 100 years will be no different. GM is committed to leading the industry in alternative fuel propulsion.
>
> **(See: http://www.company-statements-slogans.info/list-of-companies-g/general-motors.htm)**

McDonald's Vision Statement

> McDonald's vision is to be the world's best quick service restaurant experience. Being the best means providing outstanding quality, service, cleanliness, and value, so that we make every customer in every restaurant smile.
>
> **(See: http://www.specimentemplates.org/mission-statements/mcdonald-visionstatement.htm)**

As noted by Bart (1997), an organization's mission statement should contain three important elements: its key market(s), the contribution of the organization (e.g., in terms of a product/service), and the distinction of the organization's contribution.

An individual's values are often more difficult to determine than an organization's values, even for the individual himself or herself. However, by focusing on his or her values, an individual can often develop more and better alternatives and, through expansion on these values, often develop a more relevant set of attributes for a particular decision problem.

An *objective* can be thought of as a desired direction of improvement for some performance measure. Since the concept of "direction" is often associated with objective, one typically attaches one of the words "minimize," "maximize," "optimize," or "improve" to an objective. Examples of objectives would be "minimize cost," "maximize reliability," "maximize throughput," and so on. Objectives should be things that are derived from the values of an organization for any particular decision-making context. One of the things that makes decision making difficult is that most, if not all, decision problems have conflicting objectives. That is, given a set of alternative solutions, some alternatives might do well on some objectives, some might do well on other objectives, but there is typically no alternative that optimizes all of the objectives simultaneously.

Sometimes, objectives are stated in such a way that the word optimize (or maximize/minimize) is implied and not stated. For example, one might state "enforce speed limits" as an objective; the implication here is that the objective is actually "optimize enforcement of speed limits."

Another difficulty is that sometimes an objective might be confused with an alternative action. For example, "optimize enforcement of speed limits" might be thought of as an alternative with respect to an overall objective of "optimize transportation safety," whereas in reality this is an objective. In this particular situation, involving speed limit enforcement an alternative action might be related to where cameras are placed in order to detect speeding vehicles.

A *goal* is a desired level for a performance measure. An example would be "achieve a throughput of 1000 parts per day." The concept of a goal is often used in conjunction with that of "satisficing," which can be defined as selecting the first solution identified that meets a goal or a set of goals, even if additional searching might achieve better results for the respective set of objectives associated with the goals. In addition, the concept of goals is associated with *goal programming*, an optimization technique discussed in Chapter 4.

A *constraint* is a restriction placed on a performance measure value, typically by stating that the performance measure value must be greater than or equal to, equal to, or less than or equal to some specific number. An example would be that any solution chosen for plant expansion must have a cost of less than or equal to $20 million. The concepts of constraints and objectives are important ones in the field of *mathematical programming*. In addition, many of the approaches associated with *multiple objective mathematical programming*, in which multiple conflicting objectives are considered in the optimization process, are modeled by considering one of the objectives in the objective function and the other objectives through the use of respective constraints where the objectives are constrained to achieve a certain value; these values are represented as the right-hand values in the constraints. By selectively varying the right-hand sides of the constraints and exchanging one objective for another in the objective function, one can explore the Pareto optimal surface of the problem. This area will be discussed in more detail in Chapter 4.

An *attribute* can be thought of as a measure of how well an objective is achieved. For example, if an objective is to minimize cost, then an attribute associated with that objective would be cost in dollars. The term *attribute* is often used interchangeably with the phrase performance measure. Note that an attribute should be defined in such a way, as much as possible, so that there is no question as to its interpretation. One reason for this requirement is that the DMs and stakeholders will typically be answering questions concerning their trade-offs between different attribute values in a decision analysis application. Hence, there must be a clear understanding on the part of stakeholders and DMs as to the meaning associated with various levels of an attribute.

Attributes can be categorized along two dimensions as being (1) *natural attributes* or *constructed attributes* and (2) *nonproxy attributes* or *proxy attributes* (see Table 2.5). The best situation is if all attributes in a study are natural and nonproxy.

The interpretation as to the meanings of various attribute levels is fairly straightforward if the attribute in question is a natural attribute. A natural attribute is one that is quantitative in nature. Examples of natural attributes and their respective associated objectives are shown in Table 2.6. Even with natural attributes, especially in decision settings involving highly technical

TABLE 2.5
Categorization of Attributes

Attributes that are *natural* and *nonproxy*	Attributes that are *constructed* and *nonproxy*
Attributes that are *natural* and *proxy*	Attributes that are *constructed* and *proxy*

TABLE 2.6
Objectives and Associated Natural Attributes

Objective	Natural Attribute
Minimize transportation costs	Transportation costs in thousands of dollars
Minimize mean number of defective parts produced per day during the month of May	Mean number of defective parts produced per day during the month of May
Minimize cost per purchased part for vendor selected	Cost per purchased part for vendor selected

problems, the meaning may not be clear to particular DMs or stakeholders. For example, a (nontechnical) governmental DM may not be aware of the significance of "tons of SO_2 emitted" as an attribute in a decision problem involving environmental issues; hence, an analyst would need to be sure that the DM became educated in this area if this attribute were used in the problem.

Natural attributes can be contrasted with constructed attributes for which the interpretation of meaning may not be quite so clear. These constructed attributes are typically subjective in nature and also normally require the use of a scale with numbers and associated meanings assigned to the numbers. As would be expected, constructed attributes are associated with objectives that are somewhat nebulous in nature. An example of such an objective would be "optimize reputation" (of the selected vendor). The associated attribute would be vendor reputation, as measured by a scale with numbers and associated phrases, as shown in Table 2.7.

Another example of a constructed attribute for the quality of a writing assignment is shown in Table 2.8 (derived from work performed by Dr. Patricia Ralston, chair of the Department of Engineering Fundamentals at the University of Louisville). Note the large amount of detailed information shown for each of the ratings. Typically, the more detailed information that can be given for each of the numerical rating values associated with a constructed attribute, the better. Such detailed information allows for less error in interpretation.

One of the keys for the construction of the scale would be that any two reasonable DMs or stakeholders for the problem agree on the mapping for any particular outcome into the correct number for the scale. As discussed in the next section, these nebulous objectives and associated constructed attributes are more likely to occur at the higher levels of the "objectives–attributes" hierarchy than at the lower levels.

In addition to reputation and quality, other examples of constructed attributes could be comfort, fun, appearance, pain, attitude, and public acceptance. The key point in their identification is the idea that their measurement

TABLE 2.7

Scale for the Constructed Attribute of Reputation

Number	Associated Meaning
1	World-renowned reputation, achieved from receiving numerous international awards
2	National reputation, well known for high-quality work on a national basis; not well known outside of the United States
3	Locally prestigious, within the state and local area
4	Has a mixed reputation for the quality of its work
5	Associated with numerous scandals for the quality of its work

TABLE 2.8

Constructed Scale for the Quality of a Written Report

Rating	Description
4—Excellent	Clearly identifies the purpose including all complexities of relevant questions
	Accurate, complete information that is supported by relevant evidence
	Complete, fair presentation of all relevant assumptions and points of view
	Clearly articulates significant, logical implications, and consequences based on relevant evidence
3—Good	Clearly identifies the purpose including some complexities of relevant questions
	Accurate, mostly complete information that is supported by evidence
	Complete, fair presentation of some relevant assumptions and points of view
	Clearly articulates some implications and consequences based on evidence
2—Satisfactory	Identifies the purpose including irrelevant and/or insufficient questions
	Accurate but incomplete information that is not supported by evidence
	Simplistic presentation that ignores relevant assumptions and points of view
	Articulates insignificant or illogical implications and consequences that are not supported by evidence
1—Poor	Unclear purpose that does not include questions
	Inaccurate, incomplete information that is not supported by evidence
	Incomplete presentation that ignores relevant assumptions and points of view
	Fails to recognize or generate invalid implications and consequences based on irrelevant evidence

involves the use of a subjective scale. Hence, an attribute that might normally be thought of as natural could be a constructed attribute depending on the scale used; for example, an attribute like cost, which would typically be a natural attribute since it would normally be measured in units such as dollars, could be thought of as a constructed attribute if descriptions for values like "very high cost," "moderately high cost," and so on were used. In this regard, whether an attribute is natural or constructed might depend on the amount of effort put forward in the measurement activity.

Whether an attribute is proxy or nonproxy refers to how directly it measures an objective. As noted by Keeney and Raiffa (1993, p. 55), "a proxy attribute is one that reflects the degree to which an associated objective is met but does not directly measure an objective." Hence, a proxy attribute measures an objective in an indirect fashion. For example, a proxy attribute associated with the objective of "optimize comfort" might be "deviation of temperature from 70° Fahrenheit." The deviation in temperature is not a direct measurement of comfort, but it is useful as a proxy.

Another example of a proxy attribute is described in Keeney and Raiffa (1993, Chapter 2). They describe the use of the proxy attribute of "response time" in order to measure the objective of "delivering patients to the hospital in the best possible condition." Note that the proxy attribute of response time is relatively easy to measure as compared to the patient's condition when arriving to the hospital. Of course, this is the key advantage associated with the use of a proxy attribute—ease of measurement—but at the expense of not being directly related to the objective being measured.

One error often made with selection of attributes, and often with proxy attributes, has to do with using an input in order to measure the "quality" of a system; an example of this is the use of per capita student spending in order to measure the quality of a school system. The advantage in this case of course is that this information is relatively easy to obtain, but with the disadvantage being that it's probably not a good measure of quality.

Of course, in some sense, all attributes are nonproxy since no objective can be measured in a completely accurate fashion. However, in most situations, it will be clear whether an attribute should be considered as a proxy attribute or not.

2.6.2 Formation of a Hierarchy/Network of Objectives and Associated Attributes

Objectives for a decision situation or system can typically be structured as either a hierarchy or as a network. These hierarchies or networks consist of nodes (representing the objectives) and arcs connecting the nodes. An arc (line) connecting two nodes (objectives) indicates that these two objectives are closely related.

Fundamental (or "high-level") objectives are structured as hierarchies, which are shaped something like a triangle, with just one node at the top and an increasing number of nodes at each level of the hierarchy. *Means-ends* (or lower-level) objectives (or sometimes just called means objectives) are typically formed as a network. See Keeney (1992) and Clemen and Reilly (2001) for an additional discussion concerning the differences between fundamental objectives and means objectives.

The main difference between a network and a hierarchy in this case is that in a network, a lower-level objective can be connected to multiple higher-level objectives, while for a hierarchy, each objective is connected to only one other higher-level objective. An example of such a network (as opposed to a hierarchy) involving apartment choice is shown later in this section.

A particular objective could be a fundamental objective for one decision problem and yet a means objective for another problem, depending on the scope of the problem. That is, for a tactical decision problem, an objective that might be considered as fundamental could be considered as a means objective for a related strategic problem. An example of such a situation is provided by Keeney (1992, pp. 87–88) with respect to the objective of minimization of carbon monoxide emissions. Typically, attributes are associated with the objectives at the lowest level of the hierarchy or network.

The process of generating a hierarchy or network of objectives and associated attributes is subjective in nature. That is, two different reasonable DMs, or sets of DMs, could develop two different, yet reasonable, sets of objectives and associated attributes for a decision situation. Hopefully, different yet reasonable sets of objectives and associated attributes will lead to the selection of the same alternative as the best, as well as the same (or at least similar) rankings of the alternatives.

Having a good set of objectives (through the generation of a hierarchy or network) is almost a prerequisite for obtaining a good set of attributes, but the generation of each set go hand in hand with each other; however, just because one has a good set of objectives does not mean that one will necessarily have a good set of attributes.

Many techniques have been suggested for the formation of these hierarchies or networks. Most of these techniques are listed in Table 2.9. For example, MacCrimmon (1969) suggests three complementary approaches: (1) examine the relevant literature, (2) conduct an analytical study, and (3) perform causal empiricism. The first approach is especially useful when addressing systems/problems arising from the public/governmental sector. An example of the second approach would be to develop a model of the inputs, processes, and outputs of the system under study; the outputs and their categorizations could be used in the development of a hierarchy. Causal empiricism refers to talking with the DMs and stakeholders associated with the problem/system. A sensible approach for an analyst without much initial knowledge of the system would be to first examine any relevant literature and then interview DMs/stakeholders.

Buede (1986) suggests two systematic approaches: top down (or objective driven) and bottom up (or alternative driven). He notes that the top-down approach is the one that is most commonly espoused in the literature and is especially appropriate for strategic problems or decisions. This approach is

TABLE 2.9

A List of Approaches for Generating a Hierarchy or Network of Objectives

Device	References
Literature examination, analytical study, causal empiricism	MacCrimmon (1969)
Top-down versus bottom-up approaches	Buede (1986)
Listing of devices	Keeney (1992)
Specification, means-ends	Chapter 2 of Keeney and Raiffa (1993)
Categorization of stakeholders	—
Fundamental versus means objectives	Clemen and Reilly (2001)
Questions for moving upward/downward in hierarchy	Clemen and Reilly (2001)
Two-stage approach for generating objectives	Bond et al. (2010)

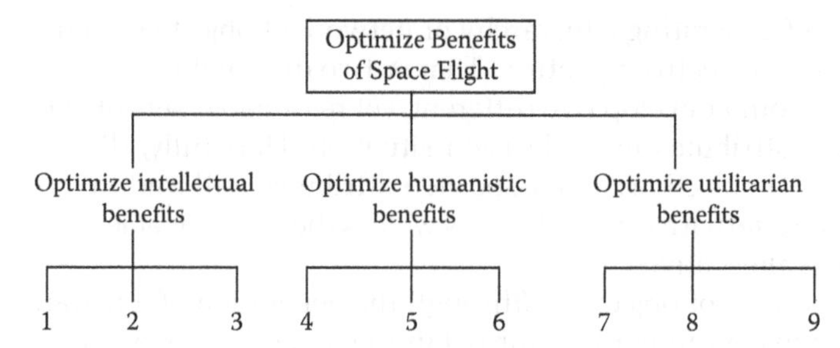

Where the numbers denote the following subobjectives:

1. Optimize mapping morphology and mineralogy of the moon, planets, and asteroids.
2. Optimize benefits associated with research beyond the solar system.
3. Optimize benefits associated with improving our understanding of astronomy, biology, geology, and meteorology.
4. Optimize benefits associated with improving domestic/global communications.
5. Optimize benefits related to understanding/controlling of the earth's climate.
6. Optimize benefits associated with understanding of the earth's upper atmosphere.
7. Optimize benefits associated with the research and development of industrial processes.
8. Optimize benefits associated with providing services such as satellite repair, maintenance, and refueling.
9. Optimize benefits associated with mining of minerals and other resources.

FIGURE 2.7
Hierarchy of objectives associated with the benefits of space flight.

also useful in the development of an objectives' hierarchy for an entire system, as opposed to for a particular decision situation.

As an example of an objective-driven hierarchy, during the 1980s, NASA utilized a database system called the Advanced Missions Information System, which contained information on various prospective missions to outer space over the next 40 years (Evans and Fairbairn, 1989). The objectives of this hierarchy could be broadly categorized into two groups: those relating to costs and those relating to benefits. Figure 2.7 shows part of the benefits hierarchy. Obviously, this hierarchy could be expanded even further.

The bottom-up (or alternative-driven) approach described by Buede is useful for tactical decisions in which the alternatives are fairly well defined. Basically, in this approach, one looks for differences between the alternatives in order to build up from the bottom of the hierarchy. For example, this approach might be useful for someone who was trying to decide which of three new cars to purchase. The cars might differ in fuel economy, maintenance costs, comfort, cargo capacity, and so on. In addition, Keeney (1992) notes that particular preselected alternatives may be used as prompts in order to generate fundamental objectives, and Butler et al. (2006) note that by reflecting on prespecified attributes, DMs can generate fundamental objectives. Each of these latter two approaches can be categorized as bottom-up procedures, as described by Buede (1986).

Keeney (1992, p. 57) lists the following devices as being useful in identifying relevant objectives:

- A wish list
- Problems and shortcomings
- Consequences
- Goals, constraints, and guidelines
- Different perspectives
- Strategic objectives
- Generic objectives
- Structuring objectives
- Quantifying objectives

For example, if one were developing a wish list associated with a finding a new job, one of the things on the list might be a nice office; this could be translated into the objective of optimize office amenities. Suppose for this same decision situation, the person's spouse is not happy living far from his or her parents. This could be translated into the objective of minimizing the distance of the job accepted from the spouse's parents' home.

In Chapter 2 of their book, Keeney and Raiffa (1993) suggest the use of *specification* and *means-ends* as devices in developing a hierarchy of objectives. Specification refers to stating in more detail what is meant by an objective and allows one to expand "downward" from an objective into several, more detailed, objectives. For example, in the hierarchy of Figure 2.7, a further specification of "humanistic benefits" would be "optimize benefits associated with improving domestic/global communications," among other objectives.

The concept of means-ends allows one to expand in either direction in a hierarchy; that is, to expand downward from an objective into several, more detailed objectives, one could ask the question: "What is the *means* by which this higher-level objective could be achieved?" To expand "upward" from an objective in the hierarchy, one might ask the question: "To what *end* do we want to achieve this objective?" Note that the answer and question for the "means" should interchange with the question and answer for the "ends" question.

As an example, in the hierarchy of Figure 2.7, the means by which one optimizes the humanistic benefits of space flight is by (1) optimizing benefits associated with improving domestic/global communications, (2) optimizing benefits related to understanding/controlling of the earth's climate, and (3) optimizing benefits associated with understanding of the earth's atmosphere. Conversely, the end associated with optimizing benefits associated with improving domestic/global communications is to optimize the humanistic benefits of space flight.

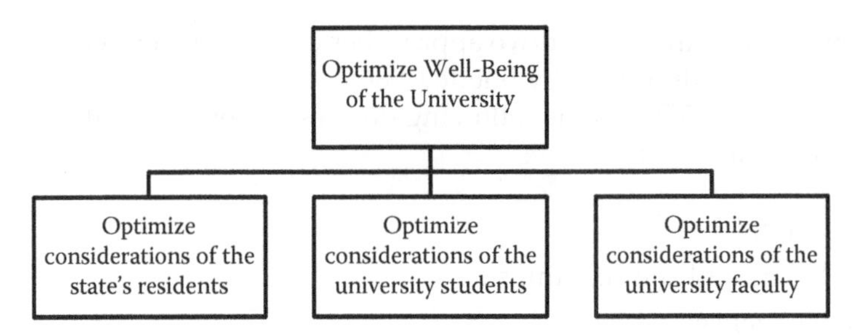

FIGURE 2.8
Partial hierarchy for considering strategic decisions at a public university.

Another way by which one can expand downward in a hierarchy is by considering the various stakeholders associated with a system. For example, in evaluating strategic decisions for a public university, one might want to consider all of its stakeholders, as shown in the partial hierarchy of Figure 2.8.

The hierarchy of Figure 2.8 might be further expanded by considering subcategories of stakeholders. For example, the state's residents might be subdivided into the taxpayers and corporate infrastructure.

In addition to the context-dependent categorization of objectives as described earlier, one could also categorize objectives according to short-term versus long-term and personal versus professional objectives (Bond et al., 2010).

Clemen and Reilly (2001) stress that fundamental objectives arise from one's values and appear toward the top of the hierarchy and that means objectives, as their name indicates, are the means by which the fundamental objectives are achieved. The attributes that are used to evaluate decisions should be directly related to the means objectives if possible. They also suggest asking the following questions for the development of the hierarchy:

To move downward in the hierarchy:

> *Ask*: "What do you mean by that?" or "How could you achieve this?"

To move upward in the hierarchy:

> *Ask*: "Of what more general objective is this an aspect?" "Why is that important?"

Note that asking these questions corresponds to the procedures discussed earlier.

Consider the problem of a college student who wants to select an apartment to rent in the city where he or she is attending college. Given that the city is of moderate size, he or she might have several alternatives from which to choose and therefore wants to develop a hierarchy of objectives and associated attributes to aid his or her choice.

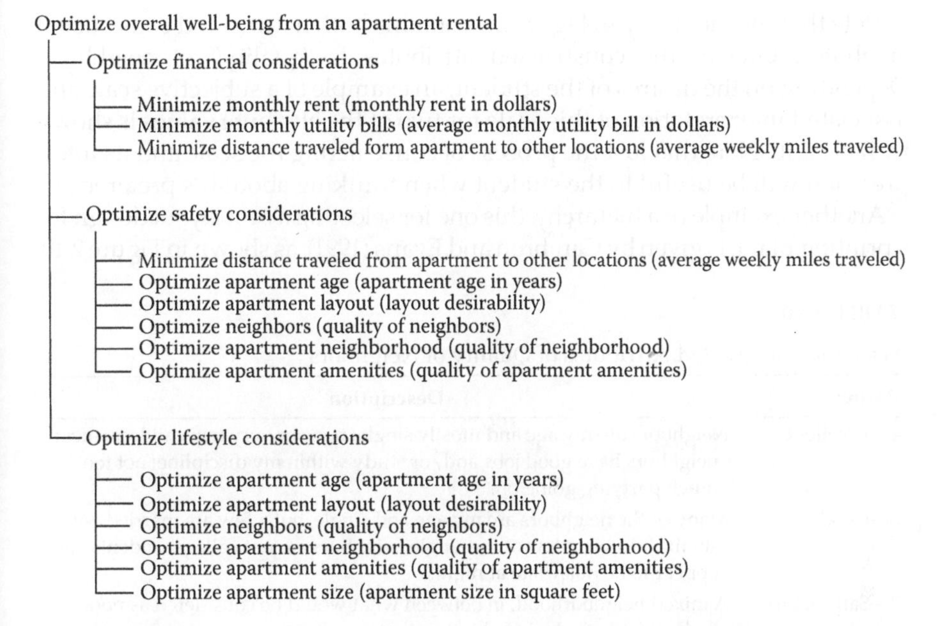

FIGURE 2.9
Hierarchy of fundamental objectives, means objectives, and attributes for an apartment rental decision.

Note that in Figure 2.9, the attributes for the means objectives are shown in parentheses. Also, note that a means objective can be associated with more than one fundamental objective. For example, the means objective "minimize distance traveled from apartment to other locations" can be associated with each of the fundamental objectives: (1) optimize financial considerations and or she (2) optimize safety considerations. The further the student travels, the more he or she has to spend on gasoline and maintenance for a private automobile or on fares for public transportation. In addition, the further he or she has to travel, the higher the chances of having an accident in that travel. In developing the hierarchy, note how the student would have given much thought as to how this means objective affected more than one fundamental objective, thereby affecting his or her thought process in the consideration of the value of this means objective.

Also, in Figure 2.9, note that the three, second-level, fundamental objectives indicate why the student is concerned about this decision situation. In addition, careful consideration of the "means objectives" in the figure can be helpful in the generation of new alternatives for consideration. For example, consideration of the means objective of "optimize neighborhood …" might cause the student to think about other neighborhoods that might have good apartments for rent.

Note that the hierarchy in Figure 2.9 contains both natural and constructed attributes. One of the constructed attributes is "quality" of neighbors. Depending on the desires of the student, an example of a subjective scale and associated interpretation of this scale for two of the attribute values is shown in Table 2.10. Note that just the process of constructing the scale and its interpretation will be useful to the student when thinking about his preferences.

Another example of a hierarchy, this one for selecting a best layout design for a printing plant, is given by Cambron and Evans (1991), as shown in Figure 2.10.

TABLE 2.10

Scale for Constructed Attributes of Quality of Neighbors

Rating	Description
4—Excellent	Neighbors are my age and mostly single; none or very few children; most neighbors have good jobs and/or study within my discipline; not too much partying going on.
3—Good	Many of the neighbors are my age and single, but a few are married with small children; a few of the single neighbors are not college students and appear to be noisy late at night.
2—Satisfactory	A mixed neighborhood, in between what would be considered as poor (below) and good (above).
1—Poor	Most neighbors are unemployed and/or not attending school; many small and noisy children in the neighborhood; much loud partying going on.

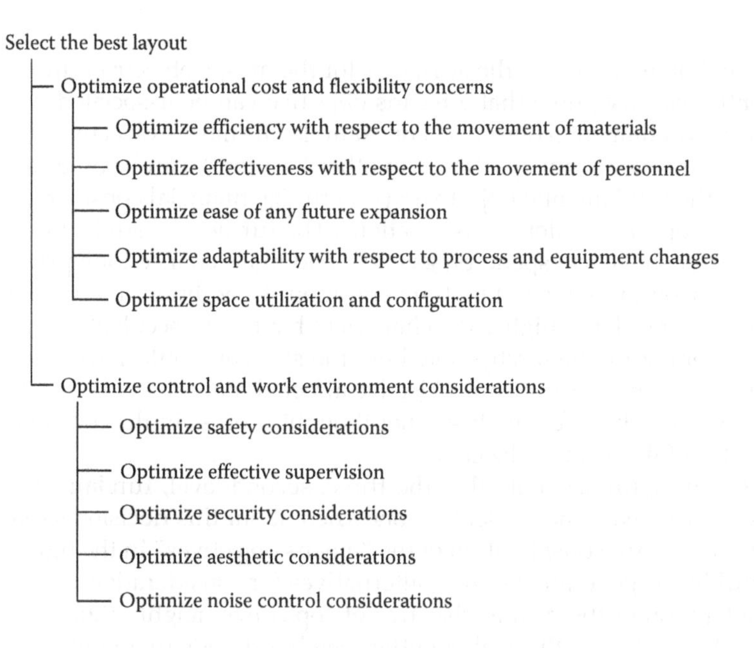

FIGURE 2.10
A hierarchy of objectives for layout design. (From Cambron, K. and Evans, G.W., *Comput. Ind. Eng.*, 20, 211, 1991.)

Note that this hierarchy could easily be expanded further, since many of the lowest-level objectives for this hierarchy are still at a fairly high level.

A good set of objectives for a specific decision situation must (1) cover all aspects of concern of the DMs and stakeholders, (2) be understandable to the DMs and stakeholders and therefore allow for careful consideration of trade-offs between these objectives, (3) have associated values (attributes) that are reasonably computable for any alternative under investigation, and (4) allow for differing outcomes for the various alternatives.

For example, selection of a staffing schedule for a fast-food restaurant that considered the cost of staffing, but not the waiting time of restaurant patrons, would not satisfy the first concern. As another example, design of an inventory policy in which one of the objectives involved minimization of average shortage cost per day may not satisfy the second concern if the DMs did not understand the concept associated with computation of an integral within a simulation model (see Kelton et al., 2015, pp. 257–270, for an example of a simulation model with this output). The third concern would not be satisfied if the construction of a model that would compute a value for the objective for any alternative under investigation required 3 months of effort and the decision needed to be made in 2 months. Finally, with respect to the fourth concern, if all alternatives resulted in the same value of an objective, then there would be no need to consider that objective.

Empirical research has indicated that without direction, DMs are not able to generate a substantial number of relevant objectives for any particular decision situation. The research also suggests two impediments to generation of most of the relevant objectives: (1) not thinking broadly enough about the possible objectives and (2) not thinking deeply enough about the situation (Bond et al., 2010). Since typically the objectives can be formed as a hierarchy (i.e., in categories), the first impediment corresponds to not thinking of entire categories of objectives, while the second impediment corresponds to not expanding the depth of the hierarchy.

As a result of their research, Bond et al. (2010) suggest a two-stage approach to the generation of objectives. The first stage involves just having the DM (or DMs) just generate a list of objectives for the decision situation. Following this stage, the analyst/facilitator would organize the objectives generated into categories (possibly a hierarchy) and then show this grouping to the DM(s) as a prelude to the second stage. The analyst/facilitator would then ask the DM(s) to add additional objectives to the list, by considering additional categories that may not have been considered in the first stage and by thinking more "deeply" about categories that had been considered. The analyst/facilitator would also set a goal of, say, about 100% more objectives for generation and also indicate the existence of research that shows that with additional effort, about this many more objectives can be added to the original list. See Keller et al. (2009) and Keeney (2012) for an additional discussion in this area.

Desirable characteristics associated with a good set of fundamental objectives are also given by Keeney (1992, pp. 82–86):

- *Essential*: Indicates outcomes in terms of the fundamental concerns of the DMs and stakeholders.
- *Controllable*: Outcomes associated with these objectives are affected only by the choice of alternatives.
- *Complete*: Covers all aspects of concern.
- *Measurable*: Define objectives precisely and specify degrees to which objectives are achieved.
- *Operational*: Values associated with the objectives can be computed within a reasonable time.
- *Decomposable*: Respective values associated with the objectives can be considered separately from one another by the DM(s).
- *Nonredundant*: Aspects associated with any objective are not included in another objective.
- *Concise*: Use as few objectives as possible.
- *Understandable*: The DM(s) implicitly understand the meanings of the objectives.

The reader should note that there are both trade-offs and complementary aspects associated with the characteristics described above. The most obvious example of a trade-off is with respect to the two characteristics of concise and complete—an analyst wants to use as few objectives as possible (i.e., be concise) while still covering all aspects of concern with respect to the decision situation (i.e., be complete). Hence, trade-offs need to be considered with respect to the set of characteristics listed earlier.

In addition to the examples presented in this section, the reader is referred to any of the following hierarchies found in the literature related to (1) cost and environmental concerns for the decisions associated with the scheduling of refueling for a nuclear power plant (Dunning et al., 2001); (2) for evaluating alternatives for the disposition of excess waste plutonium from dismantled nuclear weapons (Butler et al., 2005); (3) for the evaluation of various vision and mission statements of a software company (Keeney, 1999); (4) for investigating alternatives associated with the regional organization of the U.S. Army's Installation Management Agency (Trainor et al., 2007); (5) for selecting a geographic information system (GIS) (Ozernoy et al., 1981); (6) for ranking 542 hydroelectric projects in Norway (Wenstöp and Carlsen, 1988); (7) for relating objectives in the areas of cost, theft, environment, health, and safety with respect to the selection of a technology for the disposition of surplus weapons-grade plutonium (Dyer et al., 1998); and (8) for evaluating various types of plants for generating electricity (Keeney et al., 1986). In addition, Keller et al. (2009) discuss situations involving multiple stakeholders/DMs and objective hierarchies.

As noted earlier, attributes are used to measure the quality of an outcome that's associated with an objective. Typically, one will have one attribute associated with each of the lowest-level objectives in an objectives' hierarchy or network. In some cases, one might employ multiple attributes to measure a lowest-level objective.

Of course, all other things being equal, the best way to judge the quality of a set of objectives and attributes would be after the fact—that is by looking at the quality of the decision made. Since this is not possible prior to making the decision, one should assure that the attributes and objectives have certain characteristics.

In many cases, the attribute to be associated with an objective is an obvious choice. For example, in the design of a call center, if the objective is to "minimize the number of callers who must wait longer than 12 seconds to have their calls answered," then the obvious associated attribute would be the "number of callers who must wait longer than 12 seconds to have their calls answered." The difficulty arises when the objective is not so clear-cut such as an objective like "maximize service to callers."

Keeney and Gregory (2005) suggest that the formation of a hierarchy or network of objectives and the determination of the associated attributes be accomplished in a concurrent fashion. In particular, they suggest that a natural attribute should be used for an objective if there is an obvious choice; and in fact, if reasonable to do, one should consider several natural attributes and select the best of the alternative choices (i.e., another decision in itself).

If there is no good natural attribute for an objective, then the analyst should either (1) decompose the objective into "component objectives" and then proceed to the development of natural attributes for those component objectives or (2) consider the use of constructed attribute(s) for the objective.

In the case of the call center design, suppose that the initial objective considered is "optimize service to callers." Since there is no obvious natural attribute for this objective, one might consider dividing this into component objectives: "minimize number of callers who hang up prior to their call being answered" and "minimize waiting time for callers who reach the system." Assuming that a simulation model is being constructed to analyze different staffing schedules for the call center, then two natural attributes for these objectives might be "number of callers who hang up prior to their call being answered" and "sample mean waiting time for callers who reach the system" (see Figure 2.11). Note that even in the case for the second objective—"minimize waiting time for callers who reach the system"—the attribute to use is not necessarily obvious. For example, one might consider as an attribute "the maximum waiting time over all callers for some period of time" instead of the mean waiting time. In addition, it may be desirable to attach different values to the waiting times for different categories of callers (see version 3 of the call center model in Kelton et al. [2015, Chapter 5] for an example).

Guidance associated with the selection of attributes is given in Keeney and Raiffa (1993, Chapter 2), in Clemen and Reilly (2001), and in

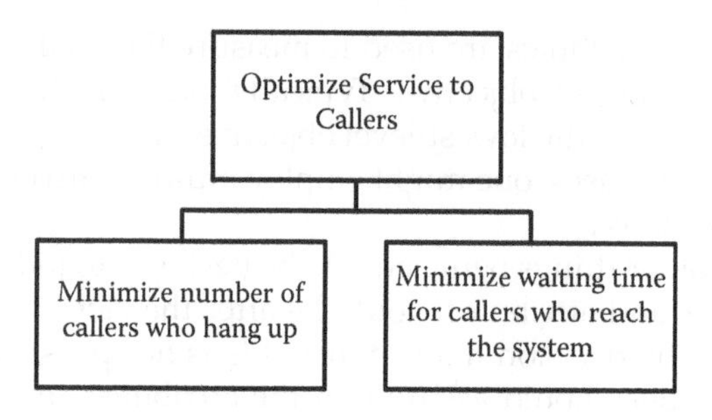

FIGURE 2.11
Sample portion of objective–attribute hierarchy for the determination of a staffing schedule for a call center.

Keeney (1992, Chapter 4). In particular, Keeney notes that if the fundamental objectives are carefully chosen, then the corresponding attributes should be *measurable*, *operational*, and *understandable*.

Additional references for the characteristics of a good set of attributes are Keeney and Raiffa (1993) and Keeney and Gregory (2005). For example, Keeney and Raiffa suggest that a set of attributes should have the following characteristics: completeness, operability, decomposability, lack of redundancy, and small size with respect to their number.

Completeness means that the set of attributes for a problem should cover all important aspects of the problem. For example, a set of attributes involving only economic factors for a decision problem involving the setting of the speed limits for interstate highways would be incomplete, since such a set would not address the safety factors involved (e.g., number of automobile accidents, number of highway fatalities).

Operability basically means that it is possible (or at least not too difficult) to compute values for the various attributes as a function of each feasible alternative. For example, consider the problem of where to locate fire stations in a metropolitan area. An attribute such as annual property damage associated with the placement of fire stations might not meet this criterion of operability because of the difficulty associated with constructing an accurate model of the relationship between annual property damage and the placement of fire stations. In addition, operability means that the attributes used are understandable (or at least easily explainable) to everyone involved in the decision-making process.

Decomposability refers to the idea that when one has a large set of attributes for a problem, the assessment of the preference structure of the DM can be decomposed over subsets of the larger set of attributes.

Nonredundancy refers to the idea that there is no overlap among the attributes so that there is no double counting of effects.

Small size refers to the fact that the number of attributes should be "as small as possible." There is usually a trade-off involved here, that is, the larger the number of attributes, the more accurate the decision-making process will be. However, the larger the number of attributes, the more effort will be required in determining attribute values for each of the alternative, as well as ranking the alternative in terms of the attribute values. The bottom line is that one should consider whether or not a different alternative will be chosen in considering whether or not to add a particular attribute to the mix. Of course, it's difficult to know the answer to this question until the analysis is finished.

Keeney and Gregory (2005) note that the attributes for a decision situation should have the following characteristics:

- *Unambiguous*: The relationship between the outcomes and their respective descriptions using the various possible values of the attributes is clear.
- *Comprehensive*: The ranges of the attributes cover the full range of outcomes, and any implied value judgments obtained from the attribute values are well understood.
- *Direct*: The various levels of the relevant attributes are sufficient to represent the outcomes for the fundamental objectives of interest.
- *Operational*: The levels of the attributes associated with the alternatives can be reasonably obtained, and value trade-offs can be made by the DM(s).
- *Understandable*: Outcomes and value trade-offs made using the attributes can be readily understood and clearly conveyed.

Obviously, there is dependence and trade-offs between and among these characteristics. Keeney and Gregory (2005) provide several examples and counterexamples for each of these characteristics. In addition, one can see a close relationship between the desirable characteristics associated with the aforementioned attributes and those desirable characteristics associated with fundamental objectives noted earlier.

2.7 Important Attributes for Specific Categories of Decision Situations

There are several areas of industrial engineering that employ the use of attributes, and these attributes are specific to those respective areas. In this section, we list and discuss some of these attributes. In some cases, we just refer the reader to other sources, while in other cases, we discuss these measures in some detail. Of course, almost all areas of decision analysis in industrial engineering involve some categories of cost, be they operating, capital, maintenance, or other categories.

A summary listing of the attributes presented in the following sections is provided in Table 2.11.

TABLE 2.11

Attributes for Specific Areas of Applications

Area of Application	Selected Attributes
Engineering Economics	Net present value
	Rate of return
	Internal rate of return
	Payback period
Location and Layout Design	Cultural aspects of country
	Construction costs
Quality Management	Acceptable quality level (AQL)
	Lot tolerance percent defective (LTPD)
	Producer's risk (α)
	Consumer's risk (β)
	Average outgoing quality (AOQ)
Project Management	Direct cost
	Indirect cost
	Project duration
	Weighted sum of the sums of squares of the resource usages
Health Care	Quality-adjusted life years (QALYs)
	Sensitivity (of a medical test)
	Specificity (of a medical test)
	Fraction of patients who leave (or left) without being seen (LWOBS)

2.7.1 Engineering Economics

Engineering economics is employed in many decision situations of industrial engineering. For example, in the area of project management, engineering economics can be used to choose a project to undertake from among several projects. Important performance measures (or attributes) used in engineering economics include *net present value, rate of return, internal rate of return,* and *payback period.* These performance measures are used to consolidate cash flows (both incoming and outgoing) over time into a single measure of performance. Each of these respective measures has their own advantages and disadvantages vis-a-vis each of the other measures. For example, the net present value calculation requires as input an interest rate that corresponds to the interest that the organization can receive from a general investment; the interest rate used will affect the net present value.

In many industrial engineering decision situations, one can map alternatives into outcomes that are defined by cash flows over time, along with attributes that may not be easily quantifiable, such as an organization's position in the marketplace, an organization's reputation, improvement in skill level of an organization's workforce, improvement in the level of safety, improvement in traffic flow, and so on. In such cases, the outcomes are multidimensional, involving a measure of economic performance, such as net present value, and constructed attributes such as one or more of those just listed.

2.7.2 Location and Layout Design

Organizations remain competitive or even enhance their competitive position through the development of new facilities. These facilities could be oriented toward production, service, or storage/warehousing among other purposes. In all cases however, decisions must be made regarding location, capacity, and layout.

With respect to location, decisions can be categorized by country, region, and site (Heizer and Render, 2006, p. 313). In each type of decision—country, region, and site—a variety of criteria can be considered, some of which would be described by natural attributes, while others by constructed attributes. For example, with respect to the decision of the country for location, one would want to consider cultural aspects of the respective countries under consideration. Such cultural aspects might best be considered through the use of a constructed attribute. On the other hand, decisions related to region and site location would depend upon land and construction costs, which would most likely be represented through the use of a natural attribute.

Once a location decision has been made, a decision must be made with respect to a facility's layout. Layout design is concerned with the placement/location of machines, equipment, and departments as well as the amount of space allocated to various departments. The objectives and criteria associated with these decisions and how these objectives are traded off between each other depend upon the type of facility (e.g., warehouse, manufacturing system, retail, office) being designed. Manufacturing systems could be laid out according to a process orientation, a product orientation, a work cell orientation, or some combination of these.

As with many complex decision problems, a combination of both natural attributes and constructed attributes would typically be employed in the decision process.

2.7.3 Quality Management

As noted by the American Society of Quality (see: http://asq.org/glossary/q.html), *quality* can have one of either two meanings: (1) "the characteristics of a product or service that bear on its ability to satisfy stated or implied needs" or (2) "a product or service free of deficiencies." Having products and services of high quality is of obvious importance to the bottom line of any organization. Hence, much effort is spent by most organizations in achieving such high quality.

Total quality management (TQM) addresses an entire organization's management of quality, from the suppliers and vendors to the customer, including any third-party logistics operators that aid the organization. There are many tools and methodologies employed in TQM, and each of these requires inherent decisions in their implementation. Examples of such tools include cause-and-effect diagrams, Pareto charts, process control charts, and acceptance sampling plans.

Most acceptance sampling plans involve *attributes* inspection, as opposed to *variables* inspection, in which some quantity (e.g., weight) of an item is measured. The most common type of attribute inspection is one in which an item is classified as either "good" or "defective." In single sample acceptance, sampling a randomly selected sample of items of size **n** (a decision variable) is inspected from a lot of items; if the number of the defective items in the sample is less than or equal to **c** (a second decision variable), then the defective items are repaired and the lot proceeds; if the number of defective items in the sample is greater than c, then the remainder of the lot is inspected and every defective item found in the lot is repaired. The number n is called the sample size, and the number c is called the acceptance level.

The operating characteristic (OC) curve associated with a particular acceptance sampling plan (i.e., specific values for n and c) is an X–Y graph with percent defective of the lot on the X-axis and probability of lot acceptance (i.e., probability of the number of defective items in the sample being less than or equal to c) on the Y-axis (see Figure 2.12).

The *acceptable quality level* (AQL) for a lot is the maximum percent defective allowed for a lot considered to be a *good lot*. On the other hand, the *lot tolerance percent defective* (LTPD) is the minimum percent defective for which a lot is considered to be a *bad lot*.

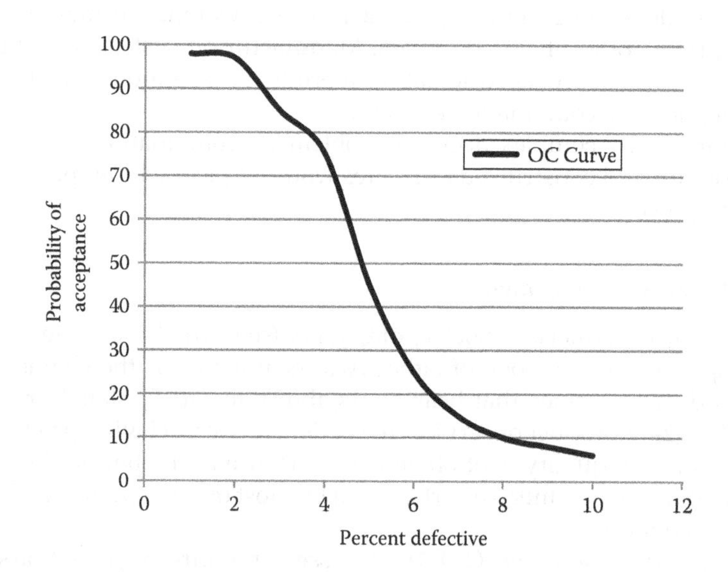

FIGURE 2.12
Operating characteristic curve for a particular sampling plan.

In acceptance sampling, there are two parties involved: the producer of the lot and the consumer of the lot. The *producer's risk* (typically denoted as α) associated with a particular plan is the probability that a *good lot* will be rejected, while the *consumer's risk* (typically denoted as β) associated with a plan is the probability that a *bad lot* will be accepted.

The *average outgoing quality* (AOQ) for a lot is the average percent defective of the lot after it has passed through the acceptance sampling plan. Keep in mind that if a lot is rejected by the plan, then its outgoing quality will be 0% (percent defective), since the entire lot will be inspected and all defective items will be repaired.

2.7.4 Project Management

A *project* can be defined as a set of activities related to the achievement of some planned objective, and "*project management* involves the coordination of group activity wherein the manager plans, organizers, staffs, directs, and controls to achieve an objective with constraints of time, cost, and performance of the end project" (Moder et al., 1995, p. 3). Related to the area of project management is *project selection* in which one or more projects are to be selected to undertake from among a group of projects.

Both project management and project selection involve many different types of decision situations. Project selection involves many aspects associated with engineering economics and therefore can employ attributes such as net present value, payback period, and internal rate of return, among others.

Two important and interrelated aspects of project management are scheduling and resource allocation. Both of these decision situations in a project involve the use of a project network, in which the individual activities of a project are represented as either arcs (for an activity-on-arc project network) or nodes (for an activity-on-node project network) in a network.

The decisions to be made in scheduling and resource allocation include (1) the amounts of various types of resources to allocate to each activity (which implicitly affects the activities' respective durations) and (2) when to start and finish each activity (which implicitly affects the resources' usages over time as well as the project's duration). The first problem is often called the *time–cost trade-off problem* in project management and is addressed by the classical critical path method. The second problem is often termed *resource-constrained project scheduling.* These decisions employ performance measures (or attributes) of *direct cost, indirect cost, project duration,* and a measure of how *level* the usages of the resources are over time.

Direct costs are those costs that can be directly associated with a respective individual activity of a project; examples of these costs would be direct labor, equipment, and material used to perform the activity in question. As such, the direct costs for an activity will be inversely proportional to the duration of that activity; that is, as an activity is shortened in duration, its direct cost will increase. For example, an activity in a project that would normally require 6 weeks of duration might be shortened to 5 weeks by increasing the overtime effort and thereby increasing its direct cost by $5000. Note that the relationship between an activity's duration and its direct cost need not, and will probably not, be linear.

Indirect costs are those costs that cannot be allocated to individual activities of a project but to the project as a whole. As such, they are proportional to the duration of the entire project—that is, the longer the duration of the project, the larger the indirect costs associated with that project. Examples of indirect costs are interest costs on the project investment, administrative and other overhead costs, penalties associated with completing the project late, and bonuses (which function as a negative cost) associated with completing the project early. In some cases, a DM may very well choose to employ project duration as a substitute attribute for indirect cost. In doing this, the DM will implicitly consider all aspects of indirect cost as well as nebulous considerations such as an organization's reputation as impacted by early/late completion of a project. See Figure 2.13 for an example of the relationships between project duration and indirect cost and project duration and direct cost.

Two types of resource-constrained scheduling problems are addressed in projects: (1) resource leveling and (2) fixed resource limit scheduling. In the first problem, the project manager attempts to schedule the activities of the project in such a way that the resource usages over time are leveled out as much as possible, subject to a constraint on the maximum project duration.

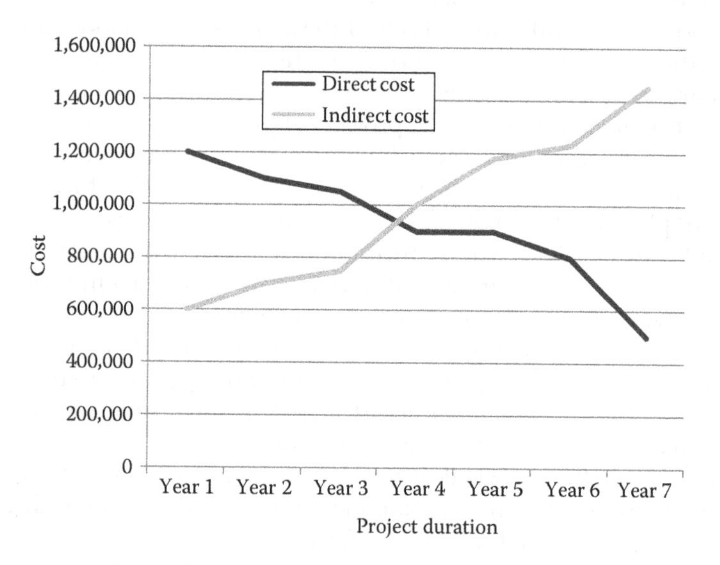

FIGURE 2.13
Relationships between project duration and direct/indirect costs of a project.

In the second problem, the project manager attempts to schedule the activities so as to minimize the project duration subject to constraints on the amounts of resources available over time.

In both problems, a level usage of the various resources employed in a project is a desirable situation, since this helps the organization in the avoidance of intermittent hiring and laying off of employees. In determining how level the resource usages are over time for a particular project schedule, a project manager might just view *resource-loading diagrams* that are graphs that depict the resource usages over time; however, if some type of algorithm is used to solve the resource leveling problem, a measure is needed for the "levelness" of the resource usage over time. One measure typically used is a weighted sum of the sum of squares of the resource usage over each time period of the project.

As a simple example to illustrate this performance measure, consider a project with duration of 3 weeks under each of two different schedules: Schedule 1 and Schedule 2. Two resources are employed in the project: Resource A and Resource B. The resource usages over each of the 3 weeks of the project for each of the two schedules are shown in Tables 2.12 and 2.13, respectively.

Note that both schedules give the same amounts of overall resource usages: 30 person-weeks for Resource A and 18 person-weeks for Resource B. This will always be the case for this type of problem.

Suppose that the "weights" attached to Resource A and Resource B are 0.8 and 0.2, respectively, indicating that Resource A is much more important to level than Resource B (note that these weights might very well be scaled by levels of usage).

TABLE 2.12

Resource Usage for Schedule 1

Week	Resource A Usage	Resource B Usage
1	10	6
2	10	6
3	10	6

TABLE 2.13

Resource Usage for Schedule 2

Week	Resource A Usage	Resource B Usage
1	8	2
2	10	10
3	12	6

Obviously, Schedule 1 provides a much more level usage of both of the resources. However, in many cases, it will be difficult to determine which schedule would provide the more level usage of resources. In addition, an algorithm for resource leveling would need to employ a single measure (attribute) of performance.

The weighted sum of the sum of squares for the resource usages is given by

$$\text{Schedule 1: } .8\,(10^2 + 10^2 + 10^2) + .2\,(6^2 + 6^2 + 6^2) = 240 + 21.6 = 261.6$$

$$\text{Schedule 2: } .8\,(8^2 + 12^2 + 10^2) + .2\,(2^2 + 10^2 + 6^2) = 246.4 + 28 = 274.4$$

Since Schedule 1 gives the smaller weighted sum of squares, it would be the preferred schedule in terms of leveling of the resource usage.

In many cases, the resources for a project might be categorized by the type of labor used. In order to simplify the analysis, these categories can be summed and the resource usage can be defined in terms of person-hours of labor for each week or month of the project.

2.7.5 Medical Decision Making and Health Care Management

There are many diverse areas in health care that have greatly benefited from the contributions of industrial engineers. These areas include medical decision making, scheduling and resource allocation in medical clinics and hospitals, design and operation of hospital supply chains, planning for pandemics, and so on. These areas employ many performance measures or attributes that are similar to those found in the classical application areas of industrial engineering. These include such measures as patient waiting times and flow times, physician utilization, nurse utilization, and so on.

Important attributes that are fairly unique to health care though are *quality-adjusted life years* (QALYs), *sensitivity* of a medical test, *specificity* of a medical test, and fraction of patients who *leave (or left) without being seen* (LWOBS). The first three attributes (QALY, sensitivity, and specificity) are used in medical decision making, while the last attribute is mainly used as a performance measure for hospital emergency departments.

The attribute of QALY is used to evaluate different outcomes for medical treatments in which these outcomes correspond to differing qualities of life. The "quality of life" is measured on a scale from 0 to 1 in which 0 corresponds to death and 1 corresponds to perfect health. A higher value on the scale corresponds to a better quality of life. Then QALY is defined as the integral over time of the quality of life. That is, if f(t) is the quality of life at time t, then the QALY from time t_0 to time t_1 where $t_1 > t_0$ is given by

$$\text{QALY}(\text{over the period from } t_0 \text{ to } t_1) = \int_{t_0}^{t_1} f(t)\,dt$$

As an example of the use of QALY, consider a patient in a wheelchair who has a choice between two alternatives:

1. Surgery that has a .9 probability of curing the patient and allowing her to live in perfect health for 30 more years and a .1 probability of resulting in death
2. No surgery that will allow the patient to stay in the wheelchair for 30 more years with a quality of life of .7

(Note that we are ignoring obvious complications in the modeling such as the possibility of a better type of surgery occurring over the next 30 years, or other things in the future that might cause death.)

The expected value of QALYs for the alternative of surgery is given by

$$.9\,(1(30)) + .1\,(0(0)) = 27$$

while the expected value of QALYs for the alternative of no surgery is given by

$$1(.7(30)) = 21$$

Hence, under the criterion of maximization of the expected value of QALY, the patient would choose the alternative of "surgery."

The quality of life associated with a particular medical condition is obviously dependent on the specific person evaluating the condition. For example, the quality of life for a person with carpal tunnel syndrome would be less for a concert pianist than it would be for someone who is not a concert pianist. Because of this subjective viewpoint associated with QALY and other reasons, QALY is rarely used for medical decision making with respect to individuals; instead, its more common use is in decision making with respect to interventions that affect entire groups of people.

Both sensitivity and specificity of a medical test refer to *conditional probabilities* associated with that test. In particular, sensitivity is the probability that the test will be positive given that the disease is present (remember that a positive medical test indicates that a disease is present in an individual). Specificity is the probability that the test will be negative given that the disease is not present. The closer that each of these conditional probabilities is to 1, the better the medical test is.

As an example, let's suppose that 1000 people are given a medical test that has a sensitivity of .96 and a specificity of .9. Suppose also that we know that 950 people of these 1000 do not have the disease for which the test is being given and that 50 people of these 1000 do have this disease. (Of course, in an actual situation we will not know the exact numbers of people who do and do not have the disease, but we may very well know the approximate values.) Then the *expected numbers* of people tested in various categories will be as follows:

- Number of people who test negative who do not have the disease = .9 * 950 = 855
- Number of people who test positive who do not have the disease = .1 * 950 = 95
- Number of people who test positive who do have the disease = .96 * 50 = 48
- Number of people who test negative who do have the disease = .04 * 50 = 2

Note that the actual numbers may very well be different from the expected values. However, the expected values may very well be enough information for decision-making purposes.

An important output to notice is that the number of people testing positive that *do not* have the disease is almost twice as many as the number testing positive that *do* have the disease. This can be explained by the large number of people in the population of 1000 tested that do not have the disease (950) versus those that do (50). See Hunink et al. (2001) for a more detailed discussion of medical decision making.

As mentioned, the number of patients (or fraction of arriving patients) who leave without being seen is an important performance measure for hospital emergency departments. Many of these cases are walk-in patients who arrive at the emergency department for nonemergency medical treatment. The LWOBS attribute may be considered with other attributes such as patient waiting time (which might be subdivided by time in the waiting room and waiting time in the actual emergency room [ER] once past the waiting room), patient flow time, physician utilization, nurse utilization, and so on.

Since LWOBS is likely to be highly correlated with patient waiting time, one might ask the reason for considering both of these attributes in the operation of emergency departments. One reason for considering these two attributes separately is that LWOBS would more likely apply to those walk-in patients arriving for nonemergency medical care; in other words, the use of both measures might be an alternative to the categorization of patients.

Patients in an ER who leave without being seen are analogous to *reneging customers* in a service system. Including this aspect of a system in a simulation model of the system may not be straightforward. For example, whether or not a waiting patient or customer leaves a system may depend upon both the number of patients/customers in front of him or her and the amount of time that he or she has already spent waiting. Both of these variables (place in waiting line and time already spent waiting) may be random in nature. Kelton et al. (2015, pp. 386–395) provide material on the simulation modeling of this aspect of a system using the Arena™ simulation software package.

Material Review Questions

2.1 Give two alternative definitions for "problem."

2.2 The "gap" referred to in one of the definitions for problem refers to a gap in what?

2.3 What are the three types of problems identified by Simon?

2.4 As noted by Ackoff, most problems encountered in "real life," as opposed to those encountered in textbooks, are of which type?

2.5 What are the terms used by Ackoff (1979), Rittel and Webber (1973), and Schon (1987) for ill-structured problems?

2.6 According to Larson, what percentage of the value of operations research lies in its contribution to the correct framing and formulation of the problem?

2.7 According to Keeney, which approach is used more often: alternative-focused thinking or value-focused thinking?

2.8 Problem structuring should occur prior to the other steps of the problem solving process (*true* or *false*).

2.9 Give a definition of "problem structuring."

2.10 What is a "problematique?"

2.11 What is "divergent thinking?"

2.12 What are characterizing features of ill-structured problems?

2.13 Explain why problem structuring methods should be "participative and interactive" in nature?

2.14 What are the four "patterns of thinking" associated with the Kepner and Tregoe method?

2.15 What are the seven core principles of "Breakthrough Thinking?"

2.16 What are the four elements associated with a "purpose" in Breakthrough Thinking?

2.17 A purpose hierarchy can allow for the generation of alternatives that are not readily apparent (*true* or *false*).

2.18 What is a "systems matrix" as described by Nadler and Hibino?

2.19 What is a difficulty that often occurs when "too much" data are available for analysis?

2.20 List two of the advantages of allowing for various perspectives in the problem structuring process (as associated with the people design principle).

2.21 What is the phrase used to start all problem statement questions within the "Why–What's Stopping" technique?

2.22 What is the main reason for using the phrase alluded to in Question 2.21?

2.23 Answers to the "why" question in the Why–What's Stopping technique will lead to a higher-level or a lower-level perspective on a problem (choose one).

2.24 Answers to the "what's stopping" question in the Why–What's Stopping technique will lead to a higher-level or a lower-level perspective on a problem (choose one).

2.25 What statements would correspond to the highest and lowest levels of a problem network in the Why–What's Stopping technique?

2.26 Briefly define the following terms:

- Criteria
- Values
- Mission statements
- Vision statements
- Objectives
- Goals
- Constraints
- Attributes

2.27 What are the three important elements that should be contained in an organization's mission statement?

2.28 Most decision situations involve conflicting objectives (*true* or *false*).

2.29 What are the two dimensions along which attributes can be categorized?

2.30 Constructed attributes are "subjective" in nature (*true* or *false*).

2.31 What would typically be the main advantage associated with the use of a proxy attribute?

2.32 At what level of an objective–attribute hierarchy is a constructed attribute more likely to occur?

2.33 What are the three complementary approaches suggested by MacCrimmon for generating a hierarchy of objectives?

2.34 What types of problems or decisions are best suited for a top-down approach with respect to the generation of a hierarchy of objectives?

2.35 What types of problems or decisions are best suited for a bottom-up approach with respect to the generation of a hierarchy of objectives?

2.36 What are the differences between a fundamental objective and a means objective?

Exercises

2.1 Recall a situation where you encountered an ill-structured problem. Describe why this problem was ill structured, as well as the solution approach you used. Did you start thinking of solutions right away or did you employ problem structuring techniques (even if you did not know them as such)? Based on the information gained in this chapter, describe how you might have approached the problem differently.

2.2 Describe at least one ill-structured problem faced by each of the following organizations. What are the characteristics of this problem that makes it ill structured?

a. The neighborhood/apartment complex in which you live
b. The metropolitan area in which you live
c. The academic department to which you belong

2.3 Consider a problem faced by many students at a commuter university: "How might we decrease the parking problems for students attending classes or other activities on campus?" Using the "Why–What's Stopping" technique, develop a problem map originating from this initial problem statement.

2.4 Consider the problem map associated with Figure 2.2. Develop several problem statements associated with answers to the question:

"What's stopping us from spending less money on reworking?"

Compare these newly generated problem statements to the ones already in Figure 2.2.

2.5 Analyze the mission statements of three different organizations. Discuss how these statements reflect the values of these organizations.

2.6 Suppose that a medical patient has a situation in which he or she can have an operation and thereby have a 90% chance of living for 10 years with a "quality of life" = .7; there is a 10% probability that he or she will die during or immediately after the operation. If he or she does not have the operation, he or she can expect to live 5 years with a "quality of life" = .7. Assuming that the patient's criterion is to maximize the expected value of QALY, compute the expected value of QALY for each decision. Which decision should be taken under this criterion?

2.7 Identify the following attributes as being either natural or constructed in nature:

a. Prestige
b. Cost
c. Appearance
d. Pain
e. Reputation

2.8 Using a "top-down approach" for the development of a hierarchy of objectives and attributes is more appropriate for what type of problem (choose one):

a. A strategic problem
b. A tactical problem

2.9 Select one of the hierarchies from the literature cited in this chapter, for example, Dunning et al. (2001), Butler et al. (2005), Keeney (1999), Trainor et al. (2007), Ozernoy et al. (1981), Wenstöp and Carlsen (1988), Dyer et al. (1998), or Keeney et al. (1986). Write a brief report relating one or more of the methodologies for generating hierarchies to the hierarchy selected from the literature.

3

Making Decisions under Conditions of Certainty with a Small Number of Alternatives

3.1 Introduction

In this chapter, we discuss various solution techniques for decision opportunities involving *no risk or uncertainty, multiple objectives, and only a few alternatives.* These problems fall into the category of *multiple criteria discrete alternative* problems (Wallenius et al., 2008) with the additional qualification that we are dealing with problems involving certainty. Of course in real life, one always has uncertainty associated with the outcome of a decision, because the decision is to be made in the future, and the results of the decision will be known in the future; in addition, and more to the point of the techniques discussed later in this book, many of the inputs and relationships associated with a model of the decision situation would be uncertain in nature.

The key here though is that we are *modeling* the situation as if there is no uncertainty. For example, it may be that the amount of uncertainty is so small that it is not worth considering, or it may be that the decision is to be made over and over again and that the variation associated with the outcome of a particular decision is not of much importance—the important aspect of the outcome is the expected values of the attributes, and these expected values can be determined with precision. The significance of having only a few alternatives to consider is that we can evaluate every single alternative under consideration—no sophisticated optimization technique is needed.

Another reason for considering the types of methods discussed in this chapter is that one might not have the time available to build a model, which considered uncertainty. Hence, a first-cut, deterministic model is employed as an aid in choosing an alternative.

The types of decision situations discussed in this chapter are important to industry (both manufacturing oriented and service oriented), government, and individuals. The discussion here applies to all situations involving multiple objectives.

Examples of relevant industrial decision situations include the following:

1. One location from among five alternatives must be chosen for a new production facility. Attributes that might be considered in making this decision might be related to distances from major suppliers, distances from customers, government incentives, quality of prospective employee base, and so on.
2. One layout must be chosen from among 10 alternatives for a new production facility. Attributes that might be considered could be related to the amount of material flow, safety, and so on.
3. A hospital administrator must choose a schedule for the personnel who work in its emergency department. Attributes could include expected waiting time for patients, personnel cost, and employee satisfaction.
4. A manufacturing firm must determine the parameters for the acceptance sampling plan for an important part for a major subassembly of one of its main products. The important attributes in this situation might include the cost associated with implementation of the plan, and the quality of an outgoing lot, as measured by the number of *good* (as opposed to defective) parts in an average lot, after the inspection process.
5. A fast-food restaurant must choose a work schedule for its personnel. Such a schedule can involve assignment of personnel to particular jobs in the restaurant as well as assignment of personnel to particular schedules (e.g., Monday through Friday, 11 a.m. to 2 p.m., or Saturday and Sunday, 9 a.m. to 5 p.m.). Important attributes to be considered in this situation might relate to factors such as waiting time of customers, personnel costs, and employee satisfaction with the work schedule.
6. In addressing its generation planning problem, which involves the number and type of electric-generating facilities to construct over time, an electric utility must consider trade-offs between cost and reliability for its future system (Moskowitz et al., 1978).

Examples of relevant situations related to governmental decision making could include the following:

1. The federal government must determine the CAFE (Corporate Average Fuel Economy) requirements for the nation's auto companies. Important attributes associated with this situation could involve the amount of petroleum used in a year and safety, as measured, for example, by the expected number of deaths on the nation's

highways. (Note that the latter consideration comes into play with the fact that stricter mandates with respect to fuel economy could imply the production of more small cars, which might be less safe in the case of a traffic accident.)

2. Related to the earlier example, decisions must be made on the maximum speeds allowed on the nation's highways. Attributes of importance here could relate to the amount of petroleum used in a year, travel times, and safety.
3. A large metropolitan area, spanning a major river, must make a decision on the location of a new bridge spanning the river. This decision would involve local, state, and federal government agencies as decision makers (DMs). Important attributes in this situation would relate to economic development, travel times, safety, relocation of homeowners, and ecological considerations.
4. The federal government must make a decision related to the location of a nuclear waste facility. Obvious considerations would include cost and safety.
5. NASA must make many decisions related to the design of lunar habitation with respect to an upcoming manned mission to the moon. Obvious considerations with respect to the development of lunar habitation would relate to safety, cost, and the development time required for the various components of the habitation.

Examples of decision making related to individuals would include the following:

1. A high school senior must choose a college to attend next year. Important considerations might include cost for tuition and living expenses, quality of the education, number of miles from home, and quality of the social life.
2. A graduate of an industrial engineering department must choose from among five different job offers. Important considerations here could involve annual salary, opportunity for advancement, fringe benefits, number of miles from home, and amount of travel required by the job.
3. An individual must choose a new car to purchase. Important considerations would be the purchase price of the car, expected annual maintenance costs, fuel economy, amount of prestige, safety, and resale value at some point in the future.

In Section 3.2, we present notations that will be used throughout the chapter. Included in this section is a simple hypothetical example involving a decision situation in which a student must choose between several job offers. In Section 3.3, we discuss concepts and techniques that do not require the elicitation of information concerning the decision maker's trade-offs between the objectives. In particular, the concepts of the *ideal*, the *negative ideal*, the *superior*, and *nondominated solutions and outcomes* are defined.

In Section 3.4, we discuss various *scales of measurement* (referring to the measurement of attributes) and *preference structures*. Starting with Section 3.5, we present methodologies that require information from the decision maker(s) about how he or she (they) "trade off" between the attributes. In particular, in Section 3.5, we present the relatively simple concept of *lexicographic ordering*.

Sections 3.6 and 3.7 are two of the most important sections of this chapter, involving basic concepts associated with *multiattribute value (MAV) functions* and their assessment, respectively. In Section 3.8, we present a simpler approach to the assessment of a multiattribute value function, entitled the "Simple Multiattribute Rating Technique." Section 3.9 provides some final discussion on MAV function assessment.

Sections 3.10 and 3.11 provide discussions of alternative approaches: the Technique for Order of Preference by Similarity to Ideal Solution (TOPSIS), and the analytic hierarchy process (AHP). Within Section 3.11, a detailed example of the use of the AHP for the design of a call center is given and Section 3.11.3 discusses criticisms of the AHP.

Finally, Section 3.12 discusses outranking techniques and Section 3.13 discusses extensions, hybrid approaches, and comparisons of the various methods of Chapter 3.

3.2 Notation

The basic problem is a special case of the problem that was defined in Chapter 1, in which there is only one state of nature. That is, we are assuming that there is a set of mutually exclusive, feasible *alternatives* (or *solutions*) denoted as $A_1, A_2, \ldots, A_n$ where n is a relatively small number. The set of *attributes* is denoted as $X_1, X_2, \ldots, X_p$ where p is the number of attributes, and $x_k(A_i)$ is the attribute value associated with attribute X_k as a function of alternative A_i where $k = 1, \ldots, p$ and $i = 1, \ldots, n$. $x(A_i)$ is defined as the *outcome* (or consequence) associated with alternative A_i. This outcome is given by the following vector:

$$x(A_i) = (x_1(A_i), x_2(A_i), \ldots, x_p(A_i)). \quad (3.1)$$

Note that when we are referring to an attribute, X_k, we use a capital letter, but when we refer to a value for that attribute, we use a lower-case letter, x_k.

The problem then is to select the best alternative, denoted as A*, or to rank the alternatives: $A_1, A_2, \ldots, A_n$ from best to worst, through the consideration of the outcomes associated with the alternatives. The assumption is that we are modeling the preferences of *one* decision maker (DM); if there is more than one DM, then we assume that they can arrive at a consensus, typically through a (question and answer) process involving an analyst (or analysts) and the DMs.

First, we denote the best value for attribute k over all alternatives as x_k^b and the worst value for attribute k over all alternatives as x_k^w. Next, we denote that one outcome, $x(A_i)$, is **preferred** to another outcome, $x(A_j)$, with the notation

$$x(A_i)\ \mathbf{P}\ x(A_j) \tag{3.2}$$

and that the DM is **indifferent** to the two outcomes, $x(A_i)$ and $x(A_j)$, with the notation

$$x(A_i)\ \mathbf{I}\ x(A_j). \tag{3.3}$$

Finally, we denote that the DM **prefers** outcome $x(A_i)$ to $x(A_j)$ or is **indifferent** to these two outcomes (i.e., either one or the other) with the notation:

$$x(A_i)\ \mathbf{PI}\ x(A_j). \tag{3.4}$$

A summary for this notation is shown in Table 3.1.

TABLE 3.1
Notations Used for Chapter 3

Quantity	Notation
Number of alternatives	n
Number of attributes	p
Alternative i	A_i
Attribute j value for alternative i	$x_j(A_i)$
Outcome for alternative i	$x(A_i) = (x_1(A_i), x_2(A_i), \ldots, x_P(A_i))$
Best and worst values for attribute k	x_k^b, x_k^w
Preferred	P
Indifference	I
Preferred or indifferent to	PI

TABLE 3.2

Attribute Values for Various Job Offers for Example 3.1

i	A_i	$x_1(A_i)$	$x_2(A_i)$	$x_3(A_i)$
1	Offer 1	70,000	5	150
2	Offer 2	60,000	15	250
3	Offer 3	75,000	7	1200
4	Offer 4	65,000	10	50
5	Offer 5	55,000	20	150
6	Offer 6	58,000	8	250
7	Offer 7	68,000	5	1200
8	Offer 8	70,000	10	250
9	Offer 9	58,000	8	50
10	Offer 10	64,000	10	300

Example 3.1: Choosing a Job

Suppose that you are the top student in your graduating class from the Department of Industrial Engineering at State University. As such, you have 10 job offers, from 10 different, highly respected, firms. You have selected several important attributes for consideration in the ranking of these job offers. The various job offers have the same values for each attribute under consideration, except for three of the attributes: annual salary in dollars (X_1), annual number of days of vacation (X_2), and number of miles from your hometown, to the nearest 50 miles (X_3). All three of the attributes are natural attributes. Attributes X_1 and X_2 are obvious choices and are both attributes, which you want to maximize. Attribute X_3 is chosen because, among other reasons, your parents are becoming older and you think that they may need your assistance in various tasks. The alternatives and associated outcomes are shown in Table 3.2.

This example will be discussed in some detail in the following sections.

3.3 Concepts Requiring No Preference Information from the Decision Maker

This section discusses techniques that allow for the elimination of alternatives without the use of any preference information (i.e., information concerning trade-offs that the DM is willing to make between the objectives). In almost all cases however, a DM will still be left with multiple alternatives following the application of these techniques. The reason for this is that typically, objectives are *conflicting* in nature—that is, one is able to improve one objective, but only at the expense of making another objective worse, in the comparison of various alternatives. For example, one may be able to purchase a higher-quality automobile, all other things being equal, but only by paying more.

In this section, we will assume, unless otherwise noted, that more of each attribute is preferred to less. This assumption is made in order to keep our discussion of the various concepts of a concise nature. If we did have an attribute for which less is preferred (e.g., cost in dollars, waiting time in a queue in minutes, number of miles from hometown in Example 3.1), then we would just convert that attribute to its negative value, since an optimal ranking of alternatives where the criterion is to minimize a function is the same ranking as one attains when the criterion is to maximize the negative of that function.

3.3.1 The Ideal, Superior, and Negative Ideal

The *ideal* is a point in the outcome space that has as its associated attribute values the best value over all the alternatives; that is, the ideal is given by

$$(x_1^b, x_2^b, \ldots, x_p^b), \tag{3.5}$$

where x_k^b is the best value for attribute k as denoted earlier, selected over all alternatives.

In Example 3.1, a best (largest in this case) value for annual salary in dollars over all of the outcomes is $75,000; a best value for annual number of days of vacation is 20 days; and a best value for number of miles from hometown is 50. Hence, the ideal is given by (75,000; 20; 50). Note that the ideal is obtained, in this case, by selecting the attribute value for annual salary from alternative 3, the attribute value for days of vacation from alternative 5, and the attribute value for miles from hometown from either alternative 4, or from alternative 9.

Normally, the ideal does not correspond to any *actual* outcome, and hence does not have an actual alternative associated with it. However, if this alternative does exist, it is called a *superior alternative* and it would simultaneously optimize each of the attributes.

The *negative ideal* is a point in the outcome space that has as its associated attribute values the worst value over all alternatives; that is, the negative ideal is given by

$$\left(x_1^w, x_2^w, \ldots, x_p^w\right), \tag{3.6}$$

where x_k^w is the worst value for attribute k, selected over all alternatives.

For Example 3.1, the negative ideal is given by (55,000; 5; 1,200), that is, the value for salary from alternative 5, the value for days of vacation from alternative 1, and the value for number of miles from home from alternative 7.

One might ask why be concerned about an ideal and a superior solution if it typically does not exist in terms of reality. One reason for being aware of these concepts is that they are associated with some of the algorithms for selecting a best alternative. In particular, the TOPSIS (*T*echnique for *O*rder of *P*reference by *S*imilarity to *I*deal *S*olution) algorithm (Hwang and Yoon, 1981), to be presented in Section 3.10, involves finding that alternative that considers the weighted distances from the positive ideal and the negative ideal. Another area of usefulness associated with the concept of the ideal is in the generation of new alternatives and outcomes. In Example 3.1, the student might negotiate with the firm that made offer 5 (the one with the largest number of vacation days) by informing the firm that there was an offer of $75,000 on the table. In a more general sense, knowing the best value for each alternative over all outcomes (and thinking about the characteristics of those alternatives) can result in the DM thinking more creatively about the generation of alternatives.

3.3.2 Nondominated (Efficient) Solutions and Outcomes, Dominance Graphs, and the Efficient Frontier

In this section, we define *deterministic dominance* and related concepts. Note that in Chapter 7, when we address decisions made under conditions of uncertainty and risk, we address the area of *stochastic dominance*. In this chapter, when we use the term *dominance* or a related term, we are referring to deterministic dominance.

An alternative A_i is said to *dominate* another alternative A_j, if and only if $x_k(A_i) \geq x_k(A_j)$ for $k = 1,\ldots, p$, and $x_k(A_i) > x_k(A_j)$ for at least one value of $k = 1,\ldots,$ p. Note that this is equivalent to saying that alternative A_i does at least as well as alternative A_j on all attributes, and better than A_j on at least one attribute. The significance of this concept is that if one alternative is dominated by another, the dominated alternative can be eliminated from further consideration, since no *rational* DM would select a dominated alternative. In a similar sense, we can say that if alternative A_i dominates alternative A_j, then we say that the outcome associated with alternative A_i dominates the outcome associated with alternative A_j.

It can be easily proven, using this definition, that if alternative A_i dominates alternative A_j and that alternative A_j dominates alternative A_k, then alternative A_i dominates alternative A_k.

Given the feasible set of alternatives, the set of all alternatives that are not dominated by any other alternative is just called (as you might expect) the "set of nondominated alternatives." (The respective outcomes associated with the set of nondominated alternatives form the *set of nondominated outcomes.*) An alternative in this set is called a "nondominated alternative" (or "efficient alternative"). Again, the significance of this set of nondominated alternatives is that one could immediately eliminate any alternative not contained in this set from further consideration. Note that this set is *specific* to a particular set

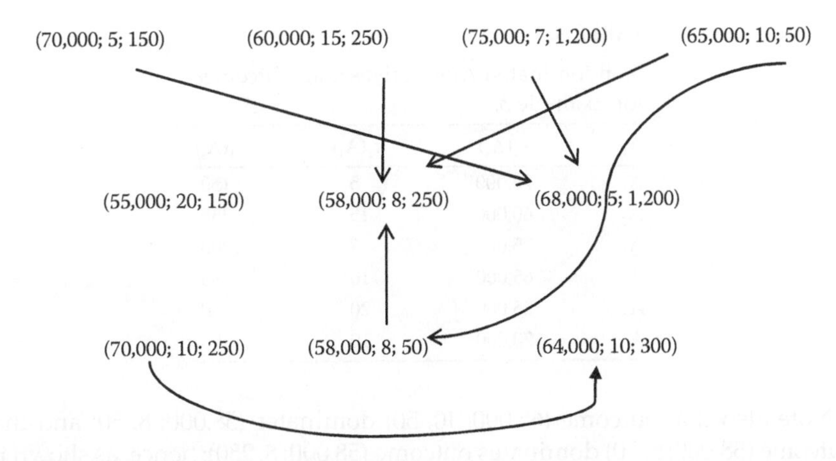

FIGURE 3.1
Dominance graph for outcomes of Example 3.1.

of feasible alternatives. That is, if an additional alternative is found to the problem, the set of nondominated alternatives could be completely changed. However, it would not be possible for a dominated alternative to become a member of the set of nondominated alternatives simply through the addition of one or more alternatives.

The *image* of the set of nondominated alternatives in the outcome space is called the *efficient frontier* or the *pareto optimal set*.

Consider Example 3.1 involving 10 job offers and 3 attributes. One can easily see that alternative 2 (job offer 2) dominates alternative 6 (job offer 6), since job offer 2 does better than job offer 6 on annual salary (60,000 > 55,000), and days of vacation (15 > 8), and as well as job offer 6 on number of miles from home (250 = 250). By considering the dominance relationships between each pair of outcomes, one can construct a *dominance graph*. Such a graph consists of nodes and arcs, where the nodes correspond to the outcomes of the situation, and an arc points from one outcome to another if the first outcome dominates the second. The set of nondominated outcomes is the collection of those outcomes, which do not have an arc pointed at them in the graph.

The dominance graph for Example 3.1 is shown in Figure 3.1. Note that the nodes in the dominance diagram correspond to the outcomes associated with their respective alternatives. An arc leading from one node to another indicates that the alternative associated with the first node dominates the alternative associated with the second node. Hence, a node, which has no arc leading into it, is associated with an alternative that is nondominated. (Hence, for this set of outcomes, there are six nondominated outcomes: [70,000; 5; 150], [60,000; 15; 250], [75,000; 7; 1,200], [65,000; 10; 50], [55,000; 20; 150], and [70,000; 10; 250].) This set of six nondominated outcomes is shown in Table 3.3.

TABLE 3.3

Nondominated Alternatives and Outcomes for Example 3.1

A_i	$x_1(A_i)$	$x_2(A_i)$	$x_3(A_i)$
A_1	70,000	5	150
A_2	60,000	15	250
A_3	75,000	7	1200
A_4	65,000	10	50
A_5	55,000	20	150
A_8	70,000	10	250

Note also that outcome (65,000; 10; 50) dominates (58,000; 8; 50) and that outcome (58,000; 8; 50) dominates outcome (58,000; 8; 250); hence, as shown in Figure 3.1, outcome (65,000; 10; 50) dominates outcome (58,000; 8; 250).

The dominance diagram of Figure 3.1 was formed through a series of pairwise comparisons. That is, outcome 1 was compared to, first, outcome 2, then outcome 3, and so on through outcome 10; then, outcome 2 was compared to outcome 3, then outcome 4, and so on through to outcome 10. This was continued until every outcome was compared to every other outcome—a total of $9 + 8 + 7 + \cdots + 1 = 45$ pairwise comparisons. Each time a comparison indicates that one outcome dominates another, an arc is drawn from the dominating outcome node to the dominated outcome node. Hence, if one has n outcomes, in order to draw a dominance graph, $(n - 1) + (n - 2) + \cdots + 1$ pairwise comparisons need to be made.

Also, if one added an alternative corresponding to an outcome of (72,000; 10; 50), then the set of nondominated outcomes would be reduced to the new outcome of (72,000; 10; 50) and a subset from the original set of nondominated outcomes: (60,000; 15; 250), (75,000; 7; 1200), and (55,000; 20; 150). In particular, the outcomes of (70,000; 5; 150), (65,000; 10; 50), and (70,000; 10; 250) would be eliminated, and the new set of nondominated outcomes would be given by: (72,000; 10; 50), (60,000; 15; 250), (75,000; 7; 1200), and (55,000; 20; 150). Note that in determining a new efficient frontier when a new alternative is added to the set of feasible alternatives, one only has to compare this new alternative to the previous set of all nondominated alternatives, not all feasible alternatives.

3.4 Scales of Measurement and Preference Structures

In the previous section, we described various concepts for which no information about the DM's preference structure was needed. For example, no information about how the DM would trade off a worse value for one objective for a better value in another objective was required. In most decision-making situations however, the objectives are *conflicting* in nature. That is, in order to obtain a better value in one objective, one must give up something in another objective. For example, a more reliable automobile typically costs more than a less reliable one, all other things being equal.

In this section, we present concepts associated with preference structures and various types of preference information. A preference structure for a DM basically defines how he or she will trade off between the objectives, as measured by the attribute values in the outcome space. A multiattribute value function, to be discussed in Sections 3.6 and 3.7, can be thought of as a mathematical representation of a DM's preference structure. Since attributes are measured on measurement scales, we first present a discussion of this concept.

3.4.1 Scales of Measurement

Attributes are measured on scales. Stevens (1946) defined four types of scales: nominal, ordinal, interval, and ratio.

Nominal scales are qualitative in nature and are basically only used to classify objects, such as biological species, languages, and so on. Objects classified according to this type of scale can be related to each other only by being "equal to" or "not equal to" each other. For this reason, this type of scale is *not* used to measure attribute values. As an example, consider the classification of countries by continent, with the coded numbers 1, 2, 3,… representing Africa, North America, Europe, and so on. Then Libya, Mexico, and Germany would have values of 1, 2, and 3, respectively.

Ordinal scales are qualitative in nature and are useful for rank ordering; however, a degree of difference between different values on the scale cannot be established with an ordinal scale. For example, with the ordinal scale associated with the constructed attribute for "quality of neighbors" in Table 2.10, one cannot determine that the difference between "poor" and "satisfactory" is the same as the difference between "good" and "excellent" even though the differences between the coded numbers for each of these ratings are the same.

Interval scales are quantitative in nature and allow for a degree of difference between items. Examples of the use of this type of scale include temperature on the Celsius or Fahrenheit scale. These scales are quantitative in nature, since mode, median, and mean can be computed from values measured on these scales. However, computation of ratios is not allowed; for example, one cannot say that 50°C is twice as hot as 25°C. One can say, however, that the difference between 30°C and 20°C is the same as the difference between 20°C and 10°C.

Ratio scales are also quantitative in nature, but in addition to allowing a degree of difference between items as with items measured on an interval scale, they allow for meaningful ratios to be formed. For example, one can say that a temperature of 200 K is twice as hot as a temperature of 100 K. As can be inferred from this discussion, ratio scales possess a meaningful value for zero, which is not arbitrary. Most scientific quantities, such as Kelvin temperature, mass, and length, are measured on this type of scale.

Constructed attributes are typically measured on ordinal scales, while natural attributes are measured on interval or ratio scales. The reason why we differentiate between the scales here is that questions used to assess utility functions and value functions may be possible to answer with attributes defined on interval/ratio scales, but not with those defined on ordinal scales (e.g., for constructed attributes). Often however, an approximation is made by constructing a utility/value function on an ordinal scale.

3.4.2 Preference Structures

According to Keeney and Raiffa (1993, p. 80), a *preference structure* is defined on an outcome space if, for any two outcomes, the DM can say that he or she prefers the first outcome to the second, the second to the first, or is indifferent to these outcomes. That is, a preference structure for a particular DM implies an "ordering" over the outcome space. More specifically, given any two outcomes, $x(A_i) = (x_1(A_i), x_2(A_i),\ldots,x_p(A_i))$ and $x(A_j) = (x_1(A_j),x_2(A_j),\ldots,x_p(A_j))$, the preference structure allows their "ranking" in one of three ways:

1. $x(A_i)$ **P** $x(A_j)$ (the DM prefers outcome $x(A_i)$ to outcome $x(A_j)$)
2. $x(A_j)$ **P** $x(A_i)$ (the DM prefers outcome $x(A_j)$ to outcome $x(A_i)$)
3. $x(A_j)$ **I** $x(A_i)$ (the DM is indifferent to outcomes $x(A_i)$ and $x(A_j)$)

The existence of a preference structure also implies that there are no intransitivities in the rankings; that is, there are no cases in which the DM would rank three outcomes, $x(A_i)$, $x(A_j)$, and $x(A_k)$ (for any three outcomes in the consequence space), as

$$x(A_i)\ \mathbf{P}\ x(A_j),\ x(A_j)\ \mathbf{P}\ x(A_k),\ \text{and}\ x(A_k)\ \mathbf{P}\ x(A_i)$$

A preference structure is a subjective concept. That is, two different, but rational, DMs, could (and probably would) have different preference structures, depending on the trade-offs that they would be willing to make among the various attributes.

3.4.3 Types of Preference Information

In order to make inferences about a DM's preference structure, information about the DM's preference structure must be provided. This information can be provided in any of several different ways. Different approaches/algorithms for solving the multiobjective problem require different types of preference information from the DM. In some cases, such as with the "scorecard approach" (where the DM or DMs are just given a list of alternatives along with the outcomes for each alternative in terms of attribute values, and then are asked to rank the outcomes, or just select the best outcome), the information is not explicit. Examples of the types of explicit information that one obtain from a DM include the following.

1. A rank ordering of the objectives/attributes, such as "annual salary is more important than the number of days of vacation, and the number of days of vacation is more important than the number of miles from home."
2. A specification of how much more important one objective/attribute is than another, such as "annual salary is moderately more important than the number of days of vacation."
3. A specification of how much better one outcome is than another on a particular attribute; an example would be, "an annual salary of $75,000 is only a little better than an annual salary of $74,000."
4. A required threshold for a particular attribute, such as "a job with less than 5 days of annual vacation to start will definitely not be acceptable."
5. Weights (typically normalized so that they sum to 1) given for the different attributes; an example would be weight assignments of .6, .3, and .1 to annual salary, number of vacation days, and miles from home, respectively.
6. A rank ordering of specific outcomes.
7. The marginal rates of substitution of one attribute for another; an example would be the amount of annual salary that the DM would give up for an extra day of vacation. (Note that this amount would likely depend on the reference point; for example, a DM would likely give up more salary for an extra day of vacation from the reference point of [$100,000 of annual salary and 20 days of vacation] than he or she would from a reference point of [$50,000 of annual salary and 20 days of vacation]).

Of these seven types of preference information, the most exacting to obtain from a DM is the seventh one, marginal rates of substitution; however, this type of information provides the most exact representation of the DM's preference structure.

3.5 Lexicographic Ordering

One approach for ranking multiobjective outcomes, which requires very little information from the DM, is *lexicographic ordering*. This approach is analogous to the ordering that one finds in the dictionary where all words that start with the letter "a" come before all words that start with the letter "b" and all words that start with the letters "aa" come before all words that start with the letters "ab," and so on.

With lexicographic ordering therefore, the DM will first rank the attributes in their order of importance. Then any outcome with a highest score on the first-ranked attribute will be in the set of the highest ranked outcomes, regardless of their scores on the other attributes. The second-ranked attribute's score is only considered if there is a tie in the first-ranked attribute's score, and so on.

In Example 3.1, suppose that the attributes are ordered in the following way, from most to least importance: annual salary (X_1), annual number of days of vacation (X_2), and number of miles from hometown (X_3). Then, in terms of lexicographic ordering, the top three job offers in this order would be: Offer 3 (75,000; 7; 1,200), Offer 8 (70,000; 10; 250), and Offer 1 (70,000; 5; 150). Note that since there is a tie in annual salary between Offer 8 and Offer 1 at $70,000, we move to the second-ranked attribute (annual number of days of vacation) to break the tie. See Table 3.4 for a complete ranking of the outcomes associated with Example 3.1 using lexicographic ordering with the ranking of the attributes, in decreasing order of importance, as X_1, X_2, and X_3.

Note that with lexicographic ordering, a dominated outcome could be ranked higher than a nondominated outcome. For example, Job Offer 5, which represents a nondominated outcome, is ranked lower than Job Offer 10.

Lexicographic ordering is a relatively simple approach to the ranking of multiobjective outcomes, but it rarely represents the way that a DM would trade off among the various objectives of a problem. For example, for the job offer situation, an offer with an annual salary of $70,001 and 2 days of vacation would be ranked above an offer with an annual salary of $70,000 with 3 weeks of vacation—an unlikely situation. Bouyssou et al. (2010, p. 191) note that lexicographic ordering should be used only in two cases: (1) when one attribute is "infinitely" more important than the other attributes and (2) as a screening process when the cost of more intricate analysis is prohibitive.

TABLE 3.4

Ranking of Job Offers via Lexicographic Ordering

Alternative	Ranking	Outcome
Offer 3	1	(75,000; 7; 1,200)
Offer 8	2	(70,000; 10; 250)
Offer 1	3	(70,000; 5; 150)
Offer 7	4	(68,000; 5; 1,200)
Offer 4	5	(65,000; 10; 50)
Offer 10	6	(64,000; 10; 300)
Offer 2	7	(60,000; 15; 250)
Offer 9	8	(58,000; 8; 50)
Offer 6	9	(58,000; 8; 250)
Offer 5	10	(55,000; 20; 150)

3.6 MAV Functions

3.6.1 Definition and Basic Concepts

Definition 3.1: A function, denoted as v, which maps from the outcome space $(X_1, X_2,\ldots, X_p)$ into a closed, continuous, interval of the real number line (usually [0,1]) is a *multiattribute value* (MAV) *function* (sometimes just called a value function) representing a particular DM's preference structure provided that

1. $x(A_i)$ **I** $x(A_j)$ if and only if $v(x(A_i)) = v(x(A_j))$
   and
2. $x(A_i)$ **P** $x(A_j)$ if and only if $v(x(A_i)) > v(x(A_j))$

In words, this means that the DM will be indifferent to two outcomes, $x(A_i)$ and $x(A_j)$, if and only if their value function values are equal, and the DM will prefer outcome $x(A_i)$ to outcome $x(A_j)$ if and only if the value function value for $x(A_i)$ is greater than the value function value for $x(A_j)$.

Note that the definition given in the previous paragraph might be considered as an "operational definition" for a multiattribute value function. In many actual situations involving evaluation with multiple objectives, a DM (or DMs) will just assign a set of (normalized) weights to the various objectives, one weight to each objective, in order to form what is thought to be an MAV function. For Example 3.1, a DM might say that salary (X_1) is twice as important as both number of days of vacation (X_2) and number of miles from hometown (X_3). Then linear "individual attribute" functions might be formed by specifying the following two points on a graph for each function of the respective attribute:

$$f_1(55{,}000) = 0, \quad f_1(75{,}000) = 1,$$

$$f_2(5) = 0, \quad f_2(20) = 1,$$

and

$$f_3(1{,}200) = 0, \; f_3(50) = 1,$$

and the multiattribute value function would be given as

$$v\,(x_1, x_2, x_3) = .5\, f_1(x_1) + .25\, f_2(x_2) + .25\, f_3(x_3).$$

Note that the slope for the third individual attribute value function is negative, owing to the fact that the DM would like to minimize this attribute value. *Typically an individual attribute value function for an attribute that is to be minimized will have a negative slope*; this is as opposed to having a negative coefficient.

Seldom will a function formed in the way described in the previous paragraph be an accurate representation of a DM's preference structure. The main difficulty is the meaning of statements like "attribute 1 is twice as important as attribute 2," from which one might infer that its weight in a function like the one earlier should be twice the weight associated with the second attribute. Many analysts contend that unless questions corresponding to the amount that the DM would give up in terms of one attribute in order to improve another attribute value are asked, one will typically not arrive at the most accurate value function for the DM.

Some of the applications from the literature involving the assessment and/or use of an MAV function include the following:

1. An MAV function over attributes involving cost and reliability for generation planning for electric utilities (Moskowitz et al., 1978)
2. An MAV function over attributes related to command and control, operational capability, and efficient use of resources for investigating alternatives for the regional organization of the U.S. Army's Installation Management Agency (Trainor et al., 2007)
3. An MAV function to evaluate alternatives for improving electric system reliability in British Columbia (Keeney et al., 1995)
4. An MAV function to investigate alternatives relative to the development of Europe's air traffic management system (Grushka-Cockayne and De Reyck, 2009)

3.6.2 Strategic Equivalence of MAV Functions

In using an MAV function, we are basically interested in how various outcomes (and their corresponding alternatives) are ranked. As it turns out, many different MAV functions can be used to represent a specific preference structure of a DM—that is, many alternative MAV functions will give the same ordering of outcomes in the outcome space. This gives rise to the concept of strategic equivalence.

Definition 3.2: A set of two or more MAV functions giving the same preference ordering of outcomes are said to be *strategically equivalent*.

As an example, consider a problem with two attributes, denoted as

X_1, with a minimum possible value of 0, and a maximum possible value of 100

X_2, with a minimum possible value of 0, and a maximum possible value of 100

Two MAV functions, v_1 and v_2, which would be strategically equivalent for this decision situation, would be

$$v_1(x_1, x_2) = 2x_1 + 3x_2 \quad \text{and} \quad v_2(x_1, x_2) = 4x_1 + 6x_2,$$

since for any (x_1, x_2) and (x_1', x_2') (where x_1, x_2, x_1', and x_2' are all within the range of 0–100),

$$v_1(x_1, x_2) > v_1(x_1', x_2') \text{ if and only if } v_2(x_1, x_2) > v_2(x_1', x_2'),$$

$$v_1(x_1, x_2) = v_1(x_1', x_2') \text{ if and only if } v_2(x_1, x_2) = v_2(x_1', x_2'),$$

and

$$v_1(x_1, x_2) < v_1(x_1', x_2') \text{ if and only if } v_2(x_1, x_2) < v_2(x_1', x_2').$$

The concept of strategic equivalence implies that a specific preference structure can be represented by many different valid MAV functions.

3.6.3 Indifference Curves

Given a particular outcome, $x(A_i)$, an *indifference curve* is the set of all points in the outcome space such that the DM is indifferent to that outcome $x(A_i)$ and all other outcomes on the curve. Note that if the number of attributes (i.e., the dimension of the outcome space) is greater than or equal to three, then the indifference "curve" is actually a surface or a hyperplane.

An indifference curve for a DM is given by the equation obtained by setting a value function for that DM equal to a constant, since every outcome point, which satisfies that equation, gives the same value function value, and therefore is equally preferable to the DM.

3.6.4 Marginal Rates of Substitution

The *marginal rate of substitution* of one attribute, say X_i, for another attribute, X_j, at a particular point in the outcome space (x_i, x_j) is roughly the amount of X_i that the DM is willing to give up for an extra unit of X_j at the point (x_i, x_j). The reason why the word *roughly* is used is that the marginal rate of substitution of one attribute for another typically changes depending upon the point in the outcome space at which it is measured. Therefore, a more exact definition is given in the following:

Definition 3.3: The *marginal rate of substitution of X_i for X_j at a particular point in the outcome space* (x_i^0, x_j^0) is given by $dx_i/dx_j \mid (x_i^0, x_j^0)$, the derivative of x_i with respect to x_j evaluated at the point (x_i^0, x_j^0), where the derivative is derived from the indifference curve at (x_i^0, x_j^0).

In fact, any time the marginal rate of substitution of any attribute for any other attribute does not depend on the reference point of the outcome space; then a linear (scaled) value function will be appropriate for this DM, as shown in (3.7):

$$v(x_1, x_2, \ldots, x_p) = \sum_{i=1}^{p} w_i(m_i x_i + b_i), \tag{3.7}$$

where the w_i, m_i, and b_i are such that

$$1.\ w_i > 0 \quad \text{for } i = 1, \ldots, p, \tag{3.8}$$

$$2.\ \sum_{i=1}^{p} w_i = 1, \tag{3.9}$$

and

$$\begin{gathered} 3.\ m_i \text{ and } b_i \text{ are given such that } m_i x_i^b + b_i = 1, \\ \text{and} \\ m_i x_i^w + b_i = 0 \text{ for } i = 1, \ldots, p. \end{gathered} \tag{3.10}$$

Let's consider only the first two attributes from Example 3.1, X_1 = annual salary and X_2 = days of vacation; suppose that we want to determine the marginal rate of substitution of X_1 for X_2 at the point x_1 = \$80,000 and x_2 = 10 days of vacation. Roughly, this would be the amount that the student would pay out of his or her \$80,000 salary for one more day of vacation. (At this point, let's not get into all of the intricacies of income taxes, etc.) Let's suppose that the student would give up \$600 for this extra day of vacation; hence, "roughly," the marginal rate of substitution of annual salary for days of vacation at the point (80,000; 10) is \$600. More specifically, the student is indifferent to the following outcomes: (80,000; 10) and (79,400; 11); that is, for *this* DM:

$$(80{,}000;\ 10)\ \mathbf{I}\ (79{,}400;\ 11). \tag{3.11}$$

As noted earlier, the marginal rate of substitution typically depends upon the point in the outcome space at which the substitution rate is determined. Let's consider the determination of the marginal rate of substitution at the point (80,000; 20). At this point, one might expect that the student would

TABLE 3.5

(Rough) Marginal Rates of Substitution of Annual Salary for Days of Vacation at Two Different Outcomes

Outcome	Rough MRS of X_1 for X_2 at Outcome	Implies
(80,000; 10)	\$600	(80,000; 10) I (79,400; 11)
(80,000; 20)	\$400	(80,000; 20) I (79,600; 21)

give up less in terms of salary for an extra day of vacation than he or she would at the point (80,000; 10), since he or she already has so many days of vacation. In particular, suppose that he or she would give up only \$400 for an extra day of vacation; hence, the marginal rate of substitution of X_1 for X_2 at the point (80,000; 20) is roughly 400, and the DM is indifferent to (80,000; 20) and (79,600; 21). The summary results associated with these trade-offs are shown in Table 3.5.

3.6.5 Independence Conditions and the Form of the MAV Function

Given certain reasonable conditions with respect to the preference structure of the DM, the analyst can assume that an appropriate MAV function for that DM will take a specific functional form.

3.6.5.1 The Case of Two Attributes

Let's first consider the case of two attributes: X_1 and X_2. As shown by Keeney and Raiffa (1993, p. 85), if the marginal rate of substitution of X_2 for X_1 does not depend on the reference point, then a valid MAV function is linear and in fact can be given by

$$v(x_1,x_2) = x_1 + cx_2, \tag{3.12}$$

where c is a constant given by the negative of the marginal rate of substitution of X_1 for X_2 (at any point in the outcome space). Note that any multiple of this function would also be a valid MAV function, since such a function would be strategically equivalent to the function given in (3.12).

Example 3.2: Illustration of a Two-Attribute Value Function, Marginal Rates of Substitution, and Indifference Curves

Suppose that a DM had a value function for the first two attributes of Example 3.1: annual salary (X_1) and days of vacation (X_2) given by

$$v(x_1,x_2) = .7v_1(x_1) + .3v_2(x_2), \tag{3.13}$$

where $v_1(x_1)$ and $v_2(x_2)$ are the individual attribute value functions for X_1 and X_2, respectively, and are given by

$$v_1(x_1) = .00005x_1 - 2.75, \quad v_2(x_2) = .0666x_2 - .3333.$$

(Note that these two individual attribute functions are linear with $v_i(x_i^b) = 1$ and $v_i(x_i^w) = 0$ for i = 1 and 2.)

Suppose that we find the value function value associated with the fourth offer for X_1 and X_2:

$$v(65,000;10) = .7(.00005(65,000) - 2.75) + .3(.0666(10) - .3333)$$
$$= .35 + .1 = .45.$$

Now, setting the value function equal to .45 will give us an equation, which relates X_1 and X_2 in terms of the DM's preferences; solving this equation for x_2 as a function of x_1 will allow us to draw a graph that would represent the indifference curve associated with the outcome for the fourth job offer:

$$v(x_1, x_2) = .45, \text{ or}$$
$$.7v_1(x_1) + .3v_2(x_2) = .45, \text{ or}$$
$$.7(.00005x_1 - 2.75) + .3(.0666x_2 - .3333) = .45, \text{ or}$$
$$x_1 = 70714.28 - 570.857x_2. \qquad (3.14)$$

Consider (3.14), which represents the indifference curve (given by x_1 as a function of x_2) associated with the value function of (3.13). The derivative of x_1 with respect to x_2 is given by

$$\frac{dx_1}{dx_2} = -570.857$$

so that the marginal rate of substitution of x_1 for x_2 for a DM with this particular value function is given by

$$\text{MRS of } x_1 \text{ for } x_2 = \frac{-dx_1}{dx_2} = -(-570.857) = 570.857.$$

Note that this marginal rate of substitution of x_1 for x_2 *for this DM* is not a function of x_1 or of x_2; in other words, the amount of salary that this DM would give up for an extra unit (day) of vacation is not a function of the current point in the outcome space. One could easily derive that the marginal rate of substitution of x_2 for x_1 is also a constant.

Substituting for the individual attribute functions, one can rewrite (3.13) as

$$v(x_1,x_2) = .7(00005x_1 - 2.75) + .3(.0666x_2 - .3333)$$
$$= .000035x_1 + .01999x_2 - 2.02499.$$

Dropping the constant term and multiplying the result by 28571.4 (in order to give x_1 a coefficient of 1), one obtains a strategically equivalent MAV function, v_{se}, as shown in (3.15):

$$v_{se}(x_1,\ x_2) = x_1 + 570.857x_2. \tag{3.15}$$

This illustrates the fact that an appropriate value function for this situation is one with a coefficient of 1 for X_1 and a coefficient of negative of the marginal rate of substitution of X_1 for X_2.

Also as shown by Keeney and Raiffa (1993), if the marginal rate of substitution of X_1 for X_2 at $(x_1^0,\ x_2^0)$ depends on x_2^0 but not on x_1^0, then a valid MAV function would be given by

$$v(x_1,x_2) = x_1 + c_2v_2(x_2), \tag{3.16}$$

where $v_2(x_2)$ is an individual attribute value function, not necessarily linear, for X_2.

Now, it is usually the case that the marginal rates of substitution are dependent on both the value of X_1 and the value of X_2 at (x_1^0,x_2^0); however, transformation of scales for $(x_1,\ x_2)$, to say $(y_1,\ y_2)$ would allow the marginal rates of substitution to be independent of the values for y_1 and y_2. In this case, a valid MAV function would have an additive (and scaled) form, as shown in (3.17).

$$v(x_1,x_2) = w_1v_1(x_1) + w_2v_2(x_2), \tag{3.17}$$

where

$$w_1 \text{ and } w_2 \text{ are constants such that } 0 < w_1 < 1,\ 0 < w_2 < 1 \text{ and } w_1 + w_2 = 1. \tag{3.18}$$

Note that the additive form of (3.17) is not the same as a linear form of (3.12) since the individual attribute functions, $v_i(x_i)$ for $i = 1, 2$, can be nonlinear. The condition that allows an additive form of (3.17) and (3.18) is termed the *corresponding trade-offs condition* (Keeney and Raiffa, 1993, pp. 90–91) and is shown to hold through a series of questions posed to the DM. Consider again Example 3.1, with only the first two attributes being employed in the formation of the MAV function.

This shows that the corresponding trade-offs condition involves asking the DM questions about the trade-offs that he or she is willing to make. More specifically, the DM would be asked to specify four indifferences, as illustrated by Example 3.3. This example corresponds to Example 3.1, where we are attempting to assess an MAV function over the first two attributes: X_1 and X_2.

Example 3.3: Illustration of Questions Asked to Show That the Corresponding Trade-Offs Condition Holds

The analyst selects a "reasonable point" in the outcome space, say (60,000; 10) (i.e., an annual salary of $60,000 with 10 days of vacation); then the analyst would ask the DM the amount of salary he or she would be willing to give up for an extra 2 days of vacation. In other words, the DM would be asked to give the value x_1 for the attribute X_1 such that he or she would be indifferent to the outcomes of (60,000; 10) and (x_1, 12).

Suppose that the DM gives a value of x_1 = $59,200. This means that the DM is indifferent to the outcomes of (60,000; 10) and (59,200; 12), or

$$(60{,}000;\ 10)\ \mathbf{I}\ (59{,}200;\ 12). \tag{3.19}$$

Then the analyst would choose a different value in the X_2 outcome space, say $x_2 = 16$, and ask the DM how much more in terms of vacation that he or she would have to receive in order to give up $800 in salary. (Note that this is the *same amount* given up for the first indifference.) Suppose that the DM indicates that he or she would need 1.5 more days of vacation at this outcome; this would mean that the DM is indifferent to the outcomes of (60,000; 16) and (59,200; 17.5), or

$$(60{,}000;\ 16)\ \mathbf{I}\ (59{,}200;\ 17.5). \tag{3.20}$$

(Note that this type of response would make intuitive sense as the DM would be expected to desire less additional vacation at an outcome where he or she already has 16 days than from the outcome associated with 10 days of vacation for the same amount of salary given up.)

Now, the analyst would choose a different value for salary: say $70,000, but the same value for days of vacation as in the first outcome shown in (3.19): 10 days. Then the DM would be asked the amount that he or she would be willing to give up in terms of annual salary in order to increase the number of days of vacation by 2; that is, the DM would be asked the value of x_1 such that he or she would be indifferent to the outcomes of (70,000; 10) and (x_1, 12). Suppose that the DM answers that the value of x_1 should be $69,000; that is, one can say that for this DM,

$$(70{,}000;\ 10)\ \mathbf{I}\ (69{,}000;\ 12). \tag{3.21}$$

Note again that this type of response, as was the case for the indifference shown for (3.20), makes intuitive sense when compared to the indifference of (3.19); that is, the DM is willing to give up more of his or her salary to go from 10 days of vacation to 12 days of vacation when he or she already has $70,000 than when he or she only has $60,000.

Finally, the analyst asks the DM to consider an outcome of (70,000; 16) the outcome associated with the X_1 value of the first outcome of (3.21) and the X_2 value associated with the first outcome of (3.20); then the DM is asked how much salary he or she would be willing to give up in order to increase his or her number of days of vacation by 1.5 days (i.e., the same amount of increase shown in the indifference of (3.20)). If the DM answers that he or she is willing to give up $1000, that is,

$$(70{,}000;\ 16)\ \mathbf{I}\ (69{,}000;\ 17.5), \tag{3.22}$$

then this is evidence that the corresponding trade-offs condition holds.

Note that for this example, four sets of indifferences are established in the questioning: (60,000; 10) **I** (59,200; 12), (60,000; 16) **I** (59,200; 17.5), (70,000; 10) **I** (69,000; 12), and (70,000; 16) **I** (69,000; 17.5) for *this* DM. Consider the first indifference, (60,000; 10) **I** (59,200; 12). The analyst selected the outcome (60,000; 10) and the attribute value of 12 days of vacation, and the DM determined the value of $59,200 such that he or she would be indifferent to these two outcomes. For the next pair of indifferent outcomes, the $60,000 and $59,200 were *set* from the previous indifference, the value of 16 days of vacation was set by the analyst, and the 17.5 days of vacation was determined by the DM. Continuing, and in summary, for each of the four sets of indifferences, the attribute values are determined as follows:

1. (60,000; 10) **I** (59,200; 12):
   a. $60,000 → set arbitrarily by the analyst
   b. 10 days → set arbitrarily by the analyst
   c. 12 days → set arbitrarily by the analyst
   d. $59,200 → determined by the DM so that indifference is satisfied
2. (60,000; 16) **I** (59,200; 17.5):
   a. $60,000 → set from the previous indifference point
   b. 16 days → set arbitrarily by the analyst
   c. 17.5 days → determined by the DM so that indifference is satisfied
   d. $59,200 → set from the previous indifference point
3. (70,000; 10) **I** (69,000; 12):
   a. $70,000 → set arbitrarily by the analyst
   b. 10 days → set from the previous indifference point
   c. $69,000 → determined by the DM so that indifference is satisfied
   d. 12 days → set from the previous indifference point
4. (70,000; 16) **I** (69,000; 17.5):
   a. $70,000 → set from the previous indifference point
   b. 16 days → set from the previous indifference point
   c. $69,000 → determined by the DM so that indifference is satisfied
   d. 17.5 days → set from the previous indifference point

Note that in determining an attribute value so that he or she is indifferent to two outcomes, the DM should *hone in* on the value. The analyst can aid the DM in this regard by asking the DM to rank order a series of pairs of outcomes. For example, in order to determine the indifference of (3.19), the analyst might ask the DM:

Which outcome do you prefer, (60,000; 10) or (58,000; 12) (i.e., do you prefer $60,000 in salary and 10 days of vacation, or $58,000 and 12 days of vacation?)?

In this case, the DM could be expected to answer $60,000 in salary and 10 days of vacation (i.e., the first outcome is preferred over the second). Now the analyst knows that an annual salary greater than $58,000 is needed in order to attain indifference on the part of the DM. Hence, the next question asked by the analyst might be:

Which outcome do you prefer, (60,000; 10) or (59,500; 12) (i.e., do you prefer $60,000 in salary and 10 days of vacation, or $59,500 and 12 days of vacation?)?

In this case, the DM could be expected to answer $59,500 in salary and 12 days of vacation (i.e., the second outcome is preferred over the first). Now the analyst knows that an annual salary of less than $59,500 is needed in order to attain indifference. Hence, the next question asked by the analyst might be:

Which outcome do you prefer, (60,000; 10) or (59,200; 12) (i.e, do you prefer $60,000 in salary and 10 days of vacation, or $59,200 and 12 days of vacation?)?

In this case, the DM could be expected to answer that he or she is indifferent to the two outcomes, as seen in the example.

One reason for asking a series of questions in which the DM provides rank orderings on two outcomes, as opposed to just asking the DM to provide an attribute value so that indifference exists, is that the rank ordering type of question is easier for the DM to answer. *Providing preference information of the rank ordering type is almost always simpler for the DM than providing an amount for an attribute so that he or she is indifferent to two outcomes.*

3.6.5.2 The Case of Three or More Attributes

Suppose that we have a situation with p attributes where $p \geq 3$: $X = (X_1, X_2,\ldots, X_p)$, as in the general case. A subset of attributes Y (with at least two attributes) contained in X is *preferentially independent* (*PI*) of its complement, denoted as Y_{comp}, if the preference structure over Y does not depend on the values of the attributes in Y_{comp}. If every subset of X with at least two attributes is PI of its complement in X, then X has the property of *mutual preferential independence* (*MPI*). Finally, if $X = (X_1, X_2,\ldots, X_p)$ has MPI, then an appropriate MAV function for the DM will be *additive*, as shown in the following equations:

$$v(x_1,x_2,\ldots,x_p) = \sum_{i=1}^{p} w_i v_i(x_i), \tag{3.23}$$

where

$$v_i(x_i^b) = 1 \quad \text{for } i = 1,\ldots,p \tag{3.24}$$

$$v_i(x_i^w) = 0 \quad \text{for } i = 1,\ldots,p \tag{3.25}$$

$$0 < w_i < 1 \quad \text{for } i = 1,\ldots,p \tag{3.26}$$

and

$$\sum_{i=1}^{p} w_i = 1 \tag{3.27}$$

As an example, in order to show MPI for a case with three attributes: X_1, X_2, and X_3, one must show that the following conditions hold:

1. (X_1, X_2) is PI of X_3
2. (X_1, X_3) is PI of X_2
3. (X_2, X_3) is PI of X_1

Obviously, as the number of attributes in the set X increases, the number of preferential independence conditions required to show mutual preferential independence increases tremendously.

Consider the situation involving Example 3.1 and suppose that we want to show that (X_1, X_2) is preferentially independent of X_3. In other words, we want to show that the preference structure over (X_1, X_2) does not depend upon the value for X_3. There is no way to do this *absolutely*. Instead, we obtain *evidence* (from the DM's answers to questions concerning his or her preference structure) that preferential independence holds. We would ask the DM questions about how he or she would rank two outcomes in the (X_1, X_2) space, in which we think the preferences would be "close" for the DM, given a particular (extreme) value for X_3. Then, we would ask the DM to consider a quite different value for X_3, and ask him or her if his or her preference ordering of the outcomes previously ranked in the (X_1, X_2) space would change.

For example, an analyst might ask the DM if given a particular job position for which $X_3 = 50$ (i.e., the job is 50 miles from his or her hometown), how would he or she rank the following outcomes given in terms of attribute values for X_1 (salary) and X_2 (days of vacation). The idea is that these two outcomes would be very close in terms of preference, thereby allowing for a possible switch in the rank orderings if the value of X_3, miles from hometown, changes:

($70,000; 10 days) and ($72,000; 7 days).

Suppose that the DM answered that he or she would prefer the first outcome to the second, given that the job was 50 miles from his or her hometown; that is, for this DM,

($70,000; 10 days; 50 miles) **P** ($72,000; 7 days; 50 miles).

Now the analyst would ask the DM if his or her rankings would change if the distance from hometown were increased to a different number, say 800 miles; that is, for this DM, would the following ranking hold?

$$(\$70,000;\ 10\ \text{days};\ 800\ \text{miles})\mathbf{P}(\$72,000;\ 7\ \text{days};\ 800\ \text{miles})?$$

If the DM's preference ordering does not shift with the change in value for X_3, then this is evidence that (X_1, X_2) is preferentially independent of X_3. The analyst may want to ask additional questions of this type in order to gain additional evidence that (X_1, X_2) is preferentially independent of X_3.

3.7 Assessment of an MAV Function

Assessment of an MAV function means defining the function for a specific DM. Remember that each DM will have his or her own function, depending upon the trade-offs he or she is willing to make between the various pairs of attributes.

Typically, the assessment process involves question and answer sessions between the analyst (or analysts) and a DM. Although technically the function is supposed to represent the preference structure of a single DM, one could conduct the process with several DMs who would arrive at a consensus with respect to their answers; in this way, one could develop an MAV function for an organization.

We will assume in this section that the DM's MAV function has an additive, scaled, form, as shown in the following equations:

$$v(x_1, x_2, \ldots, x_p) = \sum_{i=1}^{p} w_i v_i(x_i), \tag{3.28}$$

where

$$w_i > 0, \sum_{i=1}^{p} w_i = 1 \quad \text{and} \quad 0 \le v_i(x_i) \le 1 \quad \text{for } i = 1, \ldots, p. \tag{3.29}$$

In (3.28) and (3.29), the functions $v_1(x_1)$, $v_2(x_2)$,... are called *individual attribute value functions*. The constants: w_1, w_2,... are called *scaling constants*. Again, as in the two-attribute case, note that (3.28) and (3.29) do not represent a linear MAV function, since the individual attribute value functions can be, and probably will be, nonlinear in nature.

Of course, in order to assume that an additive function as shown in (3.28) and (3.29) is appropriate, one is also assuming that either the corresponding trade-offs condition holds for $p = 2$, or that mutual preferential independence holds (for $p \geq 3$).

The procedure involves several general steps. But prior to these steps, the analyst will explain the process and its purpose to the DM. In most cases, one would expect that the analyst and the DM would have had extensive interaction during the problem structuring aspect of the analysis, using some of the techniques presented in Chapter 2. Following this initial, relatively informal step, the following activities would occur:

1. The best and worst possible values for each attribute would be determined.
2. Questions would be asked to the DM to determine whether the DM's MAV function has an additive form, as shown earlier.
3. The individual attribute value functions, $v_1, v_2, \ldots, v_p$, would be determined.
4. The scaling constants, $w_1, w_2, \ldots, w_p$, would be determined.

With respect to the first step, as long as the best and worst values chosen bound the actual best and worst values over all of the alternatives, the resulting function will be useful for ranking the alternatives. In fact, one would probably choose values for best and worst that are at least slightly better and slightly worse, respectively, than the actual best and worst values in order to allow for the consideration of additional alternatives and corresponding outcomes.

Consider Example 3.1. The best and worst values over all offers for X_1 (annual salary in dollars), X_2 (vacation days), and X_3 (miles from hometown) are, respectively: \$75,000 and \$55,000, 20 and 5, and 50 and 1200. As mentioned earlier, one might choose best and worst values outside of these bounds; however, for this example, we will just use the best and worst values as derived from Table 3.2. These values are shown in Table 3.6. As mentioned earlier, we denote the best and worst values for attribute X_j as x_j^b and x_j^w, respectively.

Note that since we are using a scaled MAV function, we already know two of the points on each of the individual attribute functions:

$$v_1(75{,}000) = 1,\ v_1(55{,}000) = 0,\ v_2(20) = 1,\ v_2(5) = 0,\ v_3(50) = 1,$$

TABLE 3.6

Best and Worst Values for X_1, X_2, and X_3 of Example 3.1

Attribute	Best Value for X_j, x_j^b	Worst Value for X_j, x_j^w
X_1, annual salary in dollars	75,000	55,000
X_2, number of vacation days	20	5
X_3, miles from hometown	50	1,200

and

$$v_3(1,200) = 0.$$

Remember that, as noted earlier, the individual attribute value function for X_3 (an attribute to be minimized) will have a negative slope as opposed to a negative scaling constant. The idea here is that we want to keep all of the scaling constants positive.

As noted earlier, we are skipping the second step of the process, and just assuming that the DM's preference structure satisfies the condition required for an additive MAV function.

The next step is to determine the individual attribute value functions: $v_1(x_1)$, $v_2(x_2)$, and $v_3(x_3)$. The basic idea associated with the assessment of an individual attribute value function is to determine *points* on the function and then interpolate between those points. As noted earlier, since the individual attribute value functions are scaled, we already have the two extreme points on the graphs.

There are several techniques available for determining various points on the individual v_i. One of the most popular is the *midvalue splitting technique,* as described in Keeney and Raiffa (1993, pp. 94–96).

3.7.1 Midvalue Splitting Technique for Determining the Individual Attribute Value Functions

Suppose that we want to determine $v_1(x_1)$. In order to apply the midvalue splitting technique, we need two definitions, as given in Keeney and Raiffa (1993, p. 94).

Definition 3.4: Differentially value equivalent—The pair (x_1^a, x_1^b) is said to be *differentially value equivalent* to the pair (x_1^c, x_1^d), where $x_1^a < x_1^b < x_1^c < x_1^d$, if whenever the DM is willing to go from x_1^b to x_1^a for a given increase in X_2, the DM would be willing to go from x_1^d to x_1^c for the same increase in X_2.

The definition for differentially value equivalent can be used to define what is called a midvalue point for a particular interval of the attribute.

Definition 3.5: Midvalue point—For any interval $[x_1^a, x_1^b]$ of X_1, its *midvalue point* x_1^c is such that the pairs $[x_1^a, x_1^c]$ and $[x_1^c, x_1^b]$ are differentially value equivalent.

Let's assess the individual attribute value function for X_1 using the midvalue splitting technique. This is accomplished by first finding the midvalue point, $x_1^{.5}$, for the interval $[x_1^0,\ x_1^1]$ of X_1. By definition, this midvalue point is the one that gives a functional value of .5 for the individual attribute value function

v_1. The next steps would be to find the midvalue points for the intervals of $[x_1^{.5}, x_1^1]$ and $[x_1^0, x_1^{.5}]$. At this point, we would have five points on the individual attribute value function curve: $(x_1^0, 0)$, $(x_1^{.25}, .25)$, $(x_1^{.5}, .5)$, $(x_1^{.75}, .75)$, and $(x_1^1, 1)$. We could continue finding midvalue points, say $x_1^{.125}$ and $x_1^{.375}$, or we could stop with just the five initial points.

Finding midvalue points for a particular individual attribute value function involves setting up a hypothetical scenario in which all of the attributes except two are set at reasonable values, and then the DM is questioned about trade-offs over the two attributes that are not set at specific values.

In the situation involving Example 3.1, X_3 would be set at a reasonable value, say 200 miles. Now, the analyst would make an "educated guess" at the value for $x_1^{.5}$, say $65,000. (Note that this would be the midvalue point if v_1 were linear, since $65,000 is halfway in between the minimum [$55,000] and maximum [$75,000] values for annual salary.) Then the analyst chooses a good value for vacation days, say 18 days, and asks the DM to give the value x_2' such that the DM is indifferent to (55,000; 18) and (65,000; x_2'). Suppose that the DM determines that $x_2' = 12$, that is, the DM is indifferent to (55,000; 18) and (65,000; 12).

Now, the analyst asks the DM to rank the following two outcomes: (65,000; 18) and (75,000; 12). Suppose the DM says that he or she prefers (75,000; 12) to (65,000; 18); that is, in order for the DM to be indifferent to (65,000; 18) and (75,000, x_2''), x_2'' must be less than 12. *That is, the DM is willing to give up more in terms of vacation days to go from 65,000 to 75,000 in salary, than he or she is to go from 55,000 to 65,000, and therefore, 65,000 is not the midvalue point. The midvalue point must be larger than 65,000.*

In summary, we have determined that for this DM, (55,000; 18) **I** (65,000; 12), and (75,000;12) **P** (65,000; 18); therefore, $x_1^{.5} > 65,000$.

Suppose that the analyst now guesses that 68,000 is the midvalue point for (55,000; 75,000) for the X_1 attribute. Then the analyst chooses a good value for vacation days, say 18, and asks the DM to give the value X_2' such that the DM is indifferent to (55,000; 18) and (68,000; x_2'). Suppose that the DM determines that $x_2' = 10$, that is, the DM is indifferent to (55,000; 18) and (68,000; 10).

Now, the analyst asks the DM to rank the following two outcomes: (68,000; 18) and (75,000; 10). Suppose the DM says that he or she is indifferent to (68,000; 18) and (75,000; 10). Therefore, the DM is willing to give up the same amount in vacation days to go from 50,000 to 68,000 in annual salary as he or she is to go from 68,000 to 75,000 in annual salary, and 68,000 *is the midvalue point* for [55,000; 75,000]; that is, $v_1(68,000) = .5$. The DM is indifferent to the following pairs of outcomes:

1. (55,000; 18) and (68,000; 10)
2. (68,000; 18) and (75,000; 10)

From an algebraic perspective, let's see why $v_1(68{,}000) = .5$. We know that since the DM is indifferent to (55,000; 18) and (68,000; 10), and also between (68,000; 18) and (75,000; 10), both when $X_3 = 200$. Therefore, the value function values for these two pairs of outcomes must be equal, as shown in the following equations:

$$v(55{,}000;18;\ 200) = v(68{,}000;10;\ 200) \tag{3.30}$$

and

$$v(68{,}000;18;\ 200) = v(75{,}000;10;\ 200), \tag{3.31}$$

or, since we have an additive, scaled MAV function, we therefore have

$$\begin{aligned} &w_1v_1(55{,}000) + w_2v_2(18) + w_3v_3(200) \\ &\quad = w_1v_1(68{,}000) + w_2v_2(10) + w_3v_3(200), \end{aligned} \tag{3.32}$$

$$\begin{aligned} &w_1v_1(68{,}000) + w_2v_2(18) + w_3v_3(200) \\ &\quad = w_1v_1(75{,}000) + w_2v_2(10) + w_3v_3(200). \end{aligned} \tag{3.33}$$

Since the MAV function is scaled, we can replace $v_1(55{,}000)$ and $v_1(75{,}000)$ with 0 and 1, respectively, and therefore have the equivalent system of equations given by

$$w_2v_2(18) + w_3v_3(200) = w_1v_1(68{,}000) + w_2v_2(10) + w_3v_3(200) \tag{3.34}$$

and

$$w_1v_1(68{,}000) + w_2v_2(18) + w_3v_3(200) = w_1 + w_2v_2(10) + w_3v_3(200). \tag{3.35}$$

Now, subtracting (3.34) from (3.35), we obtain

$$w_1v_1(68{,}000) = w_1(1 - v_1(68{,}000)), \tag{3.36}$$

or dividing both sides by w_1, we obtain

$$v_1(68{,}000) = 1 - v_1(68{,}000) \tag{3.37}$$

or

$$v_1(68{,}000) = .5. \tag{3.38}$$

TABLE 3.7

Points on the Graphs for the Individual Attribute Value Functions

x_1	$v_1(x_1)$	x_2	$v_2(x_2)$	x_3	$v_3(x_3)$
55,000	0	5	0	1200	0
62,000	0.25	11	0.25	1000	0.25
68,000	0.5	15	0.5	800	0.5
72,000	0.75	18	0.75	500	0.75
75,000	1	20	1	50	1

As mentioned earlier, we can obtain other midvalue points for the X_1 attribute in order to find several points on the v_1 graph; for example, the midvalue point for x_1^0 and $x_1^{.5}$, denoted as $x_1^{.25}$, and the midvalue point for $x_1^{.5}$ and x_1^1, denoted as $x_1^{.75}$. Note that we might want to form *consistency checks* in order to assure, for example, that $x_1^{.5}$ is the midvalue point for $(x_1^{.25}, x_1^{.75})$.

Following the assessment of v_1, the analyst would want to assess the individual attribute value functions for v_2 and v_3, again using the midvalue splitting technique. Let's suppose that the points on the graphs have been assessed and are shown in Table 3.7 and that their corresponding graphs are shown in Figure 3.2.

Interpolation (either linear or something more complex) can be done to find other points on the graph.

3.7.2 Determining the Scaling Constants (Weights) for the Individual Attribute Value Functions

The next step in the assessment process involves the determination of the scaling constants: $w_1, w_2, \ldots$ Keep in mind that since we are using a scaled value function, each of these scaling constants must have respective values of between 0 and 1, and they must sum to a value of 1. There are various approaches for determining these weights, and the initial procedure described here follows Keeney and Raiffa (1993, pp. 121–123).

Just as in the assessment of the individual attribute value functions, we will glean information about the DM's preference structure from answers to questions concerning his or her trade-offs between the various pairs of attributes and then use that information to form equations about the DM's value function in order to determine the w_1, w_2, and so on. First, we need to introduce some additional notation. Let

$B_j(x_j)$ be the outcome associated with all attributes having their best values, except for attribute X_j, which has value x_j.

$W_j(x_j)$ be the outcome associated with all attributes having their worst values, except for attribute X_j, which has value x_j.

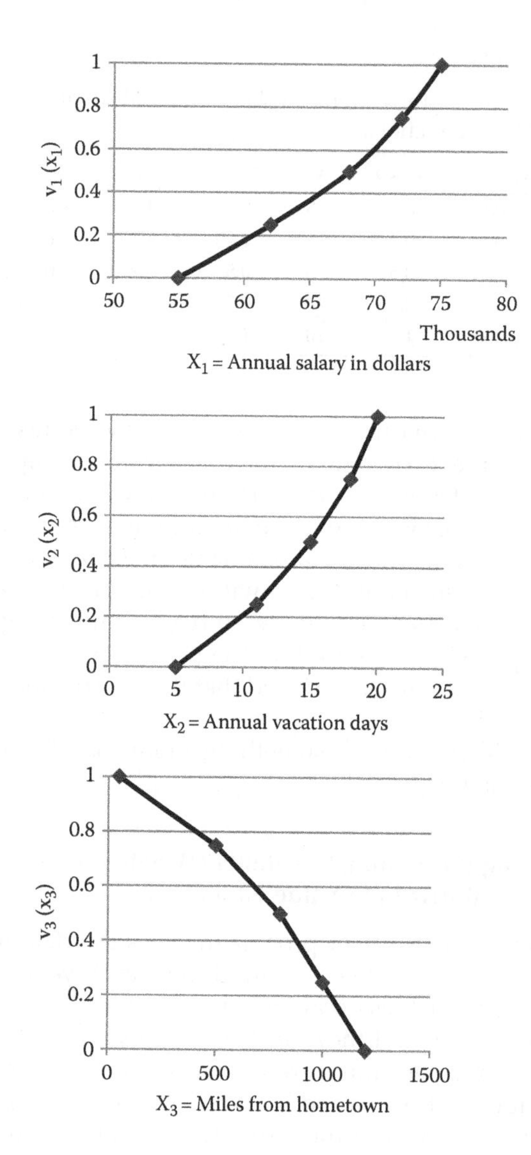

FIGURE 3.2
Graphs for the individual attribute value functions.

The steps of the process are as follows:

1. Have the DM rank the outcomes: $W_j(x_j^b)$, for j = 1,…, p. Suppose that these outcomes are ranked according to $j_1, j_2, \ldots, j_p$, where j_1 refers to the attribute number associated with the first-ranked outcome, $W_{j1}(x_{j1}^b)$; j_2 refers to the attribute number with the second-ranked outcome, $W_{j2}(x_{j2}^b)$; and so on.

2. Have the DM identify a sequence of (p − 1) values for attribute j_1 (i.e., values for $x_{j1}^2, x_{j1}^3, \ldots, x_{j1}^p$ where these numbers identify a sequence of p − 1 values for X_{j1}) such that the DM is indifferent to the following (p − 1) pairs of outcomes:

   $W_{j1}(x_{j1}^2)$ and $W_{j2}(x_{j2}^b)$, $W_{j1}(x_{j1}^3)$ and $W_{j3}(x_{j3}^b)$,…, and $W_{j1}(x_{j1}^p)$ and $W_{jp}(x_{jp}^b)$
3. Solve the system of simultaneous linear equations associated with the indifferences expressed in step 2, augmented with the normalization equation: $\sum_{i=1}^{p} w_i = 1$.

The first step in the process allows an ordering of the weights from largest to smallest; for example, if outcome $W_1(x_1^b)$ is preferred to outcome $W_2(x_2^b)$, then the value function value of outcome $W_1(x_1^b)$ must be greater than the value function value of $W_2(x_2^b)$, or

$$v(W_1(x_1^b)) > v(W_2(x_2^b)) \rightarrow w_1 > w_2.$$

Note that, for example, just because w_1 is greater than w_2, *this does not imply* that attribute 1 is *more important* than attribute 2. It does imply, however, that outcome $W_1(x_1^b)$ is preferred to outcome $W_2(x_2^b)$.

The second step of the process of determining the weights allows the analyst to set up a system of simultaneous equations. For example, if $w_1 > w_2$, and the DM is indifferent to the following two outcomes: $W_1(x_1')$ and $W_2(x_2^b)$, for a particular value for X_1, denoted as x_1', then the value function values for these two outcomes must be equal; that is,

$$v(W_1(x_1')) = v(W_2(x_2^b)),$$

or

$$w_1 v_1(x_1') = w_2 v_2(x_2^b), \text{ or since } v_2(x_2^b) = 1 \text{ because } v_2 \text{ is scaled from 0 to 1,}$$

$$w_1 v_1(x_1') = w_2.$$

Now, since we have already assessed the individual attribute value functions, we can determine the value for $v_1(x_1')$, and therefore, we have an equation in terms of w_1 and w_2.

Let's work out this process for Example 3.1, given the individual attribute value functions shown in Figure 3.2 and Table 3.7. Step 1 involves the ranking of three outcomes, since p = 3, for this example. Let's suppose that the DM ranks the relevant outcomes as follows:

$$W_1\left(x_1^b\right) \mathbf{P}\, W_3\left(x_3^b\right) \text{ and } W_3\left(x_3^b\right) \mathbf{P}\, W_2\left(x_2^b\right),$$

or using the numbers shown in Table 3.2:

$$(75{,}000;\ 5;\ 1{,}200)\ \mathbf{P}\ (55{,}000;\ 5;\ 50) \text{ and } (55{,}000;\ 5;\ 50)\ \mathbf{P}(55{,}000;\ 20;\ 1{,}200).$$

Therefore, we know that $w_1 > w_3$ and $w_3 > w_2$.

Now, for the second step, the analyst helps the DM determine two specific values for annual salary (the attribute with the largest of the three weights), denoted as x_1^2 and x_1^3 such that the DM is indifferent to two pairs of outcomes as follows:

$$W_1\left(x_1^2\right)\ \mathbf{I}\ W_3\left(x_3^b\right) \text{ and } W_1\left(x_1^3\right)\ \mathbf{I}\ W_2\left(x_2^b\right),$$

or

$$(x_1^2,\ 5,\ 1{,}200)\ \mathbf{I}\ (55{,}000,\ 5,\ 50) \text{ and } (x_1^3,\ 5,\ 1{,}200)\ \mathbf{I}\ (55{,}000,\ 20,\ 1{,}200).$$

As stated earlier with this type of questioning process, the DM would want to *hone in* on the values for x_1^2 and x_1^3. Suppose that these values turn out to be $x_1^2 = 67{,}000$ and $x_1^3 = 70{,}000$. Note that the value associated with the first indifference indicates that the DM is willing to give up $12,000 in annual salary in order to be 1,150 miles closer to his or her hometown with a job that allows for 5 days of vacation and that the second indifference indicates that the DM is willing to give up $15,000 in annual salary in order to receive 15 additional days of vacation, when he or she is at a job that is 1,200 miles from his or her hometown.

Using linear interpolation with Table 3.7, one finds that

$$v_1(67{,}000) = .458 \text{ and } v_1(70{,}000) = .625,$$

and therefore with the indifferences expressed by the DM

$$(67{,}000;\ 5;\ 1{,}200)\ \mathbf{I}\ (55{,}000;\ 5;\ 50), \text{ and}$$

$$(70{,}000;\ 5;\ 1{,}200)\ \mathbf{I}\ (55{,}000;\ 20;\ 1{,}200), \text{or}$$

$$w_1 v_1(67{,}000) = w_2,$$

and

$$w_1 v_1(70{,}000) = w_3, \tag{3.39}$$

or

$$.458 w_1 = w_2$$

TABLE 3.8

Value Function Values and Rankings for Nondominated Alternatives of Example 3.1

A_i	x_1	x_2	x_3	$v_1(x_1)$	$v_2(x_2)$	$v_3(x_3)$	$v(x)$	Ranking
1	70,000	5	150	.625	0.	.945	.586	2
2	60,000	15	250	.178	.5	.889	.462	6
3	75,000	7	1200	1.	.083	0.	.498	5
4	65,000	10	50	.375	.208	1.	.526	3
5	55,000	20	150	0.	1.	.945	.504	4
8	70,000	10	250	.625	.208	.889	.612	1

and

$$.625w_1 = w_3 \tag{3.40}$$

Combining (3.39) and (3.40) along with the normalizing equation, $w_1 + w_2 + w_3 = 1$, gives us

$$w_1 + .458w_1 + .625w_1 = 1,$$

or

$$w_1 = .48 \rightarrow w_2 = .458w_1 = .458(.48) = .22 \text{ and } w_3 = .625w_1 = .625\,(.48) = .3.$$

So, in summary, we have

$$w_1 = .48,\ w_2 = .22, \text{ and } w_3 = .3.$$

These weights, combined with the additive specification and the results for the individual attribute functions shown in Table 3.7 and Figure 3.2 complete the assessment of the MAV function for Example 3.1.

Using the assessed value function, and linear interpolation between the assessed points for the individual attribute value functions, we obtain the value function levels and ranking for the nondominated job offers of Example 3.1, as shown in Table 3.8.

3.8 An Easier Approach to Assessment: The Simple Multiattribute Rating Technique

The approach described in Section 3.7 has been found to be onerous, especially when the MAV function contains more than two or three attributes. As a result, a method called the "simple multiattribute rating technique"

(SMART) has been developed by Edwards (see Edwards, 1977; Edwards and Barron, 1994; von Winterfeldt and Edwards, 1986, pp. 278–286).

SMART has been developed through the years as an entire collection of techniques since its original development. The original procedure consisted of 10 steps as described in von Winterfeldt and Edwards, 1986, but several of these steps involve the aspect of problem structuring, which we discussed in Chapter 2. Hence, in our description, we will assume that we have already defined the attributes, alternatives, and the outcomes for those alternatives in terms of the attribute values.

SMART assumes an additive, scaled MAV function. The *first step* involves an assessment of the individual attribute value functions: $v_1, v_2, \ldots, v_p$ using a direct rating technique, which is much simpler than the midvalue splitting technique described in the previous section. This direct rating technique assigns a value of 0 to the worst attribute value over all outcomes, and a value of 100 to the best attribute value over all outcomes. (The idea is that it will be easier for the DM to think in terms of whole numbers instead of fractions when assigning intermediate values.) The DM is then asked to just assign intermediate values between 0 and 100 to the outcomes associated with the other levels for the attribute, keeping in mind the difference in value for a particular attribute level from the extremes. As an example, consider attribute 3 (number of miles from hometown) for Example 3.1; a particular DM might think in terms of days of driving for the distance in that anything between 450 and 800 miles would require an extra day of driving—hence, the difference in value between 500 and 700 miles is not nearly as great as the difference in value between 300 and 500 miles.

The *second step* of the process involves deriving the weights for the individual attribute value functions: $w_1, w_2, \ldots, w_p$, given in increasing order. As in the determination of the individual attribute value functions, the DM gives "direct numerical ratio judgments" (von Winterfeldt and Edwards, 1986, p. 281) for the weights. First, the DM rank orders the weights. Following this, the DM assigns an importance of 10 (denoted as w_1') to the least important weight, and then proceeds to assign numbers (denoted as $w_2', w_3', \ldots, w_p'$) to the other weights in order of increasing importance; there is no particular guidance in this step, and the "importance weights" have no particular bounds. These "importance weights" are then normalized by

$$w_j = \frac{w_j'}{\sum_{i=1}^{p} w_i'}.$$

The outcomes are then "scored" through the use of the following calculation:

$$\text{Outcome i score} = \sum_{j=1}^{p} w_j v_j(x_j(A_i)).$$

Note that for this calculation, the scores will range from 0 to 100 instead of from 0 to 1 as is the case for a scaled additive MAV function discussed in the previous section, since the individual attribute value functions are scaled from 0 to 100.

In order to illustrate the process, let's consider Example 3.1. Even with the same DM, one would not necessarily expect the same value function (even if one accounts for the rescaling) to result, since a different set of questions is being asked to form the function. Let's suppose that the answers from the DM give the results as shown in Tables 3.9 through 3.11 for the individual attribute value functions, importance weights, and normalized weights, respectively.

Using the results given in Tables 3.9 through 3.11, the scores for the non-dominated outcomes for Example 3.1 are shown in Table 3.12.

Note that the rankings for this particular application of SMART differ slightly from the rankings given in the previous section; namely, the rankings for the fourth and fifth ranked job offers are switched.

TABLE 3.9

Assessments for the Individual Attribute Value Functions (Scaled from 0 to 100) of Example 3.1 Using the Simple Multiattribute Rating Technique

x_1	$v_1(x_1)$	x_2	$v_2(x_2)$	x_3	$v_3(x_3)$
55,000	0	5	0	1200	0
60,000	20	7	20	250	75
65,000	40	10	35	150	85
70,000	75	15	45	50	100
75,000	100	20	100		

TABLE 3.10

Unnormalized Weights for Example 3.1 Using the Simple Multiattribute Rating Technique

j	w'_j
1	250
2	100
3	140

TABLE 3.11

Normalized Weights for Example 3.1 Using the Simple Multiattribute Rating Technique

j	w_j
1	0.51
2	0.20
3	0.29

TABLE 3.12

Scores for the Nondominated Outcomes of Example 3.1 Using the Simple Multiattribute Rating Technique

A_i	x_1	x_2	x_3	$v_1(x_1)$	$v_2(x_2)$	$v_3(x_3)$	$v(x)$	Ranking
1	70,000	5	150	75	0	85	62.9	2
2	60,000	15	250	20	45	75	40.95	6
3	75,000	7	1200	100	20	0	55	4
4	65,000	10	50	40	35	100	56.4	3
5	55,000	20	150	0	100	85	44.65	5
8	70,000	10	250	75	35	75	67	1

3.8.1 Swing Weights

An argument could be made that a DM would have fairly similar individual attribute value functions with either the midvalue splitting technique or the SMART approach. For example, one might reasonably expect the DM to employ arguments similar to the one given earlier with respect to the values attached to various levels for attribute 3 (miles from hometown) for Example 3.1.

However, the weights associated with the individual attribute value functions are a different matter. These weights cannot be inherently related quite so easily to a nebulous concept of "importance," as must be done with the approach described earlier for the SMART procedure. The weights obtained for the individual attribute value functions using the SMART approach described earlier therefore could be very inaccurate as a representation for the DM's preference structure.

The first step in determining the attribute weights using the *swing weighting method* is to rank the attributes by decreasing weight. This is accomplished by asking the DM which attribute he or she would like most to move from its worst value to its best value, with all other attributes at particular levels. (The reader will recognize that this is actually equivalent to the ranking approach used for the weighting method described in Section 3.7). This attribute is then assigned a relative weight of 100. The DM is then asked to assign a relative weight (less than 100) to the second-ranked attribute and so on. These relative weights, denoted as $r_1, r_2, \ldots, r_p$, are then normalized so that they sum to 1.

For the case of Example 3.1, the types of questions asked and example answers are shown in Figure 3.3.

With the answers given in Figure 3.3, the relative weights are given by $r_1 = 100$, $r_3 = 70$, and $r_2 = 50$. The normalized weights would be given by

$$w_1 = r_1/(r_1 + r_2 + r_3) = 100/(100 + 70 + 50) = .45$$

$$w_2 = r_2/(r_1 + r_2 + r_3) = 50/(100 + 70 + 50) = .23$$

$$w_3 = r_3/(r_1 + r_2 + r_3) = 70/(100 + 70 + 50) = .32.$$

Analyst: Given that the other attributes are at reasonable levels, which change in attribute level would you most like to make?

a. X_1 from 55,000 to 75,000

b. X_2 from 5 to 20

c. X_3 from 1,200 to 50

DM: I would rank these options in order of a, c, and then b, in order of decreasing preference.

Analyst: OK. Let's give a value of 100 to option a. In terms of degree of relative preference, assign numbers to options c and b, respectively, which indicate the amount of preference that you would have for each of these options.

DM: I would give a preference value of 70 to option c and 50 to option b.

FIGURE 3.3
Questions and answers to determine the swing weights for the case of Example 3.1.

Note that there is no particular reason to expect the weights to be the same as they were in our previous assessments. However, one could at least expect that the sets of weights would be close to each other.

Schuwirth et al. (2012) noted that an outcome in which all attribute values are at their worst levels may be difficult for the DM(s) to consider; hence, they developed a modification to the swing procedure, called the *reverse swing* procedure, which starts with all attribute values at their best levels.

Example 3.4: Resource Allocation in a Project Using an MAV Function

Consider a hypothetical project consisting of 9 activities or tasks. The activity durations, precedence relationships, early start times, early finish times, late start times, late finish times, and slack times for the activity are shown in Table 3.13.

TABLE 3.13
Early Start, Early Finish, Late Start, Late Finish, and Slack Times for Project

Activity	Immediate Predecessors	Normal Duration (weeks)	Early Start (ES)	Early Finish (EF)	Late Start (LS)	Late Finish (LF)	Slack Time (ST)
A	—	10	0	10	0	10	0
B	—	8	0	8	1	9	1
C	A	14	10	24	10	24	0
D	B	15	8	23	9	24	1
E	B	14	8	22	15	29	7
F	C,D	5	24	29	24	29	0
G	C,D	9	24	33	30	39	6
H	E,F	10	29	39	29	39	0
I	G,H	6	39	45	39	45	0

Immediate predecessor activities are those activities that must be finished before another activity can start. For example, both activities C and D must be finished before activity F can start. Calculations for the early start (ES), early finish (EF), late start (LS), late finish (LF), and slack times (ST) were accomplished through the standard program evaluation and review technique-critical path method (PERT-CPM) scheduling techniques. As is the convention, these times are relative to "time 0" that is the start time of the project.

The ES times are the earliest times that respective activities can start, while the LS times are the latest times that respective activities can start without delaying the project past its earliest finish time (45 weeks). The ST times are the amounts of times that respective activities can be delayed without delaying the project past its earliest finish time. These times *are not independent* of each other; that is, if some of an earlier activity's slack is used, then that impacts the slack of a later activity. The activities with 0 slack time (called critical activities) form the critical path(s). Any delays in these activities will result in a delay of the project past its earliest possible finish time of 45 weeks.

In this example, we are concerned with what is often called the "time-cost trade-off" problem in project management. In particular, let's suppose that the organization wants to allocate more resources to certain activities in order to complete these activities in less time than their normal durations, in order to complete the entire project sooner. The problem might be stated as "which activities should we choose to reduce in duration, and how much should we reduce their durations in order to achieve a particular project duration at a minimum cost." The cost increases for the respective activities are increases in *direct cost* (see Chapter 2). If the relationships between increases in direct cost and reductions in activity durations are linear in nature, then a linear program can be formulated to solve the problem (see Moder et al., 1995, pp. 261–264). However, such linear relationships are rarely found in practice.

An obvious approach is to allocate more resources to activities on the critical path, since the critical path is the sequence of activities which determines the project duration. What makes this approach difficult is the nonlinear relationship between the increase in the direct cost for an activity and the resulting decrease in activity duration; in addition, the critical path (and therefore the critical activities) can change as selected activity durations are reduced. Various heuristic algorithms, which iteratively select activities to reduce in duration, have been developed to address this problem; for example, see Siemens and Gooding (1975).

In this example, we are mainly concerned with an MAV function involving two attributes: project duration and the increase in direct cost as a result of reducing the durations of selected activities. In particular, we are not concerned with a large number of alternative solutions.

Let's suppose that if the project can be finished in 42 weeks, a bonus of $50,000 will be received by the organization; on the other hand, if the project is finished in 45 weeks or longer, the organization will be charged a penalty of $50,000. Given that a bonus is received or that a penalty is avoided, there is an inherent advantage, in terms of the organization's reputation, for example, of finishing the project in less time rather than more time; for example, finishing the project in 43 weeks is better than finishing the project in 44 weeks, all other things being equal, even though in both cases the penalty is avoided and the bonus is not attained.

The project team has developed three options for reducing individual activity durations. These options correspond to allocating more resources and/or working overtime on the relevant specific activities. These options will result in respective increases in direct cost for the relevant activities, as shown in Table 3.14.

The options in Table 3.14 can be combined to form independent alternatives. The various alternatives (including the "do nothing" alternative) are shown in Table 3.15, along with the resulting increase in direct cost and the resulting project duration.

Now, it is apparent from viewing Table 3.15 that some of the alternatives are dominated and can therefore be eliminated from further consideration. In particular, alternatives 3, 5, 7, and 8 can be eliminated since they correspond to respective alternatives with a lower increase in direct cost, but with the same project duration; for example, alternatives 3 and 5 have the same project duration as alternative 2, but with a larger increase in direct cost. Hence, we are left with alternatives 1, 2, 4, and 6, as indicated by the *'s in Table 3.15.

TABLE 3.14

Options for Reducing Activity Durations for the Project

Option	Description of Option	Increase in Direct Cost
1	Reduce activity A duration by 2 weeks (from 10 to 8 weeks).	$60,000
2	Reduce activity C duration by 3 weeks (from 14 to 11 weeks).	$80,000
3	Reduce activity I duration by 2 weeks (from 6 to 4 weeks).	$70,000

TABLE 3.15

Alternatives for Resource Allocation and Resulting Attribute Values

Alternative	Description	Increase in Direct Cost	Resulting Project Duration (weeks)
1*	Do nothing	$0	45
2*	Option 1	$60,000	44
3	Option 2	$80,000	44
4*	Option 3	$70,000	43
5	Options 1 and 2	$140,000	44
6*	Options 1 and 3	$130,000	42
7	Options 2 and 3	$150,000	42
8	Options 1, 2, and 3	$210,000	42

*Nondominated alternatives/outcomes.

Now, suppose that an analyst has assessed the MAV function for the project manager. What should this function look like? Let project duration in weeks be denoted by X_1 and increase in direct cost in thousands of dollars be denoted by X_2. Suppose that the analyst has determined that the MAV function is additive. Therefore, it can be written as

$$v(x_1, x_2) = w_1 v_1(x_1) + w_2 v_2(x_2).$$

The best and worst values for X1 (project duration) are 42 and 45, respectively, and the best and worst values for increase in direct cost are 0 and 130, respectively. We also know that both $v_1(x_1)$ and $v_2(x_2)$ are decreasing in x_1 and x_2, respectively, with $v_1(42) = 1$, $v_1(45) = 0$, $v_2(0) = 1$, and $v_2(130) = 0$.

Suppose that the assessment process yields an MAV function as indicated by Tables 3.16 and 3.17, with $w_1 = .6$ and $w_2 = .4$.

Using this value function, we can compute the values associated with each outcome, as shown in Table 3.18.

As can be seen from Table 3.18, alternative 6, with an outcome that provides for a project duration of 42 weeks and an increase in direct cost of $130,000, is the preferred alternative.

TABLE 3.16

Single Attribute Value Function for X_1, Project Duration in Weeks

x_1	$v_1(x_1)$
45	0
44.5	0.4
44	0.45
43.5	0.5
43	0.55
42.5	0.6
42	1.

TABLE 3.17

Single Attribute Value Function for X_2, Increase in Direct Cost, in Thousands of Dollars

x_2	$v_2(x_2)$
130	0
110	0.25
80	0.5
45	0.75
0	1.

TABLE 3.18

Value Function Values for Nondominated Outcomes

Alternative	x_1	x_2	$v_1(x_1)$	$v_2(x_2)$	$v(x_1, x_2)$
1	45	0	0	1	0.4
2	44	60	0.45	0.642	0.5268
4	43	70	0.55	0.5714	0.5585
6	42	130	1	0	0.6

Let's consider the MAV function for this example in more detail. Suppose that we are trying to find the value of x_2' for which the project manager will be indifferent to the following two outcomes:

$$(45, 0) \text{ and } (44, x_2').$$

This value must be equal to the penalty avoided by the organization from finishing the project in less than 45 weeks plus any additional value (in thousands of dollars) from having the organization finish the project in 44 weeks rather than 45 weeks. Checking the validity of this value could also be considered as a check on the validity of the DM's value function. Solving for the equation given by

$$v(45,0) = v(44,x_2')$$

for x_2' gives us

$$x_2' = 111.$$

Hence, the DM (project manager) should be indifferent to the outcomes of (45, 0) and (44, 111), or, in other words, the reduction in project duration from 45 to 44 weeks is worth $111,000, which is $61,000 plus the $50,000 associated with the averted penalty.

3.9 Final Comments on the Assessment of MAV Functions

Obviously, the assessment of an MAV function is quite an involved process, even if one is employing the SMART procedure developed by Edwards. However, if one uses the suggested approach of "honing in" on outcomes for which the DM is indifferent, then this procedure can involve only pairwise comparisons. Keeney (2002) discusses 12 common mistakes that DMs make in providing value trade-offs, along with approaches for addressing these mistakes.

Even if the DM does not use the resulting MAV function, just going through the assessment process can be helpful to the DM in thinking about his or her trade-offs.

The assessment process involves ranking pairs of outcomes that are hypothetical in nature—that is, these outcomes do not necessarily correspond to the alternatives. By not directly addressing the outcomes associated with the actual alternatives, one removes any subjective "attachments" to these alternatives from the assessment process. In addition, the reader should keep in mind that the basic idea is to obtain a function that will allow for the ranking of outcomes over the "entire" outcome space. Alternatives, and corresponding outcomes, may arise in the future, which differs from the original set of alternatives.

A different approach to assessment would involve just having the DM provide scores, of say between 0 and 100, for a large number (e.g., 100) of hypothetical outcomes; an outcome with a higher score would be preferred over an outcome with a lower score. Interpolation could be used to provide scores to any outcomes that were not scored in the original set. A disadvantage of this approach is that the DM may very well exhibit inconsistencies in providing such a large number of scores.

Finally, even if no formal approach is used to rank multidimensional outcomes, the DM will be making trade-offs between multiple objectives in an implicit way. The use of an MAV function makes these trade-offs explicit so that the decision-making process can be more easily explained to others.

The MAV function discussed in this chapter does not address the strength of preference of one outcome over another. For example, if outcomes A, B, and C have value function levels of .9, .8, and .7, respectively, one *cannot* say that the difference in preference between A and B is the same as the difference in preference between B and C (even though the differences in their value function levels are the same at .1). One can only *rank order* the outcomes. In order to consider the strength of preferences, one must employ a *measurable value function*, as presented by Dyer and Sarin (1979). Such measurable value functions "provide an interval scale of measurement for preferences under certainty" (Dyer and Sarin, 1979, p. 810) and also allow for more complex forms for the MAV function such as a multiplicative form.

3.10 TOPSIS: An Approach That Considers the Weighted Distances from the Positive and Negative Ideals

A method that involves relatively little input from the DM regarding his or her preference structure involves choosing a solution that considers weighted distances from the positive and negative ideals. Obviously, one wants to obtain an outcome as close as possible to the positive ideal and as far as possible from the negative ideal, while considering the importance of each attribute. This method, entitled the *T*echnique for *O*rder *P*reference by *S*imilarity to *I*deal *S*olution (TOPSIS), was originally proposed by

Hwang and Yoon (1981), and has since been extended by Yoon (1987) and Hwang et al. (1993), among others.

In order to simplify the following description, we will employ some additional notation:

$$x_{ij} = x_j(A_i), \text{ the value for alternative i on attribute j,} \tag{3.41}$$

where

$$J^* \text{ is the set of attributes that are to be maximized} \tag{3.42}$$

$$J' \text{ is the set of attributes to be minimized} \tag{3.43}$$

Once the alternatives and outcomes have been determined, as described in Section 3.2, the steps of the algorithm are as follows:

1. The normalized outcomes are computed by

$$r_{ij} = x_{ij} \Bigg/ \sqrt{\sum_{i'=1}^{n} x_{i'j}^2} \tag{3.44}$$

   where r_{ij} denotes the normalized outcome associated with attribute j of alternative i for i = 1,…, n and j = 1,…, p. Note that this normalization process places each of the attribute levels on basically an equal footing with respect to scale.
2. Compute the weighted values, v_{ij}, for alternative i as scored on attribute j:

$$v_{ij} = w_j r_{ij}, \tag{3.45}$$

   where

$$\sum_{j=1}^{p} w_j = 1 \tag{3.46}$$

$$0 < w_j < 1 \text{ for } j = 1,\ldots,p \tag{3.47}$$

   The weights could be computed using the method of Section 3.7 or either of the methods discussed in Section 3.8, including swing weights. (Note that if the method of Section 3.7 is used, then the individual attribute value functions would have already been computed.) Alternatively, the DM could just provide weights for each of the attributes without the assistance of any particular technique.

3. Compute a weighted positive ideal, PI, and a weighted negative ideal, NI, as given by

$$PI = \left(v_1^*, v_2^*, \ldots, v_p^*\right), \tag{3.48}$$

$$NI = (v_1', v_2', \ldots, v_p'), \tag{3.49}$$

where

v_j^* is the maximum value over $\{v_{ij}$ for $i = 1, \ldots, n\}$ for $j \in J^*$ and minimum value over $\{v_{ij}$ for $i = 1, \ldots, n\}$ for $j \in J'$

v_j' is the minimum value over $\{v_{ij}$ for $i = 1, \ldots, n\}$ for $j \in J^*$ and maximum value over $\{v_{ij}$ for $i = 1, \ldots, n\}$ for $j \in J'$

4. Compute the separation measures for each alternative, where

S_i^* = Separation measure for the positive ideal for alternative i,

$$= \sqrt{\sum_{j=1}^{p} (v_j^* - v_{ij})^2} \tag{3.50}$$

S_i' = Separation measure for the negative ideal for alternative i,

$$= \sqrt{\sum_{j=1}^{p} (v_j' - v_{ij})^2} \tag{3.51}$$

5. Compute the relative closeness to the ideal for each alternative, where

C_i is the closeness for alternative i,

$$= S_i' / (S_i^* + S_i') \tag{3.52}$$

Note that the smaller the value of S_i^*, the closer that the outcome associated with alternative i is (in a weighted sense) to the positive ideal, and hence, the better alternative i is. Alternatively, the smaller the value of S_i', the closer that the outcome associated with alternative i is (in a weighted sense) to the negative ideal, and hence, the worse alternative i is. This means that the larger is the value of the relative closeness measure, C_i, the better is the outcome associated with alternative i, and hence, the alternatives should be ranked in terms of decreasing values for the C_i.

Example 3.5: Applying TOPSIS to Example 3.1, the Job Selection Problem

Let's apply TOPSIS to Example 3.1 and its nondominated outcomes, as shown in Table 3.19. For this example, we would want to maximize the first two attributes (annual salary, $x_1(A_i)$, and days of vacation, $x_2(A_i)$) and minimize the third attribute (miles from hometown, $x_3(A_i)$); hence, we would have

$$J^* = \{1, 2\} \quad \text{and} \quad J' = \{3\}.$$

Upon normalizing the outcomes, we would have the r_{ij} values as shown in Table 3.20.

For example, r_{23} is the normalized outcome associated with the attribute 3 value of alternative 2. Its value is computed as

$$r_{23} = x_{23}/(x_{13}^2 + x_{23}^2 + x_{33}^2 + x_{43}^2 + x_{53}^2 + x_{63}^2)^{.5}$$

$$= 250/(150^2 + 250^2 + 1200^2 + 50^2 + 150^2 + 250^2)^{.5}$$

$$= 0.196875.$$

TABLE 3.19

Nondominated Outcomes (Attribute Values) for Example 3.1

i	A_i	$x_1(A_i)$	$x_2(A_i)$	$x_3(A_i)$
1	1	70,000	5	150
2	2	60,000	15	250
3	3	75,000	7	1200
4	4	65,000	10	50
5	5	55,000	20	150
6	8	70,000	10	250

TABLE 3.20

r_{ij} Values (Normalized Outcomes) Associated with the Outcomes of Table 3.19

i|j	1	2	3
1	0.431844	0.166759	0.118125
2	0.370152	0.500278	0.196875
3	0.46269	0.233463	0.944999
4	0.400998	0.333519	0.039375
5	0.339306	0.667037	0.118125
6	0.431844	0.333519	0.196875

The next step is to compute the weighted values for the normalized outputs. Let's use the swing weights computed in Section 3.8:

$$w_1 = .45,\ w_2 = .23,\ \text{and}\ w_3 = .32.$$

These weighted normalized values are shown in Table 3.21.

The positive and negative weighted ideals are given as

$$PI = (0.20821,\ 0.153419,\ 0.0126),\ \text{and}$$

$$NI = (0.152687625,\ 0.038354647,\ 0.302399705).$$

Finally, the separation measures for the positive and negative ideals, the relative closeness values, and the rankings for each of the alternatives are shown in Table 3.22.

From a cursory look at the numbers in Tables 3.21 and 3.22, it is clear that alternative 3 suffers greatly from its value for attribute 3: 1200 miles from hometown, which is much larger than the value for any other alternative. To illustrate the effect of this particular attribute value for alternative 3, the TOPSIS algorithm was rerun with changes in the weights for the attributes. In particular, the weight for attribute X_3 was reset to .2, .1, and .05 in respective runs of TOPSIS, with corresponding changes to the weights for the other two attributes. The results are shown in Table 3.23. (Note that the relative closeness values have been rounded off to the nearest two decimal places.)

TABLE 3.21

v_{ij} Values (Weighted Normalized Outcomes) Associated with the Outcomes of Table 3.19

i|j	1	2	3
1	0.19433	0.038355	0.0378
2	0.166568	0.115064	0.063
3	0.20821	0.053697	0.3024
4	0.180449	0.076709	0.0126
5	0.152688	0.153419	0.0378
6	0.19433	0.076709	0.063

TABLE 3.22

Separation Measures, Relative Closeness Values, and Rankings for the Nondominated Outcomes of Example 3.1

Alternative, i	1	2	3	4	5	6
S_i^*	0.118606	0.075798	0.306477	0.081578	0.060974	0.092829
S_i'	0.267856	0.251772	0.057603	0.293642	0.288535	0.246003
C_i	0.693098	0.768606	0.158216	0.782586	0.825544	0.726033
Ranking	5	3	6	2	1	4

TABLE 3.23
Relative Closeness Values and Rankings for Alternatives for Technique for the Order of Preference by Similarity to Ideal Solution Runs with an Alternative Sets of Weights

Run Number	w_1	w_2	w_3	Alt 1 Score (Rank)	Alt 2 Score (Rank)	Alt 3 Score (Rank)	Alt 4 Score (Rank)	Alt 5 Score (Rank)	Alt 6 Score (Rank)
Original run	.45	.23	.32	.69 (5)	.77 (3)	.16 (6)	.78 (2)	.83 (1)	.73 (4)
Run #2	.53	.27	.20	.56 (5)	.70 (2)	.24 (6)	.66 (3)	.76 (1)	.63 (4)
Run #3	.6	.3	.1	.40 (5)	.62 (2)	.33 (6)	.51 (3)	.70 (1)	.51 (3)
Run #4	.63	.32	.05	.31 (6)	.59 (2)	.36 (5)	.41 (4)	.68 (1)	.45 (3)

The TOPSIS approach and its variations (including extensions to considerations of fuzzy inputs) have been applied in a wide variety of application areas, including supplier selection (Chen et al., 2006), evaluation of preventive measure for oil spill accidents (Krohling and Campanharo, 2011), and personnel selection (Boran et al., 2011), among others. See Behzadian et al. (2012) for a survey of TOPSIS applications.

3.11 AHP

3.11.1 Introduction to the AHP

The analytic hierarchy process (AHP) was developed by Thomas L. Saaty during the 1970s (Saaty, 1980) and can be executed through the software package: Expert Choice™ (http://expertchoice.com/comparion/). AHP can be described as a "methodology for modeling unstructured problems in the economic, social, and management sciences" (Saaty, 1980). AHP can be considered as an alternative to the assessment and use of a multiattribute value function for ranking alternatives, which can be represented as respective outcomes over multiple attributes. Saaty (1986) presented a set of axioms associated with the valid use of AHP, and Harker and Vargas (1987) reviewed the theory of AHP and addressed criticisms associated with its validity.

Although there is much controversy concerning the use of AHP which is mentioned later in this section, it is generally thought to be easier for a DM to use than an approach involving the assessment and use of an MAV function. For example, one advantage associated with AHP is that it *allows* the DM to be inconsistent in answering questions concerning preferences over multiple objectives; in addition, the method only requires the DM to answer questions with respect to pairwise comparisons of two objectives, subobjectives or outcomes at a time. (Saaty notes that the original idea involving pairwise comparisons came from interaction with his grandmother as he was growing up [Palmer, 1999].)

AHP has been applied in a wide variety of areas including decision making for facility layout (Cambron and Evans, 1991), health care (Liberatore and Nydick, 2008), and diagnosis (Saaty and Vargas, 1998). Forman and Gass (2001) discussed 26 successful applications of AHP, and Zahedi (1986) and Vaida and Kumar (2006) presented surveys of applications of AHP.

3.11.2 Steps of the AHP

The process of applying AHP for ranking alternatives, which can be represented as multiattributed outcomes, is described by the steps shown in Figure 3.4.

Further discussion on each of the steps shown in Figure 3.4 is given in the text following this figure.

Step 1: Develop a hierarchy consisting of an overall goal, criteria, subcriteria, and alternatives.

Step 2: For each level and section of the hierarchy, usually start from the highest level and work downward through the hierarchy:

a. Have the DM(s) perform *pairwise comparisons* for each pair of factors in the section.

b. Form a square *influence matrix*, denoted as A, from the pairwise comparisons, where the number of rows (and columns) of the matrix is equal to the number of factors in the section of the hierarchy being evaluated.

c. Find a value for a quantity, λ_{max}, the value for λ, from solving a system of equations, λ_{max} is called the *principal eigenvalue* of the system given by (3.53).

$$|A - \lambda I| = 0 \quad (3.53)$$

d. Determine the *local weights* (sometimes called *priorities*), denoted as w_i for i = 1,...,n, for each factor in this section of the hierarchy by solving a system of equations, as given in the following equations:

$$Aw = \lambda_{max} w, \quad (3.54)$$

$$\sum_{i=1}^{n} w_i = 1, \quad (3.55)$$

where n is the number of factors in this section of the hierarchy and $w = (w_1, w_2, ..., w_n)^T$ in (3.54). These weights indicate the relative importance of a factor with respect to the next higher-level factor in the hierarchy if that factor is a criterion or subcriterion; alternatively, the weight indicates how well that factor (alternative) does with respect to the factor (subobjective) at the next higher level of the hierarchy if the factor represents an alternative. The process of finding these local weights is often termed the *Eigenvalue Method.*

Step 3: Determine the *global weight* for each alternative on each lowest-level criterion in the hierarchy, by multiplying the weights "up" the branch in the hierarchy corresponding to that alternative.

Step 4: Determine the **overall priority** of each alternative by summing the alternative's global weights over all lowest-level subcriteria.

Step 5: Rank the alternatives in terms of their decreasing values for overall priority.

FIGURE 3.4
Steps of the analytic hierarchy process.

Step 1 of the AHP: Develop a hierarchy of factors representing the decision situation.

At the highest level of the hierarchy there is one factor, which might be considered the overall (or super) objective for the system; at the lowest level of the hierarchy, there are multiple factors, or more specifically, alternatives to be evaluated.

In particular, these factors might be termed as the foci (plural of focus), criteria, subcriteria, actors, objectives, subobjectives, attributes, and alternatives. The higher levels of the hierarchy contain the fundamental objectives, while the lower levels contain the subobjectives (or means objectives as termed in Chapter 2). The procedures discussed in Chapter 2 can be useful in generating the hierarchy. The lowest level of the hierarchy contains the alternatives to be evaluated and ranked. The nodes of the hierarchy represent these factors, while arcs are used to represent connections between these nodes. In addition, each factor can be connected to only one higher-level factor (differing from means objectives network of Chapter 2), while at the lowest level of the subobjectives hierarchy, all of the alternatives are connected with each of the lowest-level. An example (generic) hierarchy is shown in Figure 3.5.

Saaty provides suggestions for the construction of hierarchies in Chapter 3, along with several examples of hierarchies in Chapter 4 of his book (Saaty, 1990a).

Step 2a: Perform "pairwise comparisons" at each level and section of the hierarchy, starting from the top section and working downward.

The comparisons are typically made for two different categories of factors:

1. *Attributes/subobjectives/objectives*: Here, the questions asked concern *how much more important* one of these factors is than another factor in the same grouping with respect to the attribute/subobjective/objective with which they are associated at the level just above.

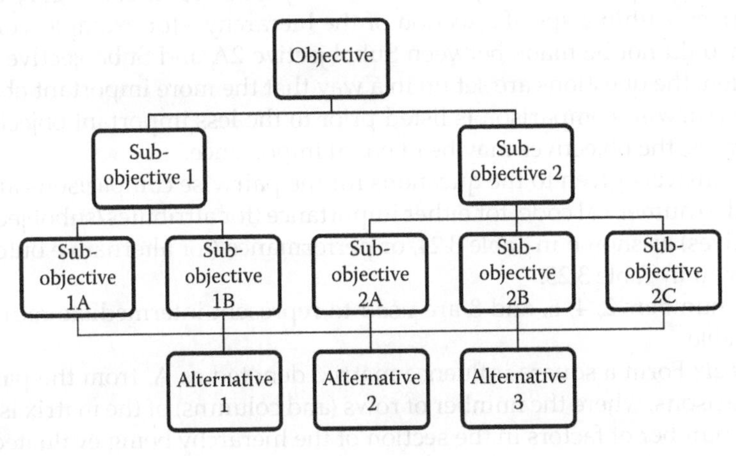

FIGURE 3.5
An example generic hierarchy for an AHP application.

2. *Alternatives*: Here, the questions asked concern *how much better* one alternative performs than another, with respect to each attribute at the next higher level in the hierarchy.

For the example hierarchy of Figure 3.5, the pairwise comparisons would be accomplished for the sections of the hierarchy as follows:

1. Subobjective 1 and Subobjective 2 with respect to the Objective
2. Subobjective 1A and Subobjective 1B with respect to Subobjective 1
3. Subobjective 2A, Subobjective 2B, and Subobjective 2C with respect to Subobjective 2
4. Alternative 1, Alternative 2, and Alternative 3 with respect to *each* of the Subobjectives 1A, 1B, 2A, 2B, and 2C

For example, with respect to the section of the example hierarchy containing Subobjectives 2A, 2B, and 2C, the questions that would be asked would include the following:

1. With respect to Objective 2, how much more important is Subobjective 2A than Subobjective 2B?
2. With respect to Objective 2, how much more important is Subobjective 2A than Subobjective 2C?
3. With respect to Objective 2, how much more important is Subobjective 2B than Subobjective 2C?

Note that the pairwise comparisons are only made for that category of subobjectives within a specific section of the hierarchy—for example, comparisons would not be made between Subobjective 2A and Subobjective 1A. In addition, the questions are set up in a way that the more important objective in the pairwise comparison is listed prior to the less important objective. In rare cases, the objectives may be of equal importance.

The answers given to the questions for the pairwise comparisons are converted to numerical code, for either importance (for attributes/subobjectives/objectives) as shown in Table 3.24, or performance (for alternative outcomes) as shown in Table 3.25.

The numbers 2, 4, 6, and 8 are used to represent intermediate ratings for each table.

Step 2b: Form a square influence matrix, denoted as A, from the pairwise comparisons, where the number of rows (and columns) of the matrix is equal to the number of factors in the section of the hierarchy being evaluated.

TABLE 3.24

Numeric Codes for Pairwise Ratings of Importance

Importance Rating	Numerical Code
Equal importance	1
Weak importance	3
Strong importance	5
Very strong importance	7
Absolute importance	9

TABLE 3.25

Numeric Codes for Pairwise Ratings of Performance

Performance Rating	Numerical Code
Equal performance	1
Weakly better	3
Strongly better	5
Very strongly better	7
Absolutely better	9

The influence matrix, $A = (a_{ij})$, is composed of the following elements:

1. For $i < j$, a_{ij} is given by the numbers corresponding to the answers provided by the DM from step 2a.
2. For $i = j$ (i.e., the diagonal of the A matrix), the value of 1.
3. For $i > j$, $1/a_{ji}$.

As an example, consider the Figure 3.5 hierarchy, and the portion of that hierarchy corresponding to Subobjective 1 and Subobjective 2. Suppose that the DM has already provided the information that Subobjective 1 is at least as important as Subobjective 2. Now, suppose that the interaction with the DM for this portion of the hierarchy is as follows:

Analyst/Software Package: How much more important is Subobjective 1 than Subobjective 2 with respect to the objective?

DM: I would say "strongly more important."

According to the numeric scale used by AHP as shown in Table 3.24, *strongly more important* (i.e., Subobjective 1 being of strong importance as compared to Subobjective 2) translates to the number 5. Hence, the A matrix would appear as

$$A = \begin{bmatrix} 1 & 5 \\ 1/5 & 1 \end{bmatrix}.$$

Step 2c: Find a value for a quantity, λ_{max}, the value for λ, from solving a system of equations: $|A - \lambda I| = 0$.

The determinant of the matrix formed by $(A - \lambda I)$ is computed. I is the identity matrix of appropriate dimension, and A is the matrix formed from the answers given by the DM with respect to the importance of one factor versus another, or the performance of one alternative versus another with respect to a particular subobjective/attribute.

The system of equations to be solved for the λ would appear as

$$\left| \begin{pmatrix} 1 & 5 \\ 1/5 & 1 \end{pmatrix} - \lambda I \right| = 0$$

or

$$\begin{vmatrix} 1-\lambda & 5 \\ 1/5 & 1-\lambda \end{vmatrix} = 0$$

or

$$1 - 2\lambda + \lambda^2 - 5(1/5) = 0$$

or

$$\lambda(\lambda - 2) = 0$$

or

$$\lambda = 2, \text{ and } \lambda = 0.$$

The maximum value found for λ, denoted as λ_{max}, is therefore given as $\lambda_{max} = 2$; this value is used in the next step of the process, as shown in Step 2d.

Step 2d: Determine the local weights, denoted as w_i for $i = 1,\ldots, n$, for each factor in this section of the hierarchy.

Determining the local weights is accomplished by solving a system of equations given by (3.54) and (3.55) in Figure 3.4. As an example, consider the A matrix and λ_{max} value found for the numeric example of Step 2c. The system of equations to be solved would appear as

$$\begin{bmatrix} 1 & 5 \\ 1/5 & 1 \end{bmatrix} w = 2w$$

$$w_1 + w_2 = 1$$

or

$$1w_1 + 5w_2 = 2w_1$$

$$1/5\, w_1 + 1w_2 = 2w_2$$

$$w_1 + w_2 = 1$$

or

$$-1w_1 + 5w_2 = 0$$

$$1/5\, w_1 - 1w_2 = 0$$

$$w_1 + w_2 = 1$$

Using the first and the third equations from this, one obtains $w_1 = 5/6$ and $w_2 = 1/6$.

Note that the numerical example shown earlier for computation of the local weights is about as simple as one can have, since there were only two factors for this section of the hierarchy. Also, when there are only two factors in the computation, one cannot have "inconsistency" on the part of the DM, since only one answer was provided by the DM in Step 2b of the process. Inconsistency can, and probably will, exist when the DM needs to provide two or more answers with respect to his or her pairwise comparisons (i.e., when the number of factors in the section of the hierarchy under consideration is greater than 2, or $n > 2$). The DM will be perfectly consistent in his or her answers if and only if λ_{max} has a value of n. In addition, the W matrix formed from the ratios of the weights will be equal to the A matrix if and only if the DM is perfectly consistent:

$$W = \begin{bmatrix} w_1/w_1 & w_1/w_2 & \cdots & w_1/w_n \\ w_2/w_1 & w_2/w_2 & & w_2/w_n \\ \vdots & \vdots & \ddots & \vdots \\ w_n/w_1 & w_n/w_2 & \cdots & w_n/w_n \end{bmatrix}$$

Let's consider a portion of a hierarchy involving three factors, as shown in Figure 3.6. In order to obtain the A matrix for this portion of the hierarchy, the DM might provide the following answers (assuming A_1 is more important than A_2 and A_3, and A_2 is equally as important as A_3):

A_1 is weakly more important than A_2, with respect to A.
A_1 is strongly more important than A_3, with respect to A.
A_2 is equal in importance to A_3, with respect to A.

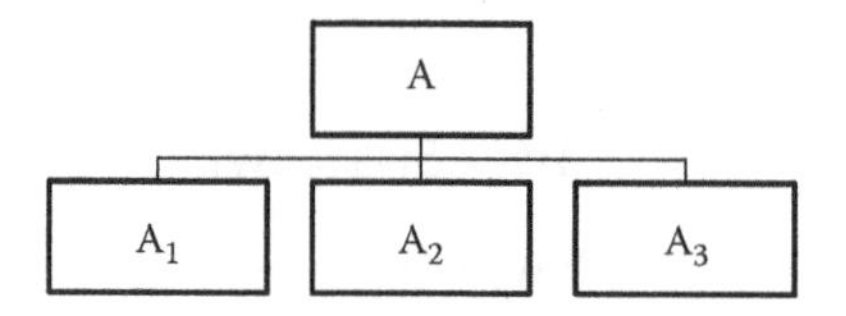

FIGURE 3.6
Portion of a hierarchy for the AHP.

$$A = \begin{bmatrix} 1 & 3 & 5 \\ 1/3 & 1 & 1 \\ 1/5 & 1 & 1 \end{bmatrix}$$

FIGURE 3.7
An A matrix resulting from inconsistent responses from the DM.

Now, looking at the first two statements, one might infer that A_2 is more important than A_3, with respect to A; hence, the third statement is inconsistent with the first two. The A matrix corresponding to this preference information is given by Figure 3.7.

Saaty (1990a, pp. 82–85) notes that almost all DMs will be inconsistent to a certain degree and a certain amount of inconsistency can be tolerated. He suggests a measure for consistency, termed the *consistency ratio* (*CR*), which should be allowed to have a maximum value of .1, or 10%. A value greater than 10% indicates that the DM should rethink his or her answers.

The consistency ratio, denoted as CR, is computed from the *consistency index*, CI, given by

$$CI = \frac{(\lambda_{max} - n)}{(n-1)}, \tag{3.56}$$

where

λ_{max} is the maximum eigenvalue associated with the solution to (3.53)
n is the number of factors (or the number of rows/columns) associated with the A matrix

(Note that if $\lambda_{max} = n$, which only occurs if the DM is perfectly consistent, then CI = 0.)

The consistency ratio, CR, is then given by

$$CR = \frac{CI}{RI}, \tag{3.57}$$

where RI denotes a "random index (RI)," which is the "average value" of CI for random matrices using the Saaty scale obtained by Forman

TABLE 3.26

Value of the Random Index as a Function of the Number of Rows of A

Number of Rows in A	RI
2	0
3	0.58
4	0.90
5	1.12
6	1.24
7	1.32
8	1.41
9	1.45
10	1.49

(Alonso and Lamata, 2006). The random index (RI) is a function of the number of rows/columns of the A matrix. The values for RI are shown in Table 3.26 (see Saaty, 1990a, p. 84).

As an example, consider the inconsistent A matrix shown in Figure 3.7. The process for finding the value for λ_{max} is shown in the following:

$$|A - \lambda I| = 0$$

or

$$\begin{vmatrix} 1-\lambda & 3 & 5 \\ 1/3 & 1-\lambda & 1 \\ 1/5 & 1 & 1-\lambda \end{vmatrix} = 0$$

or

$$-\lambda^3 + 3\lambda^2 + .26666 = 0.$$

Now, the maximum value of λ, which solves this equation, is given by λ_{max} = 3.03. (Note that $\lambda_{max} > n$.) Hence, we achieve the following:

$$CI = \frac{(\lambda_{max} - n)}{(n-1)} = \frac{(3.03-3)}{(3-1)} = .015$$

and

$$CR = \frac{CI}{RI} = \frac{.015}{.58} = .0259.$$

Since the value of CR is less than .1, good consistency is indicated.

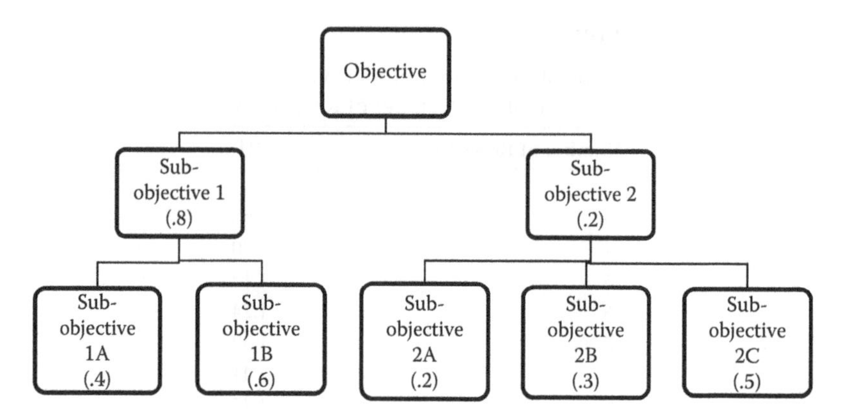

FIGURE 3.8
Hierarchy of Figure 3.5, minus the alternative factors, with local weights attached.

Saaty (1990a, p. 85) notes that one way to improve the consistency of the DM is to go ahead and compute the weights for the factors based on the original inconsistent matrix and then rank the factors based on these weights. With this ranking in mind, allow the DM to answer the pairwise comparison questions again; this should result in improved consistency.

Step 3. Determine the global weight for each alternative on each lowest-level criterion in the hierarchy, by multiplying the weights "up" the branch in the hierarchy corresponding to that alternative.

Consider the hierarchy of Figure 3.5, not including the alternatives at the lowest level. This hierarchy, with local weights shown for each respective factor, is shown in Figure 3.8.

Note that the local weights for any portion of the hierarchy sum to 1, as they always should.

Now, the *global weight* for any factor in the hierarchy is given by taking the product of weights of all of the factors associated with the relevant factor as one moves "up" the hierarchy. For example, the global weights associated with the lowest-level factors in Figure 3.8 are shown in Table 3.27.

TABLE 3.27

Global Weights for the Lowest-Level Factors of Figure 3.8

Factor	Global Weight
Subobjective 1A	.4(.8) = 0.32
Subobjective 1B	.6(.8) = 0.48
Subobjective 2A	.2(.2) = 0.04
Subobjective 2B	.3(.2) = 0.06
Subobjective 2C	.5(.2) = 0.10

TABLE 3.28
Local Weights for Alternatives of Figure 3.5

Alternative	Subobjective 1A	Subobjective 1B	Subobjective 2A	Subobjective 2B	Subobjective 2C
Alternative 1	.6	.3	.9	.6	.3
Alternative 2	.3	.5	.08	.3	.5
Alternative 3	.1	.2	.02	.1	.2

TABLE 3.29
Global Weights for All Factors of Figure 3.5

Factor	Global Weight
Sub 1	0.8
Sub 2	0.2
Sub 1A	0.32
Sub 1B	0.48
Sub 2A	0.04
Sub 2B	0.06
Sub 2C	0.10
Alt1	0.436
Alt 2	0.406
Alt 3	0.158

Note that the global weights of *all factors* at any particular *level* of the hierarchy sum to 1.

Now suppose that the local weights for the alternatives shown in Figure 3.5, relative to each of the lowest-level subobjectives, are given in Table 3.28. As seen earlier, these local weights are determined through pairwise comparisons concerning how well respective individual alternatives do with respect to other alternatives on a particular subobjective.

Note, as will always be the case, that the sum of the local weights over all alternatives for any particular subobjective is 1.

Finally, the global weights for all of the factors in the hierarchy are given in Table 3.29.

Note that the sum of the global weights over all alternatives in Table 3.29 is 1 (.436 +.406 +.158 = 1). The alternatives are then ranked according to their global weight values: Alternative 1 is preferred to alternative 2, and alternative 2 is preferred to alternative 3.

Example 3.6: Using the AHP for Resource Allocation at a Call Center

As with other service systems, the design and operation of a call center involves the consideration of multiple, conflicting performance measures related to the call center operators and staff, as well as customers.

As defined by Mehrota (1997), a call center is "any group whose principal business is talking on the telephone to customers or prospects. This group may be centralized in a single site, distributed across multiple sites or 'virtualized' with agents in individual offices."

In their excellent book, Kelton et al. (2015, Chapter 5) describe a simulation model for a generic call center that has three types of callers, classified as technical support, sales information, and order status inquiry. The technical support callers are interested in support for one of three types of products, classified as products one, two, and three.

Resources for this call center include trunk lines and staff who answer the calls and provide relevant information to the callers. These staff personnel are classified as being either of type sales support or technical support; the technical support staff personnel are further classified as being able to provide support for products of type one, two, or three (as noted in the previous paragraph) or for all three types of products for the better trained personnel. The cost per hour for staff varies by type of staff; for example, technical support staff who are able to provide support for all three types of products are more expensive than staff who can provide support for only one type of product.

Additional details concerning the system such as call arrival rates, distribution of callers by type, work schedules for the different staff, and processes associated with providing service to the callers are given by Kelton et al. (2015). Here, we are concerned with examining various policies (as defined by the number of trunk lines and numbers of various types of staff) with respect to a variety of attributes, corresponding to either (1) the cost for operating the system or (2) the level of service provided to the customers. In fact, for virtually all service systems, the attributes can be classified as being related to either cost or level of service.

A hierarchy of objectives for this call center is shown in Figure 3.9. As is noted earlier, the objectives are separated into two categories, relating to cost and service. The cost category has two subobjectives: minimization of weekly trunk line costs and minimization of weekly labor costs, both in dollars; note that a manager of the call center might very well want to consider these two types of costs separately, since the first relates to equipment and the second relates to personnel.

The service level objective has four subobjectives, relating to the waiting times for the three types of customers who access the system, and the number of callers who receive a busy signal, which occurs when all

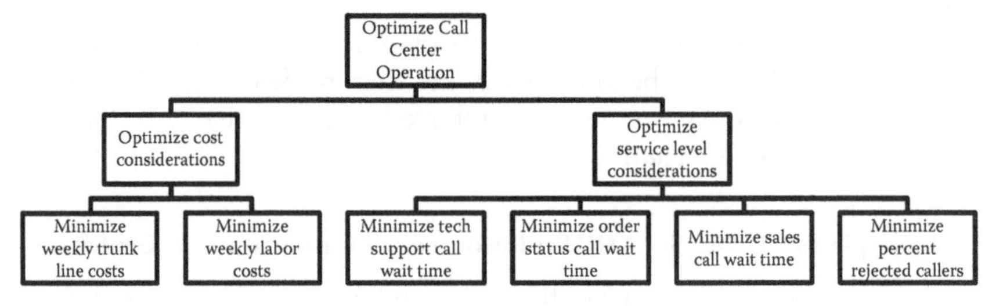

FIGURE 3.9
A hierarchy of objectives for a call center.

of the trunk lines are in use at the time of their call. These callers who receive a busy signal (and therefore are not served by the system) are called "rejected callers"; the associated relevant attribute is percent of callers who are rejected.

In summary, the attributes associated with the service level subobjectives for waiting times have threshold levels associated with them. Specifically, as described in Kelton et al. (2015), tech support callers, order status callers, and sales callers expect a certain amount of wait time, namely, 3 minutes, 1 minute, and 2 minutes, respectively; no "penalty" will be incurred if a (relevant) caller waits less than one of these times. Note that if one were thinking of these wait times in terms of individual attribute value functions, the relevant functions would have a zero slope at any value less than the threshold level.

So in summary, the attributes associated with each of the lowest-level objectives in Figure 3.9 are

1. Minimizing weekly trunk line costs: weekly trunk line cost in dollars
2. Minimizing weekly labor costs: weekly labor costs in dollars
3. Minimizing tech support call wait time: expected sum over all tech support callers of excess (over 3 minutes) wait times
4. Minimizing order status call wait time: expected sum over all order status callers of excess (over 1 minute) wait times
5. Minimizing sales call wait time: expected sum taken over all sales callers of excess (over 2 minutes) wait times
6. Minimizing percent rejected callers: percent of calls coming to the system, which are "rejected" because a trunk line is not available

The control variables for the simulation model are identified as

1. Number of trunk lines
2. Number of Tech 1 call takers
3. Number of Tech 2 call takers
4. Number of Tech 3 call takers
5. Number of Tech All call takers
6. Number of sales call takers

The Tech 1, Tech 2, and Tech 3 call takers refer to staff who can handle the three types of tech support callers, respectively (for the three types of products); the Tech All call takers refer to a type of staff who can handle tech support callers for any of the three types of products; hence, this better-trained staff is more expensive. The sales call takers refer to those call takers who handle the sales callers, as well as a small percentage of the order status callers. In terms of explaining this particular application of AHP, the actual control variables and the various solutions (in terms of values assigned to these control variables) are not that important and are only given here for illustrative purposes. We are mainly concerned with how the various outcomes for good alternatives (solutions) are evaluated, and how a DM would trade off among the various factors of the hierarchy.

A simulation model of the hypothetical system was developed, and this model was run for various optimization models in order to yield several solutions with "good outcomes." The optimization models solved to generate these solutions were of the following form:

Minimize (Weekly Call Taker Cost) + (Weekly Trunk Line Cost)

Subject to: (1) Sum of Excess Wait Times ≤ 8,000, 12,000, or 16,000, and (2) Percent of Calls Rejected ≤ 6, 8, 10, 12

An optimization model was formed by selecting a combination of two values to form the right-hand sides for the constraints. For example, one of the models solved had a right-hand side for the first constraint of 8000 minutes and a right-hand side for the second constraint of 6%. These optimization models were solved with the use of the heuristic optimization package: OptQuest™ (see Kleijnen and Wan, 2007), which is implemented as an interface with several simulation software packages.

From the solutions generated using the simulation model and OptQuest, five solutions were selected for further analysis using AHP. (Note that we are using the term *solution* in place of the term *alternative* used in describing the AHP in this example.) These five solutions, given in terms of the number of trunk lines and additional call takers of various types (over a baseline number of call takers), are given in Table 3.30. The outcomes associated with these five solutions, given in terms of the estimates of the expected attribute values as generated from the simulation, are shown in Table 3.31. All times are in minutes and all costs are in dollars in Table 3.31. As mentioned earlier, note that the "wait times" refer to excess wait times over threshold values.

Now, the answers given by a hypothetical DM to the pairwise comparison questions for the various sections of the hierarchy are shown in Figure 3.10.

Note that there are three sections of the hierarchy, which correspond to the respective set of pairwise comparison questions asked of the DM. Each of two of those sections has only two factors—hence, only one pairwise comparison is needed for these sections. The third section of the hierarchy contains four factors; hence, six pairwise comparisons are needed in order to handle all possibilities.

TABLE 3.30

Five Solutions Generated for the Call Center Example Using the Simulation Model Interfaced with OptQuest

Solution #	No. of Trunk Lines	No. of Additional Tech 1's	No. of Additional Tech 2's	No. of Additional Tech 3's	No. of Additional Tech All's	No. of Sales Call Takers
1	24	1	2	0	0	0
2	25	1	1	1	0	1
3	26	1	1	1	1	1
4	24	0	0	0	2	2
5	25	0	2	0	4	2

TABLE 3.31

Outcomes Associated with the Five Solutions in Table 3.30

Solution Number	Weekly Trunk Line Costs	Weekly Labor Costs	Tech Support Call Wait Time	Order Status Call Wait Time	Sales Call Wait Time	Percent Rejected Callers
1	2352	13,240	5203	1007	3739	9.53
2	2450	13,580	5445	689	2630	6.85
3	2548	13,940	4923	914	3024	5.26
4	2352	13,680	4957	554	2158	7.52
5	2450	15,040	4374	673	2362	4.58

Cost is equally as important as service level

Weekly labor costs are weakly more important than weekly trunk line costs

Sales call wait time is weakly more important than the number of rejected calls
Number of rejected calls is weakly more important than order status call wait time

Order status call wait time is of equal importance to tech support call wait time
Sales call wait time is strongly more important than order status call wait time
Sales call wait time is very strongly more important than tech support call wait time

Number of rejected calls is strongly more important than tech support call wait time

FIGURE 3.10
Answers to the pairwise comparison questions for the hierarchy of Figure 3.9.

TABLE 3.32

Local Weights Associated with Pairwise Comparisons of Top-Level Factors of Figure 3.9

Cost	Service Level
.5	.5

These answers were used as input to the software package Expert Choice (http://expertchoice.com/comparion/) in order to perform the computations associated with the AHP. In particular, the local weights resulting from the pairwise comparisons shown in Figure 3.10 are shown in Tables 3.32 through 3.34. Note that each set of local weights sums to 1.

The local weights associated with each of the five outcomes shown in Table 3.31 with respect to each of the six lowest-level factors in the hierarchy of Figure 3.9 are shown in Table 3.35. Note that these weights are derived from the pairwise comparisons provided by our hypothetical DM with respect to how well each of the outcomes performs on each of the lowest-level factors (objectives) of Figure 3.9. As an example, the DM

TABLE 3.33

Local Weights Associated with Pairwise Comparisons of Second-Level Cost Factors of Figure 3.9

Weekly Trunk Line Costs	Weekly Labor Costs
.25	.75

TABLE 3.34

Local Weights Associated with the Pairwise Comparisons of the Second-Level Service Level Factors of Figure 3.9

Tech Support Call Wait Time	Order Status Call Wait Time	Sales Call Wait Time	Percent Rejected Callers
0.07	0.11	0.54	0.28

TABLE 3.35

Local Weights Associated with Each of the Five Outcomes for Each of the Lowest-Level Factors in the Hierarchy

Outcome Number	Weekly Trunk Line Costs	Weekly Labor Costs	Tech Support Call Wait Time	Order Status Call Wait Time	Sales Call Wait Time	Percent Rejected Callers
Outcome 1	0.344	0.412	0.055	0.036	0.030	0.029
Outcome 2	0.129	0.243	0.038	0.200	0.147	0.154
Outcome 3	0.055	0.129	0.179	0.051	0.072	0.245
Outcome 4	0.343	0.179	0.137	0.432	0.445	0.072
Outcome 5	0.129	0.037	0.591	0.281	0.306	0.500

would be expected to provide an answer to the question of how much better a technical wait time of 5203 minutes (associated with Outcome 1) is than the technical wait time of 5445 minutes (associated with Outcome 2).

Note that the weights in Tables 3.32 through 3.34 are rounded off to one, two, and two digits, respectively, but the weights in Table 3.35 are rounded to three digits.

The global weights associated with each of the factors in the hierarchy are shown in Table 3.36. As noted previously, these global weights are computed as the product of the local weights as one moves down the hierarchy. For example, the global weight of sales call wait time is computed as the product of the local weights for sales call wait time and service level, or as .54 * .5 ≈ .2669. (Note that the rounding of the first two weights results in these two quantities not being exactly equal.)

Note that the *local weight* for a factor could be thought of as a reflection of the importance of that factor with respect to the next higher-level factor to which the relevant factor is connected; for example, the local weight of .75 for *weekly labor cost* is a reflection of the importance of that factor with respect to *cost considerations* only.

TABLE 3.36

Global Weights Associated with Each of the Factors in the Hierarchy of Figure 3.9

Factor	Global Weight
Cost	0.5
Service level	0.5
Weekly trunk line costs	0.125
Weekly labor costs	0.375
Tech support call wait time	0.0389
Order status call wait time	0.053
Sales call wait time	0.2669
Percent rejected callers	0.1412

TABLE 3.37

Global Weights for the Solutions/Outcomes

Solution/Outcome	Global Weight
1	0.23
2	0.18
3	0.11
4	0.27
5	0.21

On the other hand, the *global weight* for a factor could be thought of as a reflection of the importance of that factor with respect to the overall (top level) goal or factor of the system: optimize call center operation; for example, the global weight of .27 for *sales call wait time* is a reflection of the importance of that factor with respect to the overall goal of *optimizing the call center operation*.

The respective global weights for each of the outcomes are shown in Table 3.37. These global weights are computed as the sum (overall lowest-level objectives) of the products of an outcome's local weight associated with a lowest-level objective and that lowest-level objective's global weight. For example, the global weight for Outcome 4 (associated with Solution 4 of Table 3.37) is given by

$$0.125 * .343 + 0.375 * .179 + 0.0389 * .137 + 0.053 * .432 + 0.2669 * .445 + 0.1412 * .072 = .27.$$

As can be seen from the global weights, the solutions/outcomes, listed in order of decreasing preference, are given as 4, 1, 5, 2, and 3. It is clear that, upon examining the outcomes closely, solution number 4 is helped greatly by its excellent value (i.e., low numerical value) for sales call wait time and also by the high value associated with the global weight for sales call wait time.

In considering this example in more detail, the reader should be aware of the large amount of effort made prior to even applying the AHP. Specifically, a simulation model was developed in order to model the relationships between the control variables and the performance measures; then an optimization model was developed and solved for various parameter values in order to obtain several good alternative solutions.

3.11.3 Criticisms of the AHP, Comparison of the AHP and MAVT, and Extensions

Following his original development of the AHP, Saaty (1986) provided an axiomatic foundation for his methodology. The theory relies on the use of *ratio scale priorities* (as opposed to a nominal, ordinal, or interval scale) for the factors in the hierarchy. To quote Forman and Gass (2001),

> Any Hierarchical-based methodology must use ratio-scale priorities for elements above the lowest level of the hierarchy. This is necessary because the priorities (or weights) of the elements at any level of the hierarchy are determined by multiplying the priorities of the elements in that level by the priorities of the parent element.

As noted by Bouyssou et al. (2000, p. 121), the "models" derived in AHP and multiattribute value theory (MAVT) are the same (additive, scaled value functions), but one should not necessarily expect these respective methodologies to yield the same results with respect to evaluation and ranking of the alternatives. However, in a simulation study involving AHP, MAVT, and other methodologies, Buede and Maxwell (1995) found that AHP and MAVT gave close to the same results. Schoner et al. (1997) showed that with their two modifications of AHP (called referenced AHP and linking pin AHP), total agreement with MAVT with respect to the first-ranked alternatives would have been achieved in the Buede and Maxwell study. (See Belton [1986] for an additional discussion involving the comparison of AHP and MAVT.)

Among other difficulties, the detractors of AHP say that one of its main problems is that the concept of "importance" is not defined well enough to accurately represent a DM's preference structure (see Bouyssou et al., 2010, p. 151). In particular, consider Example 3.6 involving staffing levels for a call center. One might very well expect a DM, in comparing the importance of cost to service level, to say that they are both important and that without more specific information, it would be impossible to say that one is more important than the other.

Harker and Vargas (1987), Dyer (1990a), Harker and Vargas (1990), Saaty (1990b), and Dyer (1990b) engaged in a spirited discussion of the merits of AHP in general, and versus MAVT in particular, and the reader is directed to this series in order to gain a detailed understanding of the controversy.

In addition, Gass (2005) provides a summary of the discussion, albeit from the viewpoint of a proponent of AHP. He notes that the major criticisms of

AHP involve its measurement scale (the 0–9 scale presented earlier), rank reversal, and transitivity of preferences. As noted earlier, AHP does *not* require transitivity of preferences (e.g., if Subobjective A is more important than B and B is more important than C, then A must be [much] more important than C if transitivity is to hold); but multiattribute value theory does assume this concept as an axiom.

Rank reversal can be described by a situation in which one has several alternatives that are ranked from best to worst through the use of AHP, say A_1, A_2, and A_3, in this respective order. Then another alternative, A_4, is introduced and the AHP process is performed again with the same DM acting in a consistent fashion; the subsequent computation results in A_2 being ranked ahead of A_1 with AHP—that is, the ranking of A_1 and A_2 become reversed. (See Belton and Gear [1983] for a specific example of rank reversal.) Rank reversal can happen with AHP, but obviously not with a multiattribute value function. As noted by Gass (2005), however, rank reversal does happen in real life, but it can be considered a function of the problem structuring process.

A variation of AHP involves the use of a geometric scale to make pairwise comparisons and geometric means to perform the aggregation over the levels of the hierarchy (Lootsma, 1993, 1996). This variation, termed multiplicative AHP, allows the elimination of some of the rank reversal cases (Lootsma, 1993). Multiplicative AHP also allows for the separation of the pairwise comparisons into two separate categories of assessment: the pairwise comparisons involving all of the factors except for the alternatives and the comparisons involving only the alternatives with respect to the lowest-level attributes. Lootsma (1996) describes the general situation in which such an approach, involving two separate sets of DMs, may be advisable.

In considering the use of the multiplicative AHP by a group, the influence/power of the members of the group may be important to consider, as shown by Barzilai and Lootsma (1997); however, this use of the multiplicative AHP within a group has also raised controversy (see Korhonen, 1997; Larichev, 1997; Lootsma and Barzilai, 1997; Vargas, 1997).

Triantaphyllou (2001) describes additional cases of rank reversal, which occur with standard AHP that do not occur with multiplicative AHP. See Schoner and Wedley (1989), Schoner et al. (1993), and Barzilai and Golany (1994) for additional discussion of rank reversals.

As noted earlier, AHP was originally designed so as to apply to a hierarchy in which each factor is attached to exactly one higher-level factor in the hierarchy. This contrasts with some of the fundamental objectives–means objectives hierarchies in which a means objective can be connected to multiple fundamental objectives (e.g., see Figure 2.9). An extension of AHP, called the analytic network process (ANP), has been developed by Saaty and Vargas (2006) to address these more complex situations. Finally, Forman and Peniwati (1998) discuss how individual judgments and priorities can be aggregated within the AHP.

3.12 Outranking Methods

The so-called outranking methods originated from the work of Bernard Roy (1968). This group of techniques/methods is often termed the European or French approach, as opposed to the "American approach" as represented by multiattribute value/utility theory and the analytic hierarchy process. Since Roy's initial work in the area, many researchers have developed variations on this initial method, to the point where there currently exists a whole collection of related methods.

As noted by Rogers and Bruen (1998), the word "outranking" refers to the degree of dominance of one alternative over another. As such, these methods do not necessarily provide a complete ranking of the alternatives and also avoid strong assumptions about the true preferences of the DM(s) (Kangas et al., 2001b). As with the AHP, the outranking methods rely on pairwise comparisons, but without the use of a hierarchy; in addition, many of the outranking methods use the differences in the attribute values for pairs of alternatives as the basis for their inputs.

The original outranking method, developed by Roy (1968), was called ELECTRE for *EL*imination *E*t *C*hoix *T*raduisant la *RE*alité (ELimination and Choice Expressing REality). As noted earlier, since its initial development, a whole group of related methods have been developed, including ELECTRE I (ELECTRE One), ELECTRE II, ELECTRE III, ELECTRE IV, ELECTRE IS, and ELECTRE TRI (ELECTRE tree). (See Figuerira et al. [2005] for a review of the ELECTRE methods.)

Additional outranking methods that have been developed include the PROMETHEE (*p*reference *r*anking *o*rganization *met*hod for the *e*nrichment of *e*valuations), which include PROMETHEE I and PROMETHEE II. See Brans and Vincke (1985); Brans et al. (1986); and Mareschal and Brans (1988) for descriptions of these methods. Mareschal (1986) describes an extension of PROMETHEE (stochastic PROMETHEE) used to analyze a group of uncertain outcomes. Dubois et al. (1989) describe an application of PROMETHEE as part of an expert system for diagnosis, and Mareschal and Brans (1991) describe an application of PROMETHEE to "industrial evaluation." Finally, see Behzadian et al. (2010) for a literature review of the methodologies and applications of PROMETHEE.

The outranking methods have been applied to a variety of areas including risk assessment for natural gas pipelines (Brito et al., 2010), selection of land mine detection strategies (De Leeneer and Pastijn, 2002), selection of a solid waste management system in Finland (Hokkanen and Salminen, 1997), and ranking of extension projects for the Paris metro line (Roy and Huggonard, 1982). Leyva-López and Fernández-González (2003) developed a method for group decision support based on ELECTRE III.

The following description corresponds to the ELECTRE III method, which has been described as the most commonly used of the ELECTRE methods

(Diakoulaki et al., 2005, p. 889). The basic steps, following the identification of the p attributes, alternatives, and associated outcomes, as described in Section 3.2, are as follows (Kangas et al., 2001a):

1. Three types of thresholds are identified by the DM for each attribute: *indifference thresholds*, denoted as q_j for $j = 1,\ldots, p$; *strong preference thresholds*, denoted as r_j for $j = 1,\ldots, p$, where $r_j > q_j$ for $j = 1,\ldots, p$; and *veto thresholds*, denoted as v_j for $j = 1,\ldots, p$, where $v_j > r_j$ for $j = 1,\ldots, p$.

   Indifference thresholds identify the maximum amount that two outcomes can differ on a particular attribute for which the DM would be *indifferent*. For Example 3.1, if the DM would be indifferent to two annual salaries that differed by at most $100, then q_1 for Example 3.1 would be 100.

   Strong preference thresholds identify the minimum amount that two outcomes could differ on a particular attribute for which the decision maker would *strongly prefer* one outcome over another with respect to that attribute. Again, for Example 3.1, if the DM would strongly prefer one salary over another if they differed by at least $1000, then r_1 would be 1000.

   If the difference between two outcomes for a particular attribute is greater than q_j but less than r_j then the DM is said to *weakly prefer* one outcome to another with respect to that attribute.

   Veto thresholds allow for the possibility for an alternative with a very poor value on a particular attribute (as compared to another alternative) to not be chosen as a preferred alternative no matter how well that alternative performs on the other attributes. Just as with indifference thresholds, weak preference thresholds, and strong preference thresholds, veto thresholds refer to a *difference* in the outcomes for two alternatives with respect to a particular attribute, j.

   The closer the veto threshold is to the strong preference threshold for a particular attribute, the more important that attribute is (Roy, 1991).

   Figure 3.11 illustrates the relationships between the various thresholds.

   As noted by Kangas et al. (2001a), these indifference thresholds, weak preference thresholds, strong preference thresholds, and veto thresholds can be either constant, proportional to the attribute value, or a linear function of the attribute value. This owes to the fact that the marginal rates of substitution may depend on the attribute value at which the rate is measured, as discussed earlier. In particular for Example 3.1, q_1, the indifference threshold for annual salary, might be $100 if the threshold is applied within the range of $60,000–$65,000, but q_1 might be $200 if applied in the range of $70,000–$75,000).

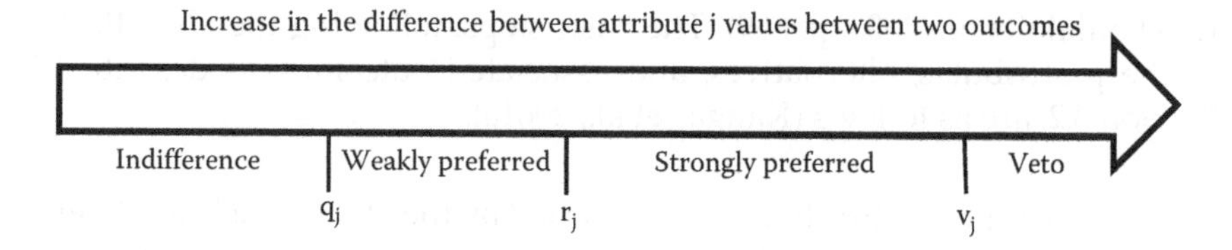

FIGURE 3.11
Relationships between the various thresholds used in ELECTRE III.

2. Weights, denoted as w_j for j = 1,…, p (where $\sum_{j=1}^{p} w_j = 1$, and $0 < w_j < 1$), are defined for the respective attributes. These weights can be computed using any of the methods already presented.
3. A *concordance index*, denoted as $C(A_i, A_\gamma)$ for i = 1,…, n; γ = 1,…, n; and i ≠ γ, is computed for each distinct pair of alternatives as follows:

$$C(A_i, A_\gamma) = \sum_{j=1}^{p} w_j c_j (A_i, A_\gamma)$$

and $c_j(A_i, A_\gamma)$ for i ≠ γ is defined as a local concordance index:

$$\begin{aligned} c_j(A_i, A_\gamma) &= 0, \text{if } x_j(A_i) + r_j \le x_j(A_\gamma), \quad \text{if attribute j is to be maximized or} \\ & \quad x_j(A_\gamma) + r_j \le x_j(A_i), \quad \text{if attribute j is to be minimized.} \\ &= 1, \text{if } x_j(A_i) + q_j \ge x_j(A_\gamma), \quad \text{if attribute j is to be maximized or} \\ & \quad x_j(A_\gamma) + q_j \ge x_j(A_i), \quad \text{if attribute j is to be minimized.} \\ &= (r_j - |x_j(A_i) - x_j(A_\gamma)|)/(r_j - q_j) \quad \text{otherwise.} \end{aligned}$$

4. A *discordance index*, denoted as $d_j(A_i, A_\gamma)$, is computed for each attribute and for each pair of alternatives for j = 1,…, p; i = 1,…, n; γ = 1,…, n; and i ≠ γ, as follows:

$$\begin{aligned} d_j(A_i, A_\gamma) &= 0, \text{if } x_j(A_\gamma) - r_j \le x_j(A_i), \quad \text{if attribute j is to be maximized or} \\ & \quad x_j(A_i) - r_j \le x_j(A_\gamma), \quad \text{if attribute j is to be minimized.} \\ &= 1, \text{if } x_j(A_\gamma) - v_j \ge x_j(A_i), \quad \text{if attribute j is to be maximized or} \\ & \quad x_j(A_i) - v_j \ge x_j(A_\gamma), \quad \text{if attribute j is to be minimized.} \\ &= (|x_j(A_i) - x_j(A_\gamma)| - r_j)/(v_j - r_j) \quad \text{otherwise.} \end{aligned}$$

Note that the discordance index is computed for *each attribute* with a veto threshold (for every pair of distinct alternatives) and is not weighted by attribute importance.

5. The degree of outranking, denoted as $O(A_i, A_\gamma)$, is computed for each pair of alternatives for $i = 1,\ldots, n$; $\gamma = 1,\ldots, n$; and $i \neq \gamma$, as follows:

$$O(A_i, A_\gamma) = C(A_i, A_\gamma), \quad \text{if } \beta(A_i, A_\gamma) = \Phi, \text{ the null set,}$$
$$= C(A_i, A_\gamma) \prod_{j \varepsilon \beta(A_i, A_\gamma)} (1 - d_j(A_i, A_\gamma))/(1 - C(A_i, A_\gamma)), \quad \text{otherwise.}$$

where $\beta(A_i, A_\gamma)$ is the set of attributes for the pair of alternatives (A_i, A_γ) for which $d_j(A_i, A_\gamma) > C(A_i, A_\gamma)$.

6. There are at least two different ways that the alternatives can be ordered, based on the values associated with the $O(A_i, A_\gamma)$ (Miettinen and Salminen, 1999). One approach is the "minimum procedure" (Pirlot, 1995) in which the alternatives are ranked according to the decreasing values of inf $\{O(A_i, A_\gamma)$ for $\gamma = 1,\ldots, p, i \neq \gamma\}$; that is, select the smallest element of each row and then rank the rows (alternatives) according to these values from largest to smallest. Note that this procedure could very easily lead to ties in the ranking.

   Note that while considering a pair of local concordance index values, $c_j(A_i, A_\gamma)$ and $c_j(A_\gamma, A_i)$ for $i \neq \gamma$, and for a particular attribute j, one could have three possibilities:

   1. $c_j(A_i, A_\gamma) = 1$ and $c_j(A_\gamma, A_i) = 1$
   2. $c_j(A_i, A_\gamma) = 1$ and $c_j(A_\gamma, A_i) = 0$

   or

   3. $c_j(A_i, A_\gamma) = 0$ and $c_j(A_\gamma, A_i) = 1$

   but not $c_j(A_i, A_\gamma) = 0$ and $c_j(A_\gamma, A_i) = 0$.

   Note also that the discordance values are useful in situations where a very poor attribute value for an alternative can eliminate that alternative from consideration, no matter what its other attribute values are.

Example 3.7: Applying ELECTRE III to Example 3.1, the Job Selection Problem

Let's return to Example 3.1 of our graduating student with the job offers, as shown in Table 3.3, and apply ELECTRE III to this problem. Suppose that the student supplies indifference thresholds, strong preference thresholds, and veto thresholds as shown in Table 3.38. In order to keep this example relatively simple, we will assume that these thresholds remain constant over the range of respective attribute values.

As noted earlier, any of the methods presented thus far could be used to determine the respective weights for the attributes/criteria. Let's employ the swing weights computed in Section 3.8, that is, $w_1 = .45$, $w_2 = .23$, and $w_3 = .32$.

The local concordance indices (the $c_j(A_i, A_\gamma)$ values) for the attributes of annual salary (X_1), annual days of vacation (X_2), and miles from

TABLE 3.38

Indifference Thresholds, Strong Preference Thresholds, and Veto Thresholds for Example 3.1

Attribute	Indifference Threshold	Strong Preference Threshold	Veto Threshold
1: Annual salary	$1000	$4000	$15,000
2: Days of vacation	3	5	15
3. Miles from hometown	100	200	600

TABLE 3.39

Local Concordance Indices for the Annual Salary (X_1) Attribute

	Alt 1	Alt 2	Alt 3	Alt 4	Alt 5	Alt 6
Alt 1	—	1	0	1	1	1
Alt 2	0	—	0	0	1	0
Alt 3	1	1	—	1	1	1
Alt 4	0	1	0	—	1	0
Alt 5	0	0	0	0	—	0
Alt 6	1	1	0	1	1	—

TABLE 3.40

Local Concordance Indices for the Annual Days of Vacation (X_2) Attribute

	Alt 1	Alt 2	Alt 3	Alt 4	Alt 5	Alt 6
Alt 1	—	0	1	0	0	0
Alt 2	1	—	1	1	0	1
Alt 3	1	0	—	1	0	1
Alt 4	1	0	1	—	0	1
Alt 5	1	1	1	1	—	1
Alt 6	1	0	1	1	0	—

hometown (X_3) are shown in Table 3.39 through Table 3.41, respectively. The first row and the first column of each of the matrices are headings for the six alternatives of the problem. So for example, the value for the local concordance index for attribute 2 (days of vacation) for the first and third alternatives ($c_2(A_1, A_3)$) is given by locating the alternative 1 row and the alternative 3 column in Table 3.40—that is, $c_2(A_1, A_3) = 1$. Note that this value of 1 is derived from the fact that $x_2(A_1) + q_2 \geq x_2(A_3)$, or $5 + 5 > 7$.

Note that these are given for each pair of distinct alternatives. Also, note that since none of the differences in given attribute values for any pair of alternatives lie between the indifference threshold and strong preference threshold values for that attribute, none of the local concordance indices has a fractional value—each of the values is either 0 or 1.

As noted earlier, when $c_j(A_i, A_\gamma) = 1$, then $c_j(A_\gamma, A_i)$ can equal either 0 or 1 for $i \neq \gamma$. For example, for the first attribute, salary, the first and

TABLE 3.41

Local Concordance Indices for the Miles from Hometown (X_3) Attribute

	Alt 1	Alt 2	Alt 3	Alt 4	Alt 5	Alt 6
Alt 1	—	1	1	1	1	1
Alt 2	1	—	1	0	1	1
Alt 3	0	0	—	0	0	0
Alt 4	1	1	1	—	1	1
Alt 5	1	1	1	1	—	1
Alt 6	1	1	1	0	1	—

TABLE 3.42

Local Discordance Indices for the Annual Salary (X_1) Attribute

	Alt 1	Alt 2	Alt 3	Alt 4	Alt 5	Alt 6
Alt 1	—	0	1/11	0	0	0
Alt 2	6/11	—	1	1/11	0	6/11
Alt 3	0	0	—	0	0	0
Alt 4	1/11	0	6/11	—	0	1/11
Alt 5	1	1/11	1	6/11	—	1
Alt 6	0	0	1/11	0	0	—

TABLE 3.43

Local Discordance Indices for the Annual Days of Vacation (X_2) Attribute

	Alt 1	Alt 2	Alt 3	Alt 4	Alt 5	Alt 6
Alt 1	—	5/10	0	0	1	0
Alt 2	0	—	0	0	0	0
Alt 3	0	3/10	—	0	8/10	0
Alt 4	0	0	0	—	5/10	0
Alt 5	0	0	0	0	—	0
Alt 6	0	0	0	0	5/10	—

sixth alternatives have the same value, \$70,000. So, in this case, both $c_1(A_1, A_6) = 1$ and $c_1(A_6, A_1) = 1$. However, consider the values for salary for the third and fifth alternatives ($x_1(A_3) = 75{,}000$ and $x_1(A_5) = 55{,}000$). In this case, $c_1(A_3, A_5) = 1$ and $c_1(A_5, A_3) = 0$. Hence, in considering the local concordance indices for any pair of alternatives i and γ, the DM should realize that both pairs of indices are considered in the overall comparison and "scoring" of the alternatives.

The local discordance indices for the attributes of annual salary (X_1), annual days of vacation (X_2), and miles from hometown (X_3) are shown in Table 3.42 through Table 3.44, respectively. As in the case of the concordance indices, these are given for each pair of distinct alternatives. The conventions used for Table 3.39 through Table 3.41, in terms of defining the values for the rows and columns, are also used for Table 3.42 through Table 3.44.

TABLE 3.44

Local Discordance Indices for the Miles from Hometown (X_3) Attribute

	Alt 1	Alt 2	Alt 3	Alt 4	Alt 5	Alt 6
Alt 1	—	0	0	0	0	0
Alt 2	0	—	0	0	0	0
Alt 3	1	1	—	1	1	1
Alt 4	0	0	0	—	0	0
Alt 5	0	0	0	0	—	0
Alt 6	0	0	0	0	0	—

The reader should note that some of the values for the local discordance indices are fractional, due to the fact that some of the differences between the attribute values for the alternatives lay between the relevant strong preference threshold and veto threshold values. For example, consider the local discordance index for attribute 1 (annual salary) for alternatives 1 and 3 ($d_1(A_1, A_3)$), which has a value of $1/11 \approx .09$. The difference in annual salary (an attribute to be maximized) between alternatives 3 and 1 is given by $5,000; that is, $x_1(A_3) - x_1(A_1) = 75{,}000 - 70{,}000 = 5{,}000$, which is a value that lies between the strong preference threshold (4,000) and the veto threshold (15,000) for this attribute. Hence, the value for the relevant discordance index represents the "distance" that the difference (5,000) is from the strong preference threshold (4,000) to the veto threshold (15,000). The relevant calculation is given by

$$\begin{aligned} d_1(A_1, A_3) &= (|x_1(A_3) - x_1(A_1)| - r_1)/(v_1 - r_1) \\ &= (|75{,}000 - 70{,}000| - 4{,}000)/(15{,}000 - 4{,}000) \\ &= 1/11. \end{aligned}$$

Given the local concordance indices, the overall (or "global") concordance indices can be computed. These values are shown in Table 3.45.

As an example, consider $C(A_1, A_3) = .55$. This value is computed as

$$\begin{aligned} C(A_1, A_3) &= \sum_{j=1}^{3} w_j c_j(A_1, A_3) \\ &= w_1 c_1(A_1, A_3) + w_2 c_2(A_1, A_3) + w_3 c_3(A_1, A_3) \\ &= .45(0) + .23(1) + .32(1) \\ &= .55. \end{aligned}$$

Using the local discordance indices as given in Tables 3.42 through 3.44 and the global concordance indices, as given in Table 3.45, the $\beta(A_i, A_\gamma)$ sets can be determined for each distinct pair of alternatives (A_i, A_γ), $i \neq \gamma$. Recall that each β set is the set of attributes for which the relevant local

TABLE 3.45

Concordance Indices for Each Pair of Distinct Alternatives

	Alt 1	Alt 2	Alt 3	Alt 4	Alt 5	Alt 6
Alt 1	—	0.77	0.55	0.77	0.77	0.77
Alt 2	0.55	—	0.55	0.23	0.77	0.55
Alt 3	0.68	0.45	—	0.68	0.45	0.68
Alt 4	0.55	0.77	0.55	—	0.77	0.55
Alt 5	0.55	0.55	0.55	0.55	—	0.55
Alt 6	1	0.77	0.55	0.68	0.77	—

TABLE 3.46

$\beta(A_i, A_\gamma)$ Sets: The Set of Attributes j for Which $d_j(A_i, A_\gamma) > C(A_i, A_\gamma)$

	Alt 1	Alt 2	Alt 3	Alt 4	Alt 5	Alt 6
Alt 1	—	Φ	Φ	Φ	{2}	Φ
Alt 2	Φ	—	{1}	Φ	Φ	Φ
Alt 3	{3}	{3}	—	{3}	{2,3}	{3}
Alt 4	Φ	Φ	Φ	—	Φ	Φ
Alt 5	{1}	Φ	{1}	Φ	—	{1}
Alt 6	Φ	Φ	Φ	Φ	Φ	—

TABLE 3.47

$O(A_i, A_\gamma)$, Degree of Outranking Values

	Alt 1	Alt 2	Alt 3	Alt 4	Alt 5	Alt 6
Alt 1	—	0.77	0.55	0.77	0	0.77
Alt 2	0.55	—	0	0.23	0.77	0.55
Alt 3	0	0	—	0	0	0
Alt 4	0.55	0.77	0.55	—	0.77	0.55
Alt 5	0	0.55	0	0.55	—	0
Alt 6	1	0.77	0.55	0.68	0.77	—

discordance value is greater than the relevant concordance value (i.e., the set of values j for which $d_j(A_i, A_\gamma) > C(A_i, A_\gamma)$). These values are shown in Table 3.46.

In Table 3.46, a "—" indicates that the set is not defined, whereas a "Φ" indicates the null (or empty) set. As an example of the computations in Table 3.46, let's consider the set $\beta(A_3, A_5)$, the set of attributes j for which $d_j(A_3, A_5) > C(A_3, A_5)$. Since $C(A_3, A_5) = .45$, $d_j(A_3, A_5) = 0$, .8, and 1 for j = 1, 2, and 3, respectively, $\beta(A_3, A_5) = \{2,3\}$.

The values for "degree of outranking," $O(A_i, A_\gamma)$, can now be computed. These values are shown in Table 3.47.

Now, finding the minimum value in each row of the matrix given in Table 3.47 yields Table 3.48.

TABLE 3.48

Minimum Values for Each of the Six Rows (Alternatives) for Table 3.43 ($\inf\{O(A_i, A_\gamma)$ for $\gamma = 1,\ldots, p, i \neq \gamma\}$)

Row/Alternative	Minimum Value in Row
1	0
2	0
3	0
4	0.55
5	0
6	0.55

Note that either the fourth or sixth job offer would be selected with this approach, since they tie for the best value (.55) in this case. Also, alternatives 1, 2, 3, and 5 are ranked equally, tied for third in this case.

The ELECTRE III procedure was not very discriminating for this decision situation. This can be attributed to two main reasons. First, all of the values for the local concordance matrices are either 0 or 1, owing to the fact that none of the differences in the attribute values (for any particular attribute) between any two alternatives was between the indifference threshold and the strong preference threshold. Second, the veto thresholds employed basically eliminated the first, second, third, and fifth alternatives from consideration. These reasons emphasize the fact that while the amount of preference information required of the DM in using the ELECTRE III method is not nearly as great as the amount required in the assessment of an MAV function, the DM still needs to be meticulous in supplying this information.

As with the other methods presented in this chapter, sensitivity analysis is important as a part of the overall process. If the DM can quickly observe the effects on the ranking of alternatives as a result of changes to the indifference, preference, and veto thresholds as well as to changes in the attribute values for specific alternatives, a better decision (as well as a decision in which the DM will have more faith) may very well result.

Finally, the reader will note that there is much basic computation as part of the ELECTRE III process. But this computation is fairly simple and can be set up as part of a relatively simple spreadsheet application.

3.13 Extensions, Hybrid Approaches, and Comparisons

Each of the methodologies presented in this chapter has its own advantages and disadvantages vis-á-vis the other methodologies. No matter which technique is chosen however, it must be remembered that problem structuring is critical to the overall success of the decision-making process. If one does not have a good and complete set of alternatives, as well an appropriate set of attributes, even the best selection methodology may not lead to a good decision. Often, just presenting a good payoff matrix to the DM (or DMs) will lead to a good decision, satisfactory to all involved.

With respect to input, the methodologies differ with respect to the amount and type of preference information required of the DM(s). For example, the methodologies, which involve MAV function assessment, require very detailed information concerning the DM's preference structure, while other methodologies such as AHP and the outranking approaches require preference information, which is not quite so onerous to provide.

These differences can affect the "quality" of the ranking of the alternatives, that is, in comparison to how the alternatives would be ranked according to the *actual* preference structure of the DM(s). In some situations, one might even have a very inaccurate representation of the preference structure, which still gives a "correct ranking" for the alternatives (or at least gives the same first-ranked alternative).

Some of the methods such as those involving an assessment of the DM's MAV function, TOPSIS, or AHP give a complete ranking of all of the alternatives, while the outranking methods do not necessarily provide a complete ranking.

There are three important advantages associated with the approaches involving an assessment of the MAV function. First, such an approach requires the DM(s) to think very hard about the trade-offs they are willing to make concerning the important criteria for the relevant decision situation. Second, if a new alternative comes into play (with a corresponding new outcome), this new alternative can be easily placed in the overall ranking with the other alternatives, just by computing its value function level. With the AHP and some other approaches, one must repeat the entire procedure with all of the other alternatives if a new alternative is introduced. Third, a process involving MAV function assessment does not involve direct consideration of the actual outcomes associated with the alternatives, only assessments over the outcome space; this helps to remove any "emotional attachment" of the DMs to the actual alternatives. Of course, some may consider this to be a disadvantage of these approaches, since the assessment process involves only hypothetical and not actual outcomes.

The selection of an approach for multicriteria decision analysis (or the ranking of approaches) is itself a multicriteria problem, and various researchers such as Gershon (1981), Evans (1984), and Deason (1984) have suggested criteria for the selection process. For example, Gershon suggested 27 different criteria including whether or not the method provides a complete ranking of the alternatives, computer time required, and interaction time required.

Of course, one should consider the decision situation in choosing a particular approach, including the number of attributes, the number of alternatives under consideration, the number of DMs, and the "sophistication" of the DM(s) with respect to their abilities to provide preference/trade-off information. Other important considerations would be whether or not public input is needed in the process and the amount of justification/explanation required once the decision is made; for example, the use of public funds for large-scale projects obviously requires much public input and justification.

If one had many attributes to consider, then a methodology, which did not require much preference information from the DM, such as TOPSIS or one of the outranking methods, might be appropriate.

In some situations, a complementary approach might be suitable. For example, in a situation involving many attributes and alternatives, TOPSIS or one of the outranking procedures might be applied, and then the 10 or 15 top-ranked alternatives might be subjected to an approach, which required a more sophisticated analysis.

Section 3.11.3 presents some of the research involving a comparison of AHP and MAVT. Other researchers have also conducted studies involving comparisons of the various methods. For example, Olson (2001) compared four methods: SMART, a centroid method, which is a variation on SMART, PROMETHEE II, and a variation on PROMTHEE II in the ranking of major league baseball teams based on their performance statistics. In this situation, the actual team standings could be compared to the results predicted by each of the four methods. Olson found that all of the methods provided value in the decision-making process and that they were similar in their predictive capabilities.

Lootsma and Schuijt (1997) compared three methods: the multiplicative AHP (referred to in Section 3.11.3), SMART, and ELECTRE in solving a problem involving the choice of a location of a nuclear power plant. (See Keeney and Nair [1977] for the original case study.) Lootsma and Schuijt concluded that there was an "encouraging degree of similarity" in the end results associated with the three methods. They also concluded that a more complete study would have also involved the problem structuring portion of the decision analysis process, but including this aspect would be impractical for comparative studies.

Opricovic developed a method called VIKOR. (The name is an acronym derived from the Serbian language.) The method is similar to TOPSIS in that the distances from the ideal associated with the various outcomes are considered in determining a ranking. One unique aspect of VIKOR is that it allows for an explicit iterative and interactive approach with the DM(s). Opricovic and Tzeng (2007) compared an extension of VIKOR with three other methods: TOPSIS, PROMTHEE, and ELECTRE, using a case involving the evaluation of six alternative hydropower systems on a river in Yugoslavia; these systems were evaluated over eight attributes. The methods provided similar rankings. In an earlier paper, Opricovic and Tzeng (2004) performed a detailed comparison of TOPSIS and VIKOR.

Al-Shemmeri et al. (1997) addressed the problem of choosing a multicriteria decision aid for the ranking of water development projects with this multicriteria approach in mind. They employed the model selection paradigm of Deason (1984) in order to eliminate some of the methodologies, and then they employed the 27 criteria of Gershon (1981) in order to evaluate the techniques. In their study, PROMETHEE was found to be the preferred choice.

As a result of the deficiencies associated with some of the specific methods, various researchers have either extended the methods discussed in this chapter or developed "hybrid methods" involving a combination of two or more of the methods. For example, Macharis et al. (2004) provided recommendations for integrating PROMETHEE with a number of features of AHP; the suggested integration should yield a better hierarchy and weight determination.

Material Review Questions

3.1 Provide two specific examples of decision situations that one would want to model as deterministic situations. What are the characteristics of these situations, which lend themselves to a deterministic analysis?

3.2 Why would one want to eliminate all dominated outcomes prior to proceeding with a more detailed multicriteria decision analysis?

3.3 Elimination of all dominated solutions (or outcomes) from a set of solutions does not require any preference information from the DM (*true* or *false*).

3.4 What is the ideal?

3.5 What is a superior solution?

3.6 What is a nondominated solution?

3.7 Suppose that one has a collection of alternatives corresponding to multidimensional outcomes. If another alternative is added to the set, is it possible for a dominated outcome in the original set to become nondominated? Is it possible for a nondominated outcome in the original set to become dominated?

3.8 Suppose that one has 20 outcomes and wants to construct a dominance graph. How many pairwise comparisons would need to be made?

3.9 What type of scale are constructed attributes typically measured on?

3.10 Give one example each of a quantity measured on each of the four scales: nominal, ordinal, interval, and ratio.

3.11 Give an example of a situation in which a preference structure would not exist for a DM over an outcome space. How would you know that the DM would not have a valid preference structure?

3.12 Of the seven types of preference information mentioned in the chapter, which type is typically the most difficult for the DM to provide?

3.13 Give a specific example of a situation in which lexicographic ordering would clearly not be appropriate.

3.14 In a typical additive multiattribute value function, what indicates that a particular attribute is to be minimized—a negative coefficient or a negative slope for the function?

3.15 Give an example of two multiattribute value functions that are strategically equivalent.

3.16 What is the significance of the concept of "strategic equivalence" of MAV functions?

3.17 Consider two attributes, denoted as X and Y, for which a DM wants more of each of these. Keeping the value of Y constant, would you expect the marginal rate of substitution of X for Y for any particular DM to increase in value or to decrease in value as X increases?

3.18 What is the form of a multiattribute value function if the marginal rate of substitution of any attribute for any other attribute does not depend on the point in the outcome space where the rate is measured?

3.19 Suppose that more of each attribute is desired in a decision problem involving two attributes: X_1 and X_2. Why is the statement "X_1 is preferentially independent of X_2" meaningless?

3.20 If one has a set of attributes for a decision problem, X_1, X_2, X_3, what is meant by the statement, "(X_1, X_2) is preferentially independent of X_3"?

3.21 In referring to a set of attributes, X_1, X_2,..., X_p, what is meant by the statement, "the set (X_1, X_2,..., X_p) is mutually preferentially independent"?

3.22 If a set of attributes for a decision problem is mutually preferentially independent, what type of multiattribute value function is appropriate? Will this function necessarily be linear?

3.23 Two different, but reasonable, DMs can have different (not even strategically equivalent) MAV functions for the same decision situation (*true* or *false*).

3.24 What was the basic purpose behind the development of the SMART approach to MAV function assessment?

3.25 The TOPSIS approach relies on the use of both the ideal and the negative ideal (*true* or *false*).

3.26 What is one of the advantages of the AHP espoused by its proponents? Why do the detractors of AHP claim this as a disadvantage of the approach?

3.27 What is meant by "rank reversal" in the AHP? Why do some of the proponents of the AHP consider rank reversal to not be a major problem in its use?

3.28 Briefly describe the preference information required of a DM in order to use the ELECTRE III outranking method.

3.29 ELECTRE III typically provides only a partial order of the multidimensional outcomes in a decision problem (*true* or *false*).

3.30 One can often arrive at a good alternative in a decision situation by using an appropriate technique for ranking alternatives even without a good problem structuring procedure (*true* or *false*).

3.31 Briefly describe the type of preference information required from a DM (or DMs) in order to implement each of the following techniques: (1) identification of all nondominated solutions, (2) lexicographic ordering, (3) assessment of an MAV function, (4) TOPSIS, (5) AHP, and (6) ELECTRE III.

3.32 What aspects (attributes) of a decision problem should one consider in choosing a technique for ranking alternatives?

Exercises

3.1 Given that a decision problem has three attributes and that more of each attribute is preferred to less, select the nondominated outcomes from among the following set of outcomes.

(6, 5, 4), (8, 2, 1), (4, 6, 7), (3, 4, 2), (7, 6, 2), (5, 7, 2), (4, 4, 4), (3, 6, 7), (3, 4, 5).

3.2 Suppose that as a graduating student, you have a net worth of $2000. You have five different job offers to evaluate. These offers are equivalent in every important attribute except for yearly salary and number of days of vacation per year. Suppose that we have the following notation

$$X = \text{No. of days of vacation per year}$$

$$Y = \text{Yearly salary}$$

and the following outcomes to evaluate:

Outcome	(X, Y)
1	(5, 55,000)
2	(7, 54,000)
3	(10, 52,000)
4	(15, 51,000)
5	(20, 49,000)

Rank order the outcomes.

Now suppose that, because you have recently won the lottery, your net worth is $200,000. Rank order the outcomes under this assumption. Has your ranking changed? Discuss why or why not.

3.3 In a decision problem with two attributes, what can one say about the preference structure of the DM when the corresponding trade-offs condition is satisfied?

3.4 Give an example of two multiattribute value functions that are strategically equivalent, but not identical.

3.5 What is the marginal rate of substitution of X for Y at the outcome point (5, 7) given that an appropriate value function is

$$v(x, y) = 3x + 7y.$$

3.6 Suppose that you have two different job offers. Each job gives you 15 days of vacation. The first job pays $80,000 per year, while the second job pays $50,000 per year. Each job gives you the option of "buying" five more days of vacation. What is the maximum amount you would "pay" from your salary for five more days of vacation from each job; that is, what are the values for x_1 and x_2 such that the following statements are true?

$$(80,000, 15) \sim (x_1, 20)$$

and

$$(50,000, 15) \sim (x_2, 20).$$

If $(80,000 - x_1)$ is not equal to $(50,000 - x_2)$, then the MRS of salary for vacation days depends on the point you are at in the outcome space.

3.7 Suppose that one is assessing a scaled, additive multiattribute value function for a situation involving two attributes:

$$v(x, y) = c_1 v_1(x) + c_2 v_2(y).$$

Each attribute has a worst value of 0 and a best value of 100. Now, suppose that the DM prefers the outcome (0, 100) to the outcome (100, 0). What can one say about the relative values of the scaling constants: c_1 and c_2?

3.8 Suppose that you have a situation with two attributes, X and Y, and that your value function is given by $v(x, y) = x + 3y^2$ (i.e., this is a case where the MRS of X for Y depends on the value for Y but not on the value of X). Draw the graphs in the X–Y space for three indifference curves, for the cases where the value function levels are given by 10, 20, and 30, respectively.

3.9 On a Saturday morning, a friend of yours asks you to pick up a friend of his at the airport. The two attributes are the amount of time that you need to spend to perform this task and the amount of money you will be paid:

$$X = \text{time required and } Y = \text{payment, in dollars,}$$

where
x^0 is the worst possible value for X = 180 minutes
x^1 is the best possible value for X = 10 minutes

and
y^0 is the worst possible value for Y = $1
y^1 is the best possible value for Y = $200

Assess your multiattribute value function for this situation and rank order the following outcomes:

Outcome:	**1**	**2**	**3**	**4**	**5**	**6**	**7**
X:	10	30	50	80	120	150	180
Y:	5	15	25	35	100	150	200

3.10 Consider a multiobjective problem in which you have five alternatives and respective corresponding outcomes. There are two attributes, X_1 and X_2, each of which you want to maximize. The outcomes are given in the following table:

Outcome #	**X_1 Value**	**X_2 Value**
1	98	22
2	57	45
3	65	37
4	22	68
5	75	31

Using the TOPSIS method, rank the five outcomes for the following two sets of weights, where w_1 refers to the weight for attribute 1 and w_2 refers to the weight for attribute 2.

Case	**Weights**	**Relative Closeness Values: C_1, C_2, C_3, C_4, C_5**	**Rankings**
1	$w_1 = .9, w_2 = .1$		
2	$w_1 = .1, w_2 = .9$		

3.11 What is one of the major advantages of the analytic hierarchy process over the use of a multiattribute value function for solving a multiobjective problem?

3.12 Consider the following A matrix for the application of AHP:

$$A = \begin{pmatrix} 1 & 7 \\ 1/7 & 1 \end{pmatrix}$$

What are the local priorities/weights for the two attributes?

3.13 Consider a hierarchy (containing only the attributes for a problem, not the alternative solutions) for the AHP, with the following local weights:

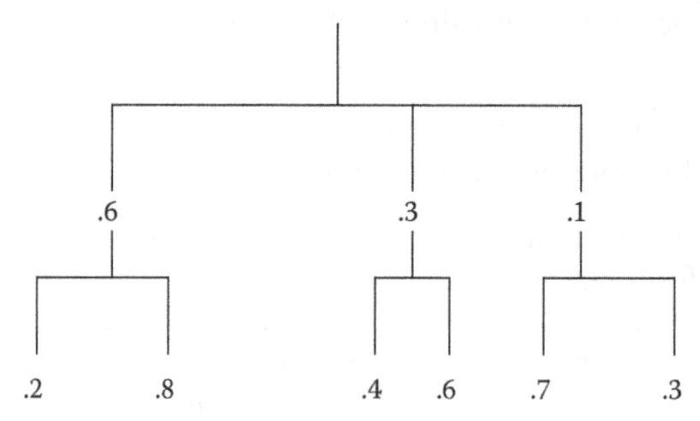

Suppose also that two alternatives are being considered, and that the local weights for each of these alternatives for each of the six lowest level attributes, from left to right, are given by

Alt. 1	.7	.4	.8	.4	.2	.2
Alt. 2	.3	.6	.2	.6	.8	.8

Rank the two alternatives by computing their respective global priorities.

3.14 Consider a simple hypothetical situation involving three alternatives and two attributes. Both of the attributes are to be maximized, and the weights are .8 for the first attribute and .2 for the second attribute. The attribute values for each of the three alternatives are given by $x_j(A_i)$, the value for attribute j of the ith alternative:

$x_j(A_i)$	j = 1	j = 2
i = 1	8	2
i = 2	2	6
i = 3	9	1

So, for example, Alternative 3 has an outcome of 9 on the first attribute and 1 on the second attribute.

Using the following threshold values (where q_j, r_j, and v_j are the indifference, strong preference, and veto thresholds, respectively, for attribute j) given in the following, determine a ranking for the three alternatives (note that ties are allowed) using the ELECTRE III procedure.

j	q_j	r_j	v_j
1	2	5	7
2	1	3	8

4

Goal Programming and Other Methodologies for Multiple Objective Optimization

4.1 Introduction

In this chapter, we present methodologies and applications for analyzing multiple objective situations with a very large (or even an infinite) number of alternatives. In addition to being multidimensional in order to represent the multiple objectives, the outcomes are also deterministic in nature.

Now the question occurs as to how a problem can have a very large or even an infinite number of alternatives. This problem characteristic occurs, because the alternatives are represented as a set of values assigned to a large number of *decision variables*, and these decision variables are classified as being either continuous or integer in nature. When the decision variables are continuous, they each can be assigned any value within a range of values; an example would be the amount of product to ship from a manufacturing facility to a distribution center. In such cases, the amount shipped could assume any of an infinite number of possible values.

In other cases, a decision variable could be integer in nature; more specifically, a decision variable could be *zero-one* or a *general integer decision variable.* A zero-one decision variable could correspond to a decision where the decision maker (DM) would either do something or *not* do something, such as *locate* a distribution center in a particular place, or *do not locate* a distribution center in that place. Other examples of such decision variables would be

1. Include, or do not include, a destination in a route
2. Include or do not include a route in a set of routes
3. Include or do not include a project in a portfolio of projects

A general integer decision variable would be one that could assume any of a large number of integer values, such as 0, 1, 2,…, 20. Examples of such general integer decision variables would be the number of items to sample from a lot for acceptance sampling purposes, or the number of call takers to assign to a particular shift at a call center. Since many of the solution methodologies for integer programming problems are designed for problems with zero-one decision variables, techniques have been developed to convert problems with general integer decision variables to those with only zero-one decision variables (e.g., see Hillier and Lieberman, 2010, pp. 478–479).

A set of values assigned to the respective decision variables in the type of problem discussed in this chapter represents an alternative. Typically, these problems can contain hundreds, thousands, or even millions of decision variables. When the problem contains only integer decision variables, the number of combinations of values that can be assigned to these variables can very easily become a very large number. For example, there are $2^{100} > 10^{30}$ various combinations of values for a problem with 100 zero-one decision variables. When there are continuous decision variables in addition to integer (or instead of integer) decision variables, the number of alternatives becomes infinite. With so many alternatives to consider, it is impossible to evaluate all of them. Hence, some type of optimization procedure, in which solution alternatives are implicitly evaluated, is required in order to determine the best solution.

Therefore, two types of processes are required in order to solve problems of the type discussed in this chapter—an optimization process and a process involving the articulation of the DM's preference structure over the multiple objectives of the problem. The relative timing of these two processes is one way to categorize the various solution methodologies discussed in this chapter (Evans, 1984): (1) prior articulation of preferences in which the articulation of preference information occurs prior to the optimization, (2) progressive articulation of preferences in which the articulation of preference information is interspersed with the optimization, and (3) a posteriori articulation of preferences in which the articulation of preference information occurs after the optimization. Some algorithms employ a combination of two of the three approaches described earlier.

In addition to the types of decision variables associated with the problem and the timing of the articulation of the preference information relative to the optimization process, a third type of categorization, which one can employ for the problems/solution methodologies discussed in this chapter, is how the relationships between the decision variables and objective/constraint functions are modeled—either as all linear, or at least some nonlinear functions. (In this chapter though, the assumption is that the relationship is always of a closed-form function.) The custom is to call the problem nonlinear if there is at least one nonlinear function out of all of the functions; otherwise, the problem is linear.

Figure 4.1 illustrates the classification scheme discussed earlier. In moving from left to right in the figure, choosing a single box out of each vertical box placement identifies the type of problem addressed.

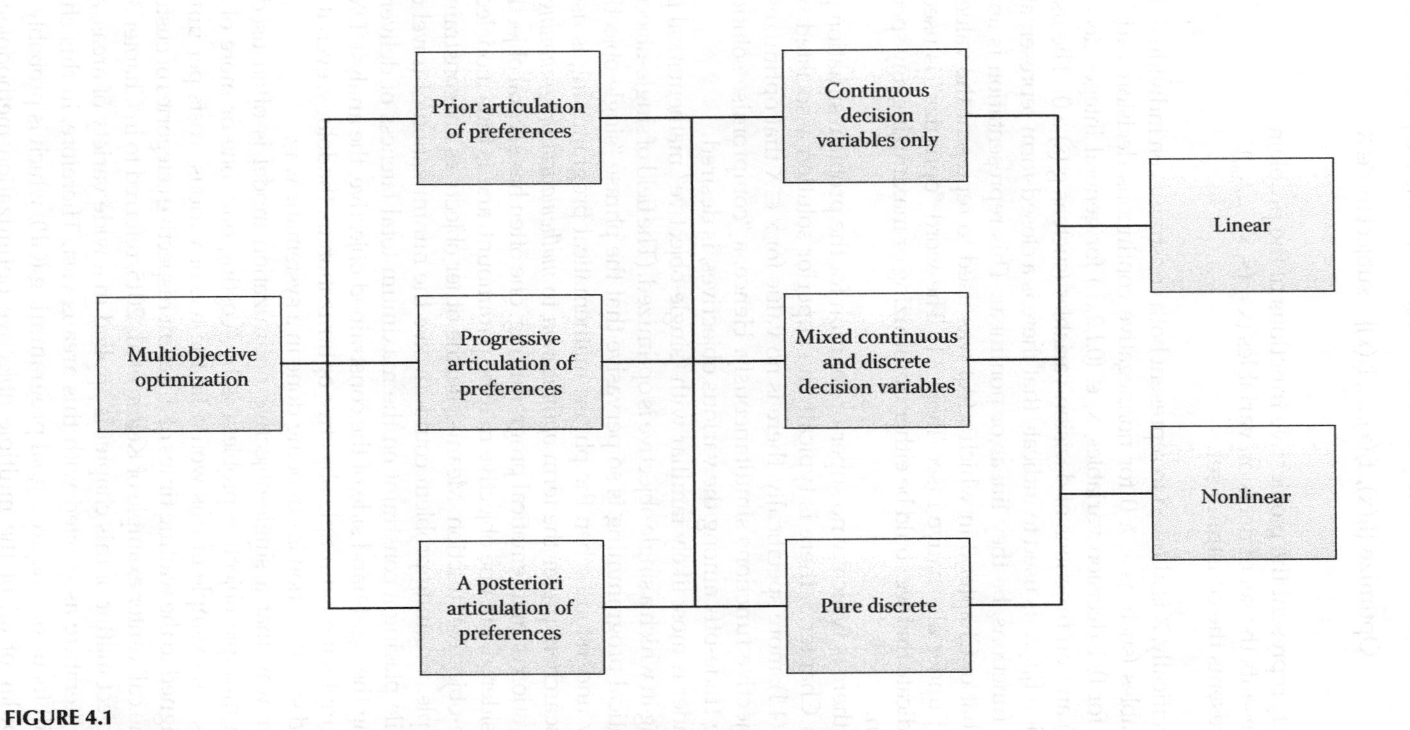

FIGURE 4.1
The classification scheme for multiple objective optimization problem/solution methodologies.

A general formulation of the problem addressed in this chapter is given by

$$\text{Optimize } \{f_1(x),\ f_2(x),\ldots,f_p(x)\} \quad \text{subject to } x \in X \tag{4.1}$$

where

$f_1,f_2,\ldots,f_p$ represent the p objective functions of the problem
x represents the set of decision variables ($x = (x_1,x_2,\ldots,x_n)$)
X represents the constraint set

More specifically, X is allowed to represent both constraints on individual decision variables (such as $x_i \geq 0$ for nonnegative continuous decision variables, $x_i \in (0,1)$ for 0/1 decision variables, $x_i \in (0,1,2,\ldots)$ for general integer decision variables) and on functions of decision variables (such as $g_j(x) \geq 0$). The use of the notation $f_1,f_2,\ldots$ is used to indicate that there is a closed-form representation for these functions, be they linear or nonlinear. This representation is analogous to that of Chapter 3 in which $x_j(A_i)$ was used to represent the value for attribute j under alternative i (see Table 3.1). The word "optimize" is used in (4.1) to indicate that we could be either minimizing or maximizing any specific f_i function.

Just as there is typically no superior solution for the problem situation presented in Chapter 3, there is typically no superior solution associated with problem (4.1); more specifically, there is no value for $x \in X$ that optimizes all of the objective functions simultaneously. Hence, a "compromise solution," involving trade-offs among the various objectives, is desired.

The reader is most likely familiar with "single-objective" mathematical programming in which a single objective is optimized. (The field of single-objective mathematical programming is so pervasive that the phrase "single objective" is usually understood when the phrase mathematical programming is used.) Many researchers prefer the term *optimization* to *mathematical programming*. In the application of mathematical programming, one often has several objectives to be considered, but one objective may be paramount, and is thus modeled as the single-objective function. Meanwhile, the other objectives are constrained. For example, a routing problem could involve the minimization of travel distance while placing a constraint on the maximum total lateness of deliveries. By varying the right-hand side of the constrained objective, the analyst/DM is actually employing a multiple objective optimization methodology, even if the right-hand side variation is not being done in a systematic way.

Another way that a single-objective optimization model is often used to address a multiple objective problem is by "costing out" one or more of the objectives. An example of this would be the use of various "costs per unit of time" assigned to the waiting times of different respective categories of customers (see the call center example of Kelton et al., 2015, referred to in Chapter 3).

The subject matter of this chapter is applied in a wide variety of areas, and thus, the literature associated with this area is vast. Therefore, in this chapter, we will focus on one area: goal programming (GP), which is probably the most popular of all of the multiple objective optimization methodologies.

For a more in-depth and comprehensive discussion of multiple objective optimization, the reader is referred to the books by Goicoechea et al. (1982), Barichard et al. (2009), Jones and Tamiz (2010), and Ehrgott (2000), among others.

In Section 4.2, we present the basic concepts of GP. Included in this section are two simple examples involving the basic product mix problem; Example 4.1 is a lexicographic goal program, and Example 4.2 is a weighted goal program. Section 4.3 discusses a decision maker's preference structure associated with a goal program, including the concepts of lexicographic redundancy and Pareto inefficiency. Section 4.4 discusses an important aspect associated with the formulation of a goal program: determination of the weights for the deviational variables.

Section 4.5 discusses the use of integer decision variables and nonlinear functions in goal programs; included in this section are respective examples involving, first, supplier selection and, second, distribution center location and transportation planning. Section 4.6 provides an introduction to interactive GP and metagoal programming. Finally, Section 4.7 briefly discusses other approaches towards multiple objective optimization.

4.2 Goal Programming: Basic Concepts

GP as a concept was originally suggested by Charnes et al. (1955) as a special type of linear program. But the term *goal programming* was first used in 1961 by Charnes and Cooper (1961). The methodology was popularized through the works of several researchers during the 1960s and 1970s, including Lee (1972) and Ignizio (1976). It remains arguably the most popular technique developed for multiple objective optimization. For example, a search of the term *goal programming* in the journal *European Journal of Operational Research* conducted on April 1, 2015, yielded 3946 entries. The methodology of goal programming has been applied in diverse areas such as portfolio management (Aouni et al., 2014), quality control (Cherif et al., 2008), design of closed-loop supply chains (Gupta and Evans, 2009), academic planning (Joiner, 1980), assignment (Lee and Schniederjans, 1983), agriculture (Minguez et al., 1988), acceptance sampling (Ravindran et al., 1986), and site location (Schniederjans et al., 1982).

Since the two are closely related, in order to understand the methodology of GP, it is useful to have a working knowledge of mathematical programming, especially linear programming; many goal programs are special cases of linear programs. In GP, the information concerning the DM's preferences over multiple objectives is embodied in goals and associated weights on those goals. In addition to decision variables, denoted as x in (4.1), a goal program has deviational variables that measure respective distances from goals. The idea of setting and achieving goals, as embodied in GP, fits nicely with Simon's concept of "satisficing" (Simon, 1960).

Referring back to Section 4.1, basic GP fits into the category of methodologies involving a prior articulation of preferences, since the DM specifies his or her goals prior to any optimization process.

As noted by Jones and Tamiz (2010, Chapter 2), there are three main variants of GP: lexicographic (also called preemptive) goal programs, weighted (also called nonpreemptive or Archimedian) goal programs, and minmax (also called Chebyshev) goal programs.

In lexicographic goal programs, the goals are placed in ordered categories of importance such that goals in the most important category must be achieved before goals in the second most important category. Goals in a particular category cannot be achieved at the expense of the goals in a more important category. In weighted goal programs, direct trade-offs between all of the objectives are allowed—that is, all goals are placed in the same category. Finally, minmax goal programs involve a minimization of the maximum deviation from any goal. This type of goal program originated through the work of Flavell (1976); it can be thought of as a solution technique where the DM seeks to achieve a balance among the various goals. Due to the popularity of the first two categories of GP, as compared to the minmax GP, the remainder of this chapter will focus on lexicographic goal programs and weighted goal programs. For additional information on minmax goal programs, see Flavell (1976) and Jones and Tamiz (2010, pp. 15–16).

A general formulation of a lexicographic goal program is given in (4.2) through (4.5):

$$\text{Lex Min } Z = \left\{ \sum_{g \in D_1} (w_{g1}p_g + w_{g2}n_g), \sum_{g \in D_2} (w_{g1}p_g + w_{g2}n_g), \ldots, \sum_{g \in D_C} (w_{g1}p_g + w_{g2}n_g) \right\} \tag{4.2}$$

subject to

$$f_g(x) + n_g - p_g = b_g \quad \text{for } g = 1, \ldots, G \tag{4.3}$$

$$x \in X \tag{4.4}$$

$$p_g, n_g \geq 0 \quad \text{for } g = 1, \ldots, G \tag{4.5}$$

In (4.2) through (4.5) the original decision variables are given as $x = (x_1, x_2, \ldots, x_n)$ and the set of original constraints are $x \in X$ as shown in (4.4) and (4.1). The parameter G denotes the number of goals. The $f_g(x)$ represent objective functions, while n_g and p_g are negative and positive deviational variables, respectively, for objective g for g = 1,…,G; the b_g represent respective target values for the goals, which are numbers determined by the DM. The goals are divided into categories: 1,2,…,C where D_i is the set of goals contained in category i for i = 1,…,C.

Each category of priorities in (4.2) through (4.5) corresponds to a weighted sum of goal deviations, $\sum_{g \in D1} (w_{g1}p_g + w_{g2}n_g)$, in the lexicographic minimization. The deviational variables, p_g and n_g in (4.2) through (4.5), represent the positive and negative deviations, respectively, from the goal target value b_g. For example, if one has a target goal of \$10,000 (i.e., $b_g = 10{,}000$) for profit, but after the optimization procedure achieves a value of only \$9,700 (i.e., $f_g(x) = 9{,}700$), then the corresponding deviational variable values will be $n_g = 300$ and $p_g = 0$.

The formulation is set up in such a way that at most, one of the two deviational variables for any goal will achieve a strictly positive value. For example in the aforementioned situation with a goal of \$10,000 for profit where an actual value of \$9,700 is achieved, the solution process will not achieve values of $n_g = 500$ and $p_g = 200$ (even though these values would still result in the satisfaction of the goal constraint $f_g(x) + n_g - p_g = b_g$ with $f_g(x) = 9{,}700$, $n_g = 500$, $p_g = 200$, and $b_g = 10{,}000$). Instead, the process would yield values of $n_g = 300$ and $p_g = 0$.

The lexicographic minimization in (4.2) through (4.5) means that a sequence of single-objective minimizations must be performed. In particular, the function associated with the first category of goals must be minimized first, followed by the minimization of the function associated with the second category of goals (without degradation of performance with respect to the first set of goals). As noted by Ignizio and Cavalier (1994), this lexicographic optimization can be accomplished by including a constraint that restricts the current optimization such that the objective function from the previous optimization is set equal to its optimal objective function value.

The constants, w_{g1} and w_{g2}, in (4.2) refer to weights assigned to the deviational variables p_g and n_g, respectively. The larger the weight, the more important that deviational variable is within its particular category. Since it may not be possible to meet all of the goals for the objectives, as given by the $b_1, b_2, \ldots, b_G$ values, the deviational variables and corresponding constraints given by (4.3) allow for some amount of deviation from these goals.

If there is only one category of goals (C = 1), then we are dealing with a weighted goal program, whereas if C > 1, we are dealing with a preemptive goal program. A general formulation for a weighted goal program is shown in (4.6) through (4.9):

$$\text{Min } Z = \sum_{g=1}^{G} (w_{g1}p_g + w_{g2}n_g), \tag{4.6}$$

subject to

$$f_g(x) + n_g - p_g = b_g \quad \text{for } g = 1, \ldots, G \tag{4.7}$$

$$x \in X \tag{4.8}$$

$$p_g, n_g \geq 0 \quad \text{for } g = 1, \ldots, G \tag{4.9}$$

The reader should note that except for the objective function, Problem (4.6) through (4.9) is the same as Problem (4.2) through (4.5). The objective function in a goal program (either Problem 4.2 or 4.6) is called an "achievement function," in order to differentiate it from an objective function in a typical mathematical program.

There are three types of goals possible, corresponding to

1. Minimization of the underachievement of a value
2. Minimization of the overachievement of a value
3. Minimization of any deviation from a value

The first case would typically correspond to a situation where one would want to maximize a value (e.g., profit) but because of other, conflicting, objectives, a target value (as opposed to an optimum value when the other objectives are ignored) is acceptable; in this case, the weight associated with the negative deviational variable, w_{g2}, would be set to 0.

The second case, involving minimization of overachievement, would correspond to minimization of a value (e.g., cost); in this case, the weight associated with the positive deviational variable, w_{g1}, would be set to 0. In the third case, both w_{g1} and w_{g2} would be set to strictly positive values. A fourth type of goal could involve achieving a value within a range (e.g., achieve a profit within a range of $30–$35 million). This situation could be formulated as two separate goals of the first and second types, for example, minimize the overachievement of $30 million and minimize the underachievement of $35 million.

Table 4.1 provides a summary of the notation given in (4.2) through (4.9).

When the concept of GP was initially developed, most of the applications involved lexicographic GP, as opposed to weighted GP. In particular, Tamiz et al. (1995) noted that about 75% of the applications prior to 1990 involved lexicographic GP, while about 25% involved weighted GP. Since 1990, there has been a shift toward the weighted approach—for example, Jones and Tamiz (2002) noted that from 1990 to 2000, about 59% of the published applications involved lexicographic GP, while 41% involved weighted GP. This shift toward the weighted GP approach and away from the lexicographic approach can be attributed to the greater flexibility (in terms of representation of preference structures) allowed by the weighted approach.

Let's consider a relatively simple hypothetical problem in Example 4.1 in order to illustrate the formulation and solution of a goal program. This example is similar to a type of example often given to illustrate the formulation of a linear program in an entry-level course in that area.

TABLE 4.1

Summary of the Notation Given by the Goal Programs of (4.2) through (4.9)

Notation	Description	Comments, If Appropriate
G	Number of goals	Typically, G will correspond to the number of objectives given in Equation 4.1.
C	Number of preemptive goal categories	C = 1 corresponds to a nonpreemptive goal program, while C > 1 corresponds to a preemptive goal program.
D_i	Set of goals contained in preemptive goal category i for i = 1,…,C	—
$x = (x_1, x_2, \ldots, x_n)$	Decision variables	—
n_g	Negative deviational variable for goal g, for g = 1,…,G	—
p_g	Positive deviational variable for goal g, for g = 1,…,G	—
w_{g1}	Weight associated with the positive deviational variable, p_g	w_{g1} is set to 0 if the goal involves minimization of an overachievement.
w_{g2}	Weight associated with the negative deviational variable, n_g	w_{g2} is set to 0 if the goal involves minimization of an underachievement.
$f_g(x)$	Functional value for goal associated with objective g, as a function of x, for g = 1,…,G	—
b_g	Target value for goal g, for g = 1,…,G	—
$x \in X$	Constraint set	

Example 4.1: A Lexicographic Goal Program for a Simple Product Mix Problem

Consider a problem involving the production of two products over a 1-month period. Suppose these two products, labeled product A and product B, result in \$50 and \$80 of profit, respectively, for each unit produced. Production requires two types of resources. Four and ten units of Resource 1 are required, respectively, for each unit of products A and B produced, and seven and eleven units of Resource 2 are required, respectively, for each unit of products A and B produced. There are 200 units of Resource 1 available and 264 units of Resource 2 available. The information regarding this production problem is summarized in Table 4.2.

As mentioned, a linear program associated with this type of problem is often found in many entry-level books in operations research. This problem can be stated as follows: determine the number of units of each product to manufacture over a 1-month period so as to maximize profit without exceeding the number of units of each resource available. If x_1 and x_2 represent the decision variables associated with the number

TABLE 4.2

Summary of Information Regarding the Production Mix Problem

Product Type	Per Unit Profit ($)	Units of Resource 1 Required per Unit of Production	Units of Resource 2 Required per Unit of Production
A	50	4	7
B	80	10	11
Units of resources available		200	264

of units of products A and B to produce, respectively, then this linear program can be stated as (4.10) through (4.13):

$$\text{Maximize} \quad Z = 50x_1 + 80x_2, \tag{4.10}$$

subject to

$$4x_1 + 10x_2 \leq 200 \tag{4.11}$$

$$7x_1 + 11x_2 \leq 264 \tag{4.12}$$

$$x_1, x_2 \geq 0 \tag{4.13}$$

In this formulation, Z stands for the objective function of profit, while the two "less than or equal to" constraints restrict the amounts of Resources 1 and 2 used according to the respective amounts available.

Now, let's consider a GP formulation of this problem. This GP formulation will allow much more flexibility in terms of considering goals in addition to profit. Suppose that the DM has four goals to be considered, according to the following three (decreasing) priority levels.

1. At least $1800 of profit should be achieved.
2. Not more than 190 units of Resource 2 should be used.
3. (a) At least 10 units of Product 1 and (b) at least 10 units of Product 2 should be manufactured.

Note that each goal corresponds to (1) the specification of an attribute/measure (e.g., profit achieved, amount of resource used, or number of units of a product manufactured), (2) a target value, and (3) a direction level ("at least," or "not more than"). For the GP formulation, the goals would be restated as a minimization of a weighted sum of deviational variables. In addition, goal constraints would be added to the formulation, as follows:

$$50x_1 + 80x_2 + n_1 - p_1 = 1800$$

$$7x_1 + 11x_2 + n_2 - p_2 = 190$$

$$x_1 + n_3 - p_3 = 10$$

$$x_2 + n_4 - p_4 = 10.$$

Note that each goal constraint has the same format: a function of the decision variables representing the measure, with a negative deviational variable added and a positive deviational variable subtracted, set equal to a target value. (Students are sometimes confused by the fact that the *negative* deviational variable is *added*, while the *positive* deviational variable is *subtracted* from the left-hand side of a goal constraint.) For example, for the first goal constraint, $50x_1 + 80x_2$ represents the profit in terms of the decision variables and 1800 is the target value.

A comment concerning the goal associated with the second-level priority (not more than 190 units of Resource 2 should be used) is in order. The reader might see this as being in contrast to the hard constraint associated with this resource (only 264 units of this resource are available for use). The idea here is that the DM would like to have some of this resource available for other uses; however, this might be considered as a "soft constraint" especially in light of the fact that it is only of second-level priority.

The complete lexicographic goal program is given as in (4.14) through (4.21):

$$\text{Lex Min } Z = \{n_1, p_2, n_3 + n_4\} \tag{4.14}$$

subject to

$$50x_1 + 80x_2 + n_1 - p_1 = 1800 \tag{4.15}$$

$$7x_1 + 11x_2 + n_2 - p_2 = 190 \tag{4.16}$$

$$x_1 + n_3 - p_3 = 10 \tag{4.17}$$

$$x_2 + n_4 - p_4 = 10 \tag{4.18}$$

$$4x_1 + 10x_2 \le 200 \tag{4.19}$$

$$7x_1 + 11x_2 \le 264 \tag{4.20}$$

$$x_1,\ x_2 \ge 0 \tag{4.21}$$

Note that the constraint set contains two categories of constraints: the goal constraints and the "hard constraints" from the original linear programming formulation. The lexicographic minimization consists of one category for each priority level. For example, in the first priority level, we are attempting to minimize the negative deviation from a profit of $1800 by minimizing n_1; if in solving this optimization problem, the optimal value of n_1 turned out to be $25, then the profit achieved from the solution would be $1775, or $1800 – $25.

Similarly, for the second priority level, we are trying to minimize the positive deviation from 190 units of Resource 2 used by minimizing p_2. The third priority level contains two goals. By just minimizing the sum of these of the two deviational variables, we are implicitly assigning equal weights to these two goals within the third priority level. Any time multiple goals are contained within a single priority level, individual weights need to be assigned to the goals. Various approaches for assigning weights to the deviational variables within a goal category will be presented later in this chapter.

As mentioned earlier, solving a lexicographic goal program such as Problem (4.14) through (4.21) involves solving a sequence of linear programs. The first linear program for solving Problem (4.14) through (4.21) involves optimizing the first objective in (4.14) subject to the constraints as shown in (4.15) through (4.21). Using the software package LINGO™, the following optimal solution was obtained for this linear program:

$$x_1^* = 36, x_2^* = 0, p_2^* = 62, p_3^* = 26, \text{and } n_4^* = 10,$$

with the optimal values for all of the other deviational variables being 0. This solution, since it has $n_1^* = 0$, meets the goal associated with the first-level priority.

The second linear program in the sequence involves optimizing the second objective in the lexicographic minimization of (4.14), subject to the set of constraints, (4.15) through (4.21), along with the additional constraint: $n_1 = 0$. The optimal solution obtained from this linear program is given as

$$x_1^* = 11.11\ldots, x_2^* = 15.55\ldots, p_2^* = 58.88\ldots, p_3^* = 1.11\ldots, \text{and } p_4^* = 5.55\ldots,$$

with all optimal values for all of the other deviational variables being 0. The ellipsis is used to indicate a transcendental number. Note that since $p_2^* > 0$, the second goal of not using more than 190 units of Resource 2 was not achieved.

Finally, the third linear program associated with solving this lexicographic GP is given by an objective function given by the third objective in (4.14) subject to the constraints given by (4.15) through (4.21), along with the two additional constraints: $n_1 = 0$ and $p_2 = 58.88$. This linear program has the same optimal solution as the second linear program:

$$x_1^* = 11.11\ldots, x_2^* = 15.55\ldots, p_2^* = 58.88\ldots, p_3^* = 1.11\ldots, \text{and } p_4^* = 5.55\ldots$$

TABLE 4.3

Summary of Results for the Three Sequential Linear Programs Solved for GP (Equations 4.14 through 4.21)

Linear Program #	Goal 1 (Profit) Value	Goal 2 (Units of Resource 2 Used) Value	Goal 3 (Units of Product 1) Value	Goal 4 (Units of Product 2) Value
1	1800	252.	36	0
2	1800	248.88...	11.11...	15.55...
3	1800	248.88...	11.11...	15.55...

This third solution is the solution to the lexicographic goal program given by (4.14) through (4.21). A summary of the results associated with the solutions found from solving the three linear programs associated with the goal program shown in (4.14) through (4.21) is shown in Table 4.3.

Looking at the results shown in Table 4.3 for Example 4.1, the goal associated with the first-level priority (achieve a profit of at least $1800) and the goals associated with the third-level priority (manufacture at least 10 units of each product) were met; however, the goal associated with the second-level priority (do not use more than 190 units of Resource 2) was not met. This illustrates the fact that *in a lexicographic goal program, it is possible for lower-level priority goals to be met (third-level priority in this case), while higher-level priority goals (second-level priority in this case) are not met.* The key fact here is that more than 190 units of Resource 2 needed to be used (thereby not achieving the second-level priority goal) in order to achieve a profit of at least $1800 (the first-level priority goal).

Each linear program (following the first one) in the sequence of linear programs solved for a lexicographic goal program differs from the previous linear program in two ways:

1. The objective function is different.
2. The feasible region includes one more constraint.

This relatively small difference from one linear program to the next indicates that these linear programs do not have to be solved "from scratch." For example, the solution process for each linear program can start at the optimal solution found from the previous linear program, which allows for a decrease in computational effort as opposed to starting from scratch; this starting solution is guaranteed to be feasible, since it automatically satisfies the additional constraint added for the new linear program.

In addition to starting the process at the previous optimal solution, Ignizio has suggested other procedures for reducing computational effort in solving the sequence of linear programs. For example, Ignizio (1982) notes that any variables with positive reduced cost at the optimal solution for the previous problem can be restricted from assuming a value above their lower bound (normally 0) in this and following linear programs in the sequence. A further improvement was suggested by Ignizio (1985) when he noted that it is often desirable to solve the dual rather than the primal problem.

An alternate approach that is often used to solve a lexicographic goal program involves solving one formulation of the problem in which the respective objective functions associated with the priority levels are multiplied by a series of decreasing constants (e.g., 1000, 100, 10) and then summed to form the overall achievement function. For example, for the problem given in (4.14) through (4.21) the achievement function would be given by

$$\text{Minimize } 1000 * n_1 + 100 * p_2 + 10 * (n_3 + n_4). \tag{4.22}$$

Solving (4.22) subject to the constraints given in (4.15) through (4.21) yields the same solution as that given by using the sequential procedure shown previously; however, as shown, the solution is obtained by solving just one linear program instead of three. The idea is that by assigning decreasing values for the constants (i.e., 1000, 100, and 10 in this case), the optimization will attempt to achieve the first objective prior to the second and so on.

This "single pass" procedure for solving a lexicographic goal program is often called the "Big P" approach, since the objective function of Problem (4.2) through (4.5) could be represented in an alternative way as

$$P_1 \sum_{g \in D_1} (w_{g1} p_g + w_{g2} n_g) + P_2 \sum_{g \in D_2} (w_{g1} p_g + w_{g2} n_g) + \cdots + P_C \sum_{g \in D_C} (w_{g1} p_g + w_{g2} n_g), \tag{4.23}$$

where $P_1, P_2, \ldots, P_C$ are constants such that $P_1 \gg P_2 \gg \cdots \gg P_C$ (where $P_1 \gg P_2$ means P_1 is much bigger than P_2, and so on). However, as noted by Jones and Tamiz, this approach/notation is not preferred, since it expresses the lexicographic minimization as a summation, when it actually is not (see Jones and Tamiz, 2010, p. 33). In addition, one may very well run into scaling problems with this "single pass" approach.

As mentioned, a second main category of goal programs is the weighted goal program, in which all goals are placed in the same priority level. An example of this type of goal program is shown in Example 4.2.

Example 4.2: A Weighted Goal Program for a Simple Product Mix Problem

Let's consider a weighted goal program corresponding to the product mix problem of Example 4.1. Let the weights for the four deviational variables be given as

1. .4 for the goal of achieving at least $1800 of profit
2. .3 for the goal of not using more than 190 units of Resource 2
3. .15 for the goal of producing at least 10 units of Product 1
4. .15 for the goal of producing at least 10 units of Product 2

Hence, the weighted goal program for this problem is given as in the following equations:

$$\text{Minimize } Z = .4 * n_1 + .3 * p_2 + .15 * n_3 + .15 * n_4, \tag{4.24}$$

$$\text{Subject to constraints (4.15) through (4.21).} \tag{4.25}$$

The solution to the linear program of (4.24) and (4.25) is given by

$$x_1^* = 11.11\ldots, x_2^* = 15.55\ldots, p_2^* = 58.88\ldots, p_3^* = 1.11\ldots, \text{and } p_4^* = 5.55\ldots,$$

with all other variable values being 0. The reader will note that this solution is the same solution obtained from solving the lexicographic goal program of (4.14) through (4.21). The weighted goal program of (4.24) and (4.25) was modified by the use of two other sets of weights for the deviational variables. The results associated with solving these linear programs are shown in Table 4.4. The reader will note that the notation used for the weights in Table 4.4 is the same as that used in Table 4.1. For example, w_{32} is the weight for the negative deviational variable for the third goal: produce at least 10 units of Product 1.

Comments regarding the solutions shown in Table 4.4 are shown in Table 4.5.

4.3 Preference Structures Associated with Goal Programs, Lexicographic Redundancy, and Pareto Inefficiency

The preference structure associated with a lexicographic goal program is not necessarily the same as that associated with lexicographic ordering approach discussed in Chapter 3. In particular, the target(s) set for the respective goal(s) at the first-level priority for a lexicographic goal program are typically not set at the optimal value(s) for the relevant objective function(s); hence, this allows for some flexibility with respect to the achievement of second-level and lower-level priority goals.

As an example, in Example 4.1, the first priority goal was set at a target value of something less than the optimal value for a profit of $1904.40 obtained from solving the linear program given in (4.10) through (4.13). Setting the goal for

TABLE 4.4

Results for the Weighted Goal Program with Alternative Sets of Weights

Weight Set #	w_{12}	w_{21}	w_{32}	w_{42}	x_1^*	x_2^*	n_1^*	p_1^*	n_2^*	p_2^*	n_3^*	p_3^*	n_4^*	p_4^*	Profit ($)
1	0.4	0.3	0.15	0.15	11.1…	15.5…	0	0	0	58.8…	0	1.1…	0	5.5…	1800
2	0.1	0.8	0.05	0.05	0	17.27…	418.18…	0	0	0	10	0	0	7.27…	1382
3	0.1	0.1	0.4	0.4	11.1…	15.5…	0	0	0	58.8…	0	1.1…	0	5.5…	1800
4	1	0	0	0	36	0	0	0	0	62	0	26	10	0	1800
5	0	1	0	0	0	0	1800	0	190	0	10	0	10	0	0
6	0	0	1	0	37.714	0	0	85.714	0	74	0	27.714	10	0	1886
7	0	0	0	1	0	20	200	0	0	30	10	0	0	10	1600
8	0	0	0.5	0.5	10	10	500	0	10	0	0	0	0	0	1300

TABLE 4.5

Comments Regarding the Solutions Shown in Table 4.4

Weight Set #	Comments with Respect to the Solution for This Set of Weights
1	Goals 1, 3, and 4 are satisfied
2	Goals 2 and 4 are satisfied
3	Goals 1, 3, and 4 are satisfied
4	Goals 1 and 3 are satisfied
5	Goal 2 is satisfied
6	Goals 1, 3, and 4 are satisfied
7	Goal 4 is satisfied
8	Goals 2, 3, and 4 are satisfied

profit at $1800 allows for flexibility in achieving the other goals. Setting the goal for profit at its ideal level of $1904.40 could lead to what is called *lexicographic redundancy* (Romero, 1991); in effect, this leads to the feasible region associated with the subsequent sequential linear programs consisting of a single point—that is, no optimization is actually accomplished in these subsequent linear programs, since there is only one feasible solution. In addition to setting of unreasonable target levels for goals, another reason for lexicographic redundancy is having too many priority levels; the consensus is that a lexicographic goal program should have no more than five priority levels (Jones and Tamiz, 2010, p. 33).

The problem of lexicographic redundancy can be contrasted with that of *Pareto inefficiency*. That is, a solution generated from either a lexicographic goal program or a weighted goal program may be a solution for which at least one of the objectives can be improved on without degrading any of the other objectives.

As an example, consider a modification of the lexicographic goal program given by (4.14) through (4.21) in which the target value for the second goal is set to 255 units of Resource 2; that is, instead of a goal of "not more than 190 units of Resource 2 should be used," the goal would be "not more than 255 units of Resource 2 should be used." The goal program of (4.14) through (4.21) would be modified so that the right-hand side of constraint (4.16) would be 255 rather than 190. Solving this new lexicographic goal program would yield an optimal solution:

$$x_1^* = 11.1\ldots, x_2^* = 15.5\ldots, n_2^* = 6.1\ldots, p_3^* = 1.1\ldots, p_4^* = 5.5\ldots,$$

with all other variable values at 0. This solution satisfies all of the new goals, since the optimal values found for the deviational variables, namely, n_1, p_2, n_3, and n_4, are all 0.

Now, if the DM set a different target value for a profit of $1805 instead of $1800 and solved this second modified lexicographic goal program,

TABLE 4.6

Results from Solving Two Modified Versions of the Lexicographic Goal Program (Equations 4.14 through 4.21)

Problem Solved	Profit Attained ($)	Goals Achieved	Comments
First modification to lexicographic goal program 4.5 (target goal values of 1800, 255, 10, 10)	1800	All four goals achieved	Solution obtained is Pareto inefficient.
Second modification to lexicographic goal program 4.5 (target goal values of 1805, 255, 10, 10)	1805	All four goals achieved	—

the resulting solution would also satisfy all of the goals and thereby yield an outcome, which dominates the outcome from the solution to the original modification of (4.14) through (4.21). These two solutions and associated outcomes are shown in Table 4.6.

For an example of Pareto inefficiency with a weighted goal program, see Jones and Tamiz (2010, pp. 95–96). For methods to detect Pareto inefficiency and to generate a subsequent Pareto efficient solution, see Jones and Tamiz (2010, Chapter 6).

Even with the "relaxed values" for targets, thereby avoiding the problem of lexicographic redundancy, a lexicographic goal program may not accurately represent a DM's preference structure; however, the representation may very well be accurate enough for a good decision. In particular, the DM may feel much more comfortable with the ideas of setting and satisfying goals than with the interviewing procedure associated with the assessment of a multiattribute value (MAV) function.

In general, solving a weighted goal program corresponds to optimizing a multiattribute value function in which that function consists of attributes defined as respective deviations from target values for goals; moreover, the function itself is additive and linear.

4.4 Determining Weights for the Deviational Variables in a Goal Program

An important aspect associated with accurately representing the DM's preference structure in a goal program is the determination of the weights for the deviational variables in the achievement function. As previously noted with respect to Table 4.1, the type of goal determines whether the weight for a deviational variable associated with that goal in the achievement function will be zero or strictly positive. In addition, in a lexicographic goal program, for priority categories which contain only one goal, the weight for the relevant deviational variable can be set to a value of 1. Therefore, the problem of setting nonzero values for weights comes into play in two situations:

1. For the deviational variables associated with multiple goals within any particular priority level in a lexicographic goal program
2. For each of the deviational variables contained in the achievement function for a weighted goal program

One of the important considerations in setting a weight for a goal is the target value for that goal. This can be accomplished through what is called *percentage normalization* (Jones and Tamiz, 2010, p. 34). This simple approach involves setting the weights so that the deviational variable values are in effect normalized. Consider problem (4.24) and (4.25), as shown in the following equations, but with the appropriate symbols for the weights:

$$\text{Minimize } Z = w_{12}n_1 + w_{21}p_2 + w_{32}n_3 + w_{42}n_4 \tag{4.26}$$

subject to

$$50x_1 + 80x_2 + n_1 - p_1 = 1800 \tag{4.27}$$

$$7x_1 + 11x_2 + n_2 - p_2 = 190 \tag{4.28}$$

$$x_1 + n_3 - p_3 = 10 \tag{4.29}$$

$$x_2 + n_4 - p_4 = 10 \tag{4.30}$$

$$4x_1 + 10x_2 \le 200 \tag{4.31}$$

$$7x_1 + 11x_2 \le 264 \tag{4.32}$$

$$x_1, x_2 \ge 0 \tag{4.33}$$

Now, if we set the weights so that they are equal to the reciprocal of the target value associated with the corresponding goal, we would have the following:

$$w_{12} = 1/1800 = .000\overline{5}, w_{21} = 1/190 = .016129, w_{32} = 1/10 = .1,$$
$$\text{and } w_{42} = 1/10 = .1.$$

Then a particular percentage deviation from the goal, represented as a particular value for the deviational variable, would be measured the same over all goals by the achievement function. For example, a 10% deviation from the four goals would be measured by $n_1 = 180, p_2 = 19, n_3 = 1$, and $n_4 = 1$, respectively. These deviational variable values would contribute the same amount, .1, to the achievement function value; for example, $w_{12}n_1 = (1/1800)*180 = .1$.

The large differences in the weights shown indicate that it is usually good practice to scale the target values for the goals so that they have values that do not differ too much from each other. For example, for the profit goal, one might state the target value in terms of hundreds of dollars of profit, thereby making the target 18, instead of 1800. Similarly, for the goal associated with units of Resource 2, the target value might be stated in terms of 10s of units of the resource, making the target value 19 instead of 190. This approach will allow the weights to be closer in value to each other and thereby allow the DM to think about them in a more rational fashion.

While using the respective reciprocals of the target values for the weights may be intuitively appealing because of its simplicity, it also may not accurately represent the preference structure of the DM.

Therefore, a second approach would be to assess the MAV function over the deviational variables contained in the achievement function. Of course in using this approach, we are assuming that the deviational variables not in the achievement function are not relevant to the DM; for example, with respect to problem (4.26) through (4.33), this would mean that the DM places no value on achieving more than \$1800 in profit, since the p_1 deviational variable is not in the achievement function. In addition, we are assuming that the individual attribute value functions for the deviational variables are linear and that the function itself is additive.

To employ this MAV function approach to determining the weights, we would first need to determine the best and worst possible values for the deviational variables so that we can scale the implicit individual attribute value functions properly from 0 to 1. This is illustrated in Figure 4.2, for the deviational variable p_2 from problem (4.26) through (4.33), which has a best possible value of 0 and a worst possible value of 62.

In order to find these best and worst possible values for the deviational variables, we need to solve a series of single-objective optimization problems, one for each deviational variable in the achievement function for a weighted GP, or one for each deviational variable in a lexicographic category for a lexicographic GP.

For example, to find weights for the deviational variables for the GP given by (4.26) through (4.33), we would solve the sequence of problems given by

Minimize $\{n_1\}$, Minimize $\{p_2\}$, Minimize $\{n_3\}$, and Minimize $\{n_4\}$, each with the set of constraints given by (4.27) through (4.33).

Solving these four problems gives four sets of respective values for the deviational variables: n_1, p_2, n_3, and n_4, along with values for the other variables of the problem. Selecting the best (minimum) and worst (maximum) values for each deviational variable, we obtain the values shown in Table 4.7.

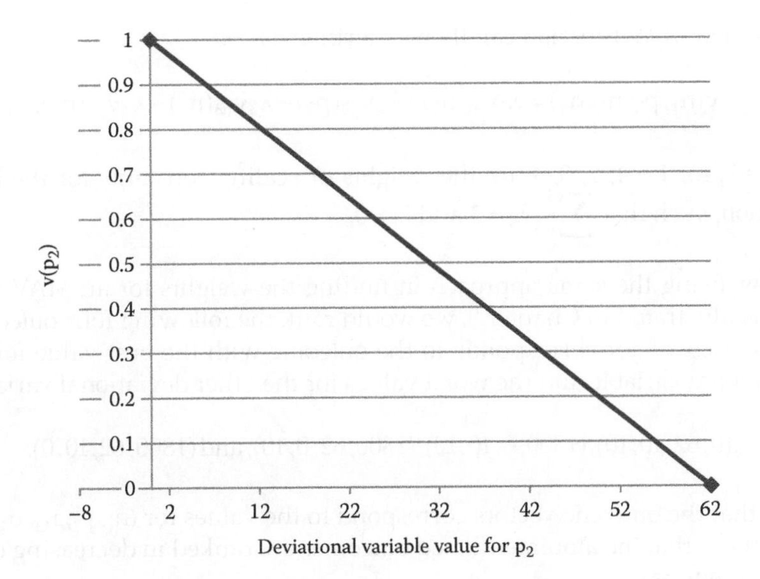

FIGURE 4.2
Implicit individual attribute value function for p_2 from problems (4.24) and (4.25).

TABLE 4.7
Worst and Best Values Found for the Deviational Variables by Solving Four Single-Objective Optimization Problems

Deviational Variable	Worst Value	Best Value
n_1	1800	0
p_2	62	0
n_3	10	0
n_4	10	0

The implicit scaled (from 0 to 1) single attribute value functions for the four deviational variables are given by the linear functions, which give a value of 0 for the worst attribute value and a value of 1 for the best attribute value. More specifically, these functions are

$$v_{n1}(n_1) = -(1/1800)n_1 + 1$$

$$v_{p2}(p_2) = -(1/62)p_2 + 1$$

$$v_{n3}(n_3) = -(1/10)n_3 + 1$$

$$v_{n4}(n_4) = -(1/10)n_4 + 1.$$

The scaled MAV function can then be written as

$$v(n_1, p_2, n_3, n_4) = \lambda_1 v_{n1}(n_1) + \lambda_2 v_{p2}(p_2) + \lambda_3 v_{n3}(n_3) + \lambda_4 v_{n4}(n_4),$$

where λ_i for i = 1, 2, 3, 4 are the weights or scaling constants for the MAV function, such that $\sum_{i=1}^{4} \lambda_i = 1$ and $\lambda_i > 0$.

Now, using the same approach in finding the weights for an MAV function as illustrated in Chapter 3, we would rank the following four outcomes, where each vector corresponds to the outcome with the best value for one deviational variable and the worst values for the other deviational variables:

$$(0,62,10,10), (1800,0,10,10), (1800,62,0,10), \text{and} (1800,62,10,0).$$

Note that the outcome vectors correspond to the values for (n_1, p_2, n_3, n_4).

Suppose that the aforementioned outcomes are ranked in decreasing order of preference as

$$(0,62,10,10), (1800,0,10,10), (1800,62,0,10), \text{and} (1800,62,10,0).$$

Then we know that $w_{12} > w_{21} > w_{32} > w_{42}$ (where w_{12} is the weight for n_1,w_{21} is the weight for p_2,w_{32} is the weight for n_3, and w_{42} is the weight for n_4). Again, following the procedure described in Chapter 3, we would find the values for n_1' such that the DM is indifferent between the following outcomes:

$$(n_1', 62, 10, 10) \quad \text{and} \quad (1800, 0, 10, 10).$$

Suppose this value is $n_1' = 200$. Similarly, we would find the values for n_1'', n_1''' such that the DM is indifferent between

$$(n_1'', 62, 10, 10) \quad \text{and} \quad (1800, 62, 0, 10)$$

and also between

$$(n_1''', 62, 10, 10) \quad \text{and} \quad (1800, 62, 10, 0).$$

Suppose that the values found for the unknowns are $n_1' = 200, n_1'' = 900$, and $n_1''' = 1350$, which allows us to set up the following equations in terms of the value function weights:

$$.\bar{8}\lambda_1 = \lambda_2, .5\lambda_1 = \lambda_3, \text{and} .25\lambda_1 = \lambda_4.$$

(Note that these equations were formed from the set of indifferences expressed by the DM from earlier.) Solving these three equations along with the normalizing equation $\left(\sum_{i=1}^{4} \lambda_i = 1\right)$ gives us the values for the value function weights, as approximately:

$$\lambda_1 = .38,\ \lambda_2 = .34,\ \lambda_3 = .19,\ \text{and}\ \lambda_4 = .09.$$

Therefore, the optimization in terms of the MAV function would be

$$\text{Maximize } v(n_1, p_2, n_3, n_4) \text{ or}$$
$$\text{Maximize}\{\lambda_1 v_{n1}(n_1) + \lambda_2 v_{p2}(p_2) + \lambda_3 v_{n3}(n_3) + \lambda_4 v_{n4}(n_4)\} \text{ or}$$
$$\text{Maximize}\{.38(-(1/1800)n_1 + 1) + .34(-(1/62)p_2 + 1) + .19(-(1/10)n_3 + 1) + .09(-(1/10)n_4 + 1)\} \text{ or}$$
$$\text{Maximize}\{-.0002\overline{1}n_1 - .005484p_2 - .019n_3 - .009n_4 + 1\}.$$

Now, in an optimization problem, a constant in the objective function can be eliminated without affecting the resulting optimal values for the decision variables. In addition, a maximization type problem can be transformed to an equivalent minimization type problem by multiplying the objective function by –1. These two steps give us the following objective:

$$\text{Minimize}\{.0002\overline{1}n_1 + .005484p_2 + .019n_3 + .009n_4\},$$

which gives us our achievement function for our goal program. Hence, the weights for the four respective deviational variables should be $.0002\overline{1}$, .005484, .019, and .009.

It is interesting in this case to compare the weights obtained from the percentage normalization procedure to the weights obtained by the MAV function procedure, respectively: $(.000\overline{5}, .016129, .1, .1)$ versus $(0002\overline{1}, .005484, .019, .009)$. The weights obtained through the percentage normalization procedure certainly required much less effort, but the MAV function procedure leads to a more accurate representation of the DM's preferences.

For a lexicographic goal program, the reader should note that, as was mentioned earlier, weights would only need to be determined for those deviational variables contained in a goal category with multiple goals. But if an MAV function approach is used to determine these weights, a separate MAV function would be needed for each category with multiple goals.

The question arises as to why not just optimize an MAV function instead of using a goal programming approach involving an achievement function. First, the preference information associated with formulating a goal program involves the specification of goals, which differs from the type of preference information required in the assessment of an MAV function. DMs often feel more comfortable in specifying goals than in specifying amounts of attributes for trade-offs, as required for the assessment of an MAV function. Second, an MAV function approach for determining the weights for the deviational variables in a goal program involves the use of the deviational variables as attributes. If one wanted to develop an MAV function for the same situation, the attributes used would correspond to the objectives of the problem. For example, for the situation involved with model (4.26) through (4.33), the attributes used would most likely be profit, amount of Resource 2 used, number of units of Product 1 produced, and number of units of Product 2 produced.

There are other methods, such as the analytic hierarchy process, that have been suggested for determining the weights for the deviational variables in an achievement function for a GP. However, in many cases, an interactive approach is used in which the DM would try different sets of weights, based upon the solution obtained for the goal program with the previous iteration's set of weights. As noted in Jones and Tamiz (2010, p. 39),

> It is important to regard weight determination as a process of interaction with the decision maker(s) rather than a single a priori declaration of a weighting scheme.

4.5 Integer Decision Variables and Nonlinear Functions in a Goal Program

So far in Chapter 4, the examples have involved continuous decision variables, linear objective functions, and linear constraint functions. However, many optimization problems encountered by industrial engineers involve either or both integer decision variables and nonlinear objective/constraint functions. From the perspective of a goal programming formulation, such problems do not cause any particular difficulty with respect to their solution process. Typically, the same software that is used to solve single-objective mathematical programs with integer decision variables and/or nonlinear functions can also be used to solve goal programs with those same features.

In this section, we present examples of goal programs, which require integer decision variables in the formulation.

Example 4.3: A Goal Program for Supplier Selection*

The selection of suppliers and the subsequent negotiation with the same are important activities for many production organizations. Examples of optimization models used in these types of activities are given by

* This problem is derived from a single-objective optimization problem described in an exercise from Shapiro (2007, p. 156).

TABLE 4.8

Data for the Supplier Problem

Supplier Number	Fixed Cost	Higher Cost (per Unit)	Break Point (Number of Units)	Cheaper Cost per Unit	Maximum Number to Sell
1	0	150	300	108	900
2	8000	120	180	105	1100
3	6000	115	240	80	600
4	9000	110	200	100	1000

Note: All costs are in dollars.

Metty et al. (2005) and Sandholm et al. (2006), among others. Multiple objective optimization is an important tool for these activities.

Consider the AAA company, which purchases an important subassembly from various suppliers for the production of their laptop computer. Four different suppliers can provide the subassembly, but the terms of purchase vary among the suppliers. In particular, the cost for purchasing from a supplier depends upon the number purchased. Each supplier (except 1) has a fixed cost if any of the subassembly is purchased, a cost per unit up to a maximum number purchased, and then a less expensive cost per unit after a particular number of units is purchased. The data for the problem are shown in Table 4.8.

As an example of the meaning of the data in Table 4.8, consider Supplier 2. If any of the item is purchased from this supplier, there will be an initial cost (fixed cost) of $8000, and the per unit cost up to 180 units (the break point) will be $120. Any units purchased from Supplier 2 beyond the break point will cost $105 per unit. So, if 500 units of the product were purchased from Supplier 2, the cost would be $8,000 + (180 units) ($120 per unit) + ((500 − 180) units)($105 per unit) = $8,000 + $21,600 + $33,600 = $63,200.

Now, the problem is to determine the amount that AAA should purchase from each supplier in order to satisfy its demand over the next year of 5000 units. Three goals are to be considered:

1. Achieve a cost of at most $160,000.
2. Purchase at least 100 units from Supplier 1.
3. Purchase at least 100 units from Supplier 2.

Suppliers 1 and 2 are companies with which AAA would like to maintain a close relationship because of other services and products that they can provide.

In order to simplify our formulation of this problem as a goal program, we will use some notation to represent some of the parameters. In particular, let

f_i be the fixed cost in thousands of dollars, associated with supplier i

b_i be the break point for supplier i

m_i be the maximum number of units that can be supplied by supplier i
c_i^h be the higher cost per unit, in thousands of dollars, associated with supplier i
c_i^l be the lower cost per unit, in thousands of dollars, associated with supplier i

Note that all costs are given in *thousands of dollars* instead of in dollars. The purpose of this is to make the right-hand side of the goal constraint for cost ($160,000.) close in value to the right-hand sides of the other two constraints (100 units). As mentioned earlier, this is a typical approach in the formulation of goal programs in order to make the weights for the deviational variables relatively close to one another.

Now, the difficulty in this optimization problem is that the cost associated with purchasing the item from any particular supplier is not a linear function of the number of units purchased from that supplier. In particular, we have a fixed cost for Suppliers 2, 3, and 4 if any units are purchased from these suppliers, and for all four suppliers, we have a break point at which the cost per unit becomes cheaper once a certain number of units have been purchased. These conditions can be handled with 0/1 decision variables. For this problem, therefore, our decision variables can be defined as the following:

x_i^h is the number of units to purchase from supplier i at the higher per unit cost, for i = 1,..., 4.
x_i^l is the number of units to purchase from supplier i at the lower per unit cost, for i = 1,..., 4.
y_i^h = 1, if any units are purchased from supplier i.
0, otherwise, for i = 1,..., 4.
y_i^l = 1, if any units are purchased from supplier i at the lower per unit cost.
= 0, otherwise, for i = 1,..., 4.

Our goal program can be written as follows:

$$\text{Minimize } Z = w_{12}p_1 + w_{21}n_2 + w_{31}n_3 \tag{4.34}$$

subject to

$$\sum_{i=1}^{4}(c_i^h x_i^h + c_i^l x_i^l + f_i y_i^h) + n_1 - p_1 = 160 \tag{4.35}$$

$$x_1^h + x_1^l + n_2 - p_2 = 100 \tag{4.36}$$

$$x_2^h + x_2^l + n_3 - p_3 = 100 \tag{4.37}$$

$$\sum_{i=1}^{4}(x_i^l + x_i^h) = 1500 \tag{4.38}$$

$$x_i^h \le b_i y_i^h, \quad \text{for } i = 1,\ldots,4 \tag{4.39}$$

$$x_i^l \le (m_i - b_i) y_i^l, \quad \text{for } i = 1,\ldots,4 \tag{4.40}$$

$$b_i y_i^l \le x_i^h, \quad \text{for } i = 1,\ldots,4 \tag{4.41}$$

$$n_i, p_i \ge 0, \quad \text{for } i = 1,\ldots,3 \tag{4.42}$$

$$x_i^h, x_i^l \ge 0 \quad \text{for } i = 1,\ldots,4 \tag{4.43}$$

$$y_i^h, y_i^l \text{ are } 0,1 \quad \text{for } i = 1,\ldots,4 \tag{4.44}$$

The achievement function (4.34) is a weighted sum of three deviational variables, one for each goal. In this function, we want to penalize positive deviations from the target goal of $160,000 of cost and also penalize negative deviations from 100 units of the subassembly ordered from both Suppliers 1 and 2; hence, the deviational variables p_1, n_2, and n_3 are included in the achievement function. The weights for these respective variables are denoted as w_{12}, w_{21}, and w_{31}, respectively.

The three goal constraints (4.35) through (4.37) are each shown as an objective function of the decision variables, plus a negative deviational variable, minus a positive deviational variable, equal to a target value.

Constraint (4.38) assures that 1500 units will be purchased, while constraint (4.39) assures that the number purchased at the higher price from each supplier will be 0 if none are purchased, and less than or equal to the break point if any are purchased, according to the value of the y_i^h variable.

Constraint (4.40) assures that the number of units purchased at the cheaper price from any supplier will be equal to 0 if none are purchased at the cheaper price (if $y_i^l = 0$ for a particular supplier i) or less than or equal to the maximum that the supplier is able to sell minus the break point.

Constraint (4.41) assures that the number purchased from a supplier at the higher price will be greater than equal to 0 if none are purchased at the cheaper price, or greater than or equal to the break point (which means that the value will be equal to the break point when combined with constraint (4.39)) if any are purchased at the cheaper price.

Constraints (4.42) through (4.44) are either nonnegativity constraints or 0/1 constraints on the individual decision variables.

Our initial problem will involve setting the weights according to percentage normalization method described in Section 4.3. Hence, we set the weights equal to the reciprocal of the target value for the respective goal: $w_{12} = 1/160 = .00625$, $w_{21} = 1/100 = .01$, and $w_{31} = 1/100 = .01$. The problem was solved using the LINGO software package, with the solution for the deviational variable values given by

$$p_1 = 12.1, n_2 = 0, n_3 = 0, p_3 = 700$$

and all other deviational variable values at 0.

The solution dictated the purchase of 100 units, 800 units, 600 units, and 0 units from Suppliers 1, 2, 3, and 4, respectively. Hence, 620 units and 360 units were to be purchased from Suppliers 2 and 3, respectively, at the cheaper price.

In viewing the values for the deviational variables, it can be seen that Goal 1, not to exceed a cost of \$160,000, was not met, since $p_1 = 12.1$; in other words, the cost for this solution is \$172,100. Goals 2 and 3 were met, since both n_2 and n_3 each achieved a value of 0; and in fact, since the solution dictated the purchase of 800 units from Supplier 2 ($p_3 = 700$), we are exceeding the third goal of purchasing 100 units from Supplier 2 by 700 units.

If the DM were unsatisfied with this solution, then he or she could vary the weights on the deviational variables in the achievement function and/or change the target values. As it turns out for this problem, if one considered cost as the only objective, the best that one could achieve would be a cost of \$163,400; hence, achieving the target value of \$160,000 is not possible as long as 1,500 units are to be purchased.

Example 4.4: A Goal Program for Distribution Center Location and Transportation Planning

The design and operation of supply chains and physical distribution systems have been a fertile area for the application of optimization models. Many of these applications are described in various journal publications and books—for example, see any of several publications in the journal, *Interfaces*, or in books such as Bramel and Simchi-Levi (1997) or Shapiro (2007).

Decisions to be made in these types of systems include locations and capacities of manufacturing facilities and distribution centers, determination of which items to produce, transportation plans, and production schedules. The decisions can involve time frames ranging from daily decisions, such as routings for daily deliveries, to long-range decisions over several years such as those involving locations of production facilities.

The example problem described in this section is one derived from a single-objective optimization given in Shapiro (2007, Chapter 4). The problem involves the determination of the locations and capacities of distribution centers, as well as the amounts to transport from the distribution centers to existing markets in order to satisfy the market demands. Initially, we will provide a general formulation for the problem, followed by a specific example, with numbers.

Our problem situation involves S locations for which we want to make one of the three possible decisions: (1) locate a large distribution center at the location, (2) locate a small distribution center at the location, or (3) do not locate a distribution center at the location. There is a fixed cost for locating a large/small distribution center at location i, denoted as fl_i/fs_i, for i = 1,…,S. In addition, there is a variable cost associated for each unit of product (where a unit is defined as a truckload in this case) shipped from a distribution center at location i, denoted as vl_i/vs_i for a large/small distribution center, for $i = 1,\ldots,S$. Each potential distribution center at a location has a capacity in terms of the number of units of product that can be shipped from the center over the time period in question, depending on whether the distribution center is large or small; for location i, the capacity for a large/small distribution center is denoted as cl_i/cs_i.

The distribution centers need to supply markets at D locations, where the demand at demand location j is denoted as b_j for $j = 1,\ldots,D$. Therefore, the other type of decision that needs to be made is the amount of product to ship from each distribution center to each market. Hence, in addition to the fixed cost associated with locating distribution centers and the variable costs of shipping product from these distribution centers, we are also interested in the transportation costs for shipping the product from the distribution centers to the demand locations or markets. This transportation cost for shipping the product from a location i to a location j is computed as a constant, ctl (cost per mile per unit of product) multiplied by the number of units shipped from i to j times the distance from i to j (denoted as d_{ij}), for $i = 1,\ldots,S$ and $j = 1,\ldots,D$.

The decision variables for this problem are denoted as follows:

yl_i = 1, if a large distribution center is located at location i. 0, otherwise.
ys_i = 1, if a small distribution center is located at location i. 0, otherwise.
xl_{ij} is the number of units of product shipped from a large distribution center at location i to market at location j.
xs_{ij} is the number of units of product shipped from a small distribution center at location i to market at location j.

A summary of the notation for this problem is shown in Table 4.9.

The decision makers for the problem have three goals to consider related to (in decreasing priority order): total cost, transportation cost, and total number of distribution centers to locate. Specifically, they would like to meet goals of C, TC, and DC Number for these three measures. This gives us a goal program as shown in (4.45) through (4.54):

$$\text{Lex Min } Z = \{p_1, p_2, p_3\} \tag{4.45}$$

subject to

$$\sum_{i=1}^{S}(fs_i ys_i + fl_i yl_i) + \sum_{i=1}^{S} vs_i \sum_{j=1}^{D} xs_{ij} + \sum_{i=1}^{S} vl_i \sum_{j=1}^{D} xl_{ij} + ctl\left(\sum_{i=1}^{S}\sum_{j=1}^{D} d_{ij}(xs_{ij} + xl_{ij})\right) + n_1 - p_1 = C \tag{4.46}$$

$$ctl\left(\sum_{i=1}^{S}\sum_{j=1}^{D} d_{ij}(xs_{ij} + xl_{ij})\right) + n_2 - p_2 = TC \tag{4.47}$$

TABLE 4.9

Summary of the Notations for the Distribution Center Location and Transportation Planning Problem

Notation	Definition
S	Number of possible locations for distribution centers
D	Number of markets
fl_i	Fixed cost for locating a large distribution center at location i
fs_i	Fixed cost for locating a small distribution center at location i
vl_i	Variable cost per unit shipped for a large distribution center at location i
vs_i	Variable cost per unit shipped for a small distribution center at location i
cl_i	Capacity in terms of number of units to ship from a large distribution center at location i
cs_i	Capacity in terms of number of units to ship from a small distribution center at location i
b_j	Demand at market j
d_{ij}	Distance in miles from location i to location j
ctl	Cost per unit mile to ship product
yl_i	0–1 variable indicating whether a large distribution center is located at i
ys_i	0–1 variable indicating whether a small distribution center is located at i
xl_{ij}	Number of units of product to ship from a large distribution center at location i to market at j
xs_{ij}	Number of units of product to ship from a small distribution center at location i to market at j

$$\sum_{i=1}^{S}(ys_i + yl_i) + n_3 - p_3 = \text{DC Number} \quad (4.48)$$

$$ys_i + yl_i \le 1, \quad \text{for } i = 1,\ldots,S \quad (4.49)$$

$$\sum_{i=1}^{S}(xs_{ij} + xl_{ij}) = b_j \quad \text{for } j = 1,\ldots,D \quad (4.50)$$

$$\sum_{j=1}^{D} xs_{ij} \le cs_i ys_i \quad \text{for } i = 1,\ldots,S \quad (4.51)$$

$$\sum_{j=1}^{D} xl_{ij} \le cl_i yl_i \quad \text{for } i = 1,\ldots,S \quad (4.52)$$

$$xl_{ij}, xs_{ij} \ge 0 \quad \text{for } i = 1,\ldots,S \text{ and } j = 1,\ldots,D \quad (4.53)$$

$$ys_i \text{ and } yl_i \text{ are } 0,1 \quad \text{for } i = 1,\ldots,S \quad (4.54)$$

The achievement function (4.45) is a lexicographic minimization of the three deviation variables, in order: p_1, p_2, p_3. Note that these are the variables for which we want values of 0; for example, we do not want the total cost to be greater than C. The goal constraints are given by (4.46) through (4.48) for total cost, transportation cost, and total number of distribution centers located. Constraint (4.49) assures that at most, one distribution center will be located at each location. Constraint (4.50) assures that demand will be met at each market. Constraints (4.51) and (4.52) assure that the amount shipped from any location will not exceed the capacity for that location; note in each case that if a distribution center is not located at a particular location, then nothing can be shipped from that location. Finally, constraints (4.53) and (4.54) are the nonnegativity constraints and zero-one constraints, respectively, on the appropriate decision variables.

Let's attach some data to a hypothetical problem of this type. Suppose that we have six potential locations for distribution centers and 15 markets. The fixed costs and the variable shipping costs for locating a small distribution center or a large distribution center for each of the locations are shown in Table 4.10. In addition, the respective capacities for small and large distribution centers at each location are given.

The demands at the markets are shown in Table 4.11, and the distances in miles from each distribution center to each market are shown in Table 4.12.

The final piece of data required for the problem is the value for ctl, the cost per unit/truckload-mile, which was set at a value of \$.75. As an example of this parameter, if 400 units are transported from DC Location 1 to Market 1, the transportation cost associated with this shipment would be .75 (dollars per truckload-mile) * 400 truckloads * 277 miles = \$83,100.

The goals associated with our example problem are \$3.2 million for total cost, \$.9 million for transportation cost, and four distribution centers located. Note that we are setting the goals related to the costs in "millions of dollars" (instead of dollars) so that these will be relatively close in numerical value to the goal for the number of distribution centers located.

The lexicographic goal program associated with this problem was solved using the "Big P" method discussed earlier through the use of the LINGO software package. A listing of the file associated with the model is shown in Figure 4.3.

The solution found to the goal program resulted in $p_1 = 0, p_2 = 0, p_3 = 1$. This means that the goals associated with a total cost of \$3.2 million and of transportation cost of \$.9 million were met, but the third priority goal of locating at most four distribution centers was not met—that is, the solution dictates the location of five distribution centers. Also, since we had values for the negative deviational variables of $n_1 = .024, n_2 = 0$, and $n_3 = 0$, the total cost goal was actually "overachieved" by \$.024 million, or \$24,000. The detailed solution is shown in Tables 4.13 through 4.15.

The total cost for the solution was, as mentioned earlier, \$3.176 million, with a fixed cost (for locating DCs) of \$1.06 million, a variable cost (for shipping from the DCs) of \$1.22 million, and a transportation cost of \$.9 million. (Note that because of round-off, the latter three costs do not quite add up to the \$3.176 million total cost.)

TABLE 4.10

Data for Hypothetical Distribution Center Location Problem

Distribution Center Location	Fixed Cost for Locating a Small DC (Millions of Dollars)	Fixed Cost for Locating a Large DC (Millions of Dollars)	Variable Cost for Shipping from a Small DC (Dollars per Unit)	Variable Cost for Shipping from a Large DC (Dollars per Unit)	Capacity of a Small DC (Number of Shipped Units)	Capacity of a Large DC (Number of Shipped Units)
A	0.175	0.300	103.	77.	2500	4000
B	0.100	0.250	110.	80.	1750	3500
C	0.150	0.325	101.	86.	2000	3200
D	0.195	0.225	96.	75.	2000	4000
E	0.160	0.275	120.	91.	2400	3800
F	0.185	0.350	108.	82.	2000	3500

TABLE 4.11

Demand at Each of the 15 Markets

Market	Demand (Units)
1	960
2	850
3	750
4	822
5	1450
6	1120
7	652
8	1650
9	490
10	690
11	570
12	534
13	587
14	605
15	1150

TABLE 4.12

Distances (in Miles) from Each Potential Distribution Center Location to Each Market[a]

277	365	315	308	320	190	298	25	98	146	126	212	215	363	309
292	334	203	127	129	22	140	207	292	336	223	213	385	491	261
368	381	221	90	28	124	107	324	410	449	320	277	490	577	284
726	753	595	463	344	442	294	575	646	769	474	368	620	598	271
530	589	462	360	266	265	184	317	375	352	196	91	336	341	19
101	25	139	271	385	305	440	347	391	476	463	504	560	716	570

[a] Row i refers to potential DC location i and column j is for market j, for i = 1…6 and j = 1…15.

4.6 Additional Approaches for Goal Programming: Interactive and Metagoal Programming

Until now, we have discussed or at least mentioned three variants of goal programming: lexicographic (or preemptive) goal programming, weighted (or nonpreemptive) goal programming, and minmax (or Chebyshev) goal programs. In addition to linear goal programs of the first two types, we have also presented examples involving the use of integer decision variables.

```
Model:
!Example 4 _ 4;
Sets:
Dist/A, B, C, D, E,
F/:Fixedcostlarge,Fixedcostsmall,Varcostlarge,Varcostsmall,
      Caplarge,Capsmall,Xlarge,Xsmall;
      Market/1,2,3,4,5,6,7,8,9,10,11,12,13,14,15/:Demand;
      Link(Dist,Market):Distance,Ylargeship,Ysmallship;
ENDSETS
DATA:
     ! attribute values;
      Fixedcostlarge = 300000, 250000, 325000, 225000, 275000, 350000;
      Varcostlarge   = 77, 80, 86, 75, 91, 82;
      Fixedcostsmall = 175000, 100000, 150000, 195000, 160000, 185000;
      Varcostsmall   = 103, 110, 101, 96, 120, 108;
      Caplarge    = 4000, 3500, 3200, 4000, 3800, 3500;
      Capsmall    = 2500, 1750, 2000, 2000, 2400, 2000;
      Demand      = 960, 850, 750, 822, 1450, 1120, 652, 1670, 490, 690, 570,
534, 587, 605, 1150;
      CTL         = .75;
Distance   =   277  365  315 308  320   190  298  25 98  146  126  212  215
363  309  292  334  203 127 129  22   140  207 292 336 223  213 385  491
261  368  381  221  90  28  124  107  324 410 449  320 277 490  577  284
726  753  595  463  344 442 294  575 646 769  474  368 620 598  271
530  589  462  360  266 265 184  317 375 352  196  91  336 341  19
101  25   139  271  385 305 440  347 391 476  463  504 560 716  570;
ENDDATA
!Achievement Function;
     Min = 1000*p1 + 100*p2 + 1*p3;
!Goal Constraints;
 (@SUM(Dist(i): Fixedcostlarge(i)*Xlarge(i) + Fixedcostsmall(i)*Xsmall(i) ) +
         @SUM(Link(i,j): Varcostlarge(i)*Ylargeship(i,j) +
Varcostsmall(i)*Ysmallship(i,j) ) +
          @SUM(Link(i,j): CTL*Distance(i,j)*Ylargeship(i,j) +
CTL*Distance(i,j)*Ysmallship(i,j) ))/1000000 + n1 - p1 = 3.2;
          (@SUM(Link(i,j): CTL*Distance(i,j)*Ylargeship(i,j) +
CTL*Distance(i,j)*Ysmallship(i,j) ))/1000000 + n2 - p2 = .9;
          @SUM(Dist(i):Xlarge(i) + Xsmall(i) ) + n3 - p3 = 4;
!Constraints;
      @For(Dist(i):Xlarge(i)  + Xsmall(i)  <= 1);
      @For(Market(j): [Demand _ Constraints]
          @SUM(Dist(i):Ylargeship(i,j)  + Ysmallship(i,j) )  = Demand(j) );
      @For(Dist(i): [Supply _ Large _ Constraints]
          @SUM(Market(j):Ylargeship(i,j) ) <= Caplarge(i)  * Xlarge(i) );
      @For(Dist(i): [Supply _ Small _ Constraints]
          @SUM(Market(j):Ysmallship(i,j) ) <= Capsmall(i)  * Xsmall(i) );
!Zero-one constraints;
      @For(Dist(i):@BIN(Xlarge(i) ));
      @For(Dist(i):@BIN(Xsmall(i) ));
End
```

FIGURE 4.3
Listing of the LINGO model for the goal program.

TABLE 4.13
The Location and Capacity Decisions for the DC Location and Transportation Planning Problem

Distribution Center Location	Location Solution
1	Locate 1 large DC
2	Locate 1 small DC
3	Locate 1 small DC
4	—
5	Locate 1 small DC
6	Locate 1 large DC

TABLE 4.14
Shipments to Market 1 through Market 8 for the DC Location and Transportation Planning Problem

Distribution Center Location	Amount to Ship to Market 1	Amount to Ship to Market 2	Amount to Ship to Market 3	Amount to Ship to Market 4	Amount to Ship to Market 5	Amount to Ship to Market 6	Amount to Ship to Market 7	Amount to Ship to Market 8
1								1670
2						1120	630	
3				680	1320			
4								
5				82			22	
6	960	850	750	60	130			

TABLE 4.15
Shipments to Market 9 through Market 15 for the DC Location and Transportation Planning Problem

Distribution Center Location	Amount to Ship to Market 9	Amount to Ship to Market 10	Amount to Ship to Market 11	Amount to Ship to Market 12	Amount to Ship to Market 13	Amount to Ship to Market 14	Amount to Ship to Market 15
1	490	683	570		587		
2							
3							
4							
5		7		534		605	1150
6							

The need for interaction with the decision maker was apparent in the discussion of the examples. This interaction should not just occur in the formulation of the initial goal program, that is, in the determination of the preemptive categories of goals, the target values for the goals, and the weights for the deviational variables in the achievement function, but it should also occur in the "reformulations" after the initial solution and after subsequent solutions of reformulated problems.

Tamiz and Jones (1997) discuss various approaches for decision maker interaction in the goal programming solution procedure. For example, one very simple and straightforward approach is to just solve a sequence of goal programs with varying weights and target levels, present the various solutions to the DM(s), and allow them to choose the most preferred. The choice of each set of weights and target levels for each respective goal program would involve little, if any, particular thought.

A more sophisticated approach involves allowing some thoughtful process by the DM between iterations. For example, the DM might not be satisfied with the value for a particular deviational variable. By increasing the weight for that deviational variable in the achievement function, increased importance is placed on that deviation at the next iteration of the interactive algorithm.

Additional information, for example, sensitivity analysis information, can also be provided to the DM as an aid in changing target values. For example, the right-hand side ranges for the goal constraints can aid the DM in setting new target values for the goals. In Section 4.4, we discussed providing the DM with the minimum and maximum values for the deviational variables as an aid in setting the weights for these variables in the achievement function.

For additional information in the area of interactive goal programming, the reader is directed to the Tamiz and Jones (1997) paper, which discusses several of the publications in this area.

Finally, Rodriguez et al. (2002) developed the concept of metagoal programming. This approach allows the DM to set secondary goals, derived from the original set of goals, within an interactive process of the DM.

For a discussion of the merging of interactive and metagoal program, see Caballero et al. (2006). In addition, see Caballero et al. (2005) for information concerning a specialized software package for solving goal programs.

For future research areas of goal programming, see Caballero et al. (2009).

4.7 Additional Approaches for Multiple Objective Optimization

As mentioned in Section 4.1, there are many approaches in addition to goal programming for multiple objective optimization. These approaches can be classified according to Figure 4.1. Whereas the classical approaches involving goal programming employ a prior articulation of preferences, many of the earlier approaches for multiple objective optimization involved a progressive articulation of preferences, which typically require an interactive procedure. Examples of these algorithms include the GDF algorithm of Geoffrion et al. (1972) and the algorithms of Zionts and Wallenius (1983), Korhonen and Laasko (1986), and Korhonen et al. (1986).

The Step method (Benayoun et al., 1971), or STEM, is another one of the early interactive multiple objective methods. The basic concept can be applied to both linear and nonlinear problems with continuous and/or integer decision variables. The method involves, first, solving the p single-objective problems (corresponding to the p objective functions in (4.1)) to form a "payoff table"; this payoff table gives an optimal solution and corresponding p objective function values for each of the p single-objective problems solved. Following this initial step, an iterative sequence of single-objective problems is solved. These problems each involve finding a nondominated solution, which minimizes a maximum weighted distance to the ideal found in the initial step of the algorithm; in addition to the initial problem constraints, the DM is allowed to impose additional constraints on the objective functions of the problem. The weights imposed at this step are determined automatically and do not necessarily relate to the DM's preferences.

More specifically, the initial step of STEM involves solving the p problems given by

$$\text{Maximize } z_i = f_i(x), \quad \text{subject to } x \in X \text{ for } i = 1, \ldots, p \tag{4.55}$$

If the objective function given by f_i initially involves minimization, that function is replaced by its negative in order to allow an equivalent maximization problem. The payoff table associated with solving these p problems is illustrated in Table 4.16. This table will have (p + 1) columns and p rows. The first column contains the optimal solution found from solving each of the p problems given in (4.55). Note that these single objective problems, which are initially of the minimization type, will correspond to negative objective function values in Table 4.16 since the minimization problems are transformed to ones where the negative of the objective function is to be maximized.

TABLE 4.16

Payoff Table Associated with the Initial Step of the Step Method

Optimal Solution for Problem i, i = 1,..., p	Objective Function Value for Objective 1	Objective Function Value for Objective 2	...	Objective Function Value for Objective p
x^1	$f_1(x^1)$	$f_2(x^1)$	...	$f_p(x^1)$
x^2	$f_1(x^2)$	$f_2(x^2)$	...	$f_p(x^2)$
⋮	⋮	⋮	⋮	⋮
x^p	$f_1(x^p)$	$f_2(x^p)$		$f_p(x^p)$

The diagonal entries of the matrix given in Table 4.16 (with the first column and first row deleted) $f_1(x^1)$, $f_2(x^2)$,..., $f_p(x^p)$ form the ideal, that is, the point in the outcome space that represents the optimal value for each of the objectives individually. Even considering the possibility of multiple optimal solutions for the individual single-objective problems given in (4.55), only very rarely will there be a single solution that optimizes each of the individual objective functions $f_1(x)$, $f_2(x)$,..., $f_p(x)$ simultaneously.

Following this initial step of the Step method, the DM might be satisfied with one of the p solutions $x^1, x^2, \ldots, x^p$ given in the payoff table. If so, the algorithm terminates with this "best compromise" solution. Otherwise, problem (4.56)–(4.58), involving the minimization of a maximum weighted deviation from the ideal, as restricted by the original set of constraints and augmented by the DM's additional restrictions on the objective function values, is solved. The weights associated with the deviations from the ideal, as well as the DM's restrictions on the objective function values, can change from one iteration to the next. The idea here is that the DM will learn about his or her preference structure as iterations continue:

$$\text{Minimize } d \tag{4.56}$$

subject to

$$w_{i\iota}(z_i^M - f_i(x)) \le d \quad \text{for } i = 1,\ldots,p \tag{4.57}$$

$$x \in X^{\iota} \tag{4.58}$$

where

ι represents the iteration number, which takes on sequential values of 1, 2,...

$w_{i\iota}$ is the weight associated with objective i and iteration ι

The expression $x \in X^{\iota}$ represents the set of constraints on x associated with iteration ι. For the initial iteration, iteration 1, this would just be the initial set of constraints, $x \in X$, but for subsequent iterations, this initial set of constraints is augmented by any additional restrictions, which the DM places on the objective function values. For example, suppose that objective 1 in a problem is "cost" and that at a particular iteration of the algorithm a cost of $150,000 is achieved; in order to achieve a better value on some objective other than cost, the DM would accept a 15% increase in cost. Then, for the next iteration, the constraints would be augmented with the following restriction on the cost objective function $f_1(x) \le 1.15(150{,}000)$, thereby allowing a relaxation from the current objective function value for cost by 15%.

For a more in-depth discussion of STEM, the reader is directed to Steuer (1986, pp. 362–367).

Material Review Questions

4.1 Briefly explain how a problem can have a very large or even an infinite number of alternatives. Give an example of this type of problem.
4.2 Give an example of a decision variable.
4.3 What are the various categories of decision variables?
4.4 What are the two types of solution processes required to solve problems of the type discussed in this chapter?
4.5 In addition to decision variables, what other types of variables are contained in a goal program?
4.6 In basic goal programming, what is the timing associated with the articulation of preference information as compared to the optimization process?
4.7 What are the three main categories/variants of goal programming?
4.8 Of the three main variants of goal programming mentioned in Question 4.7, which two are the most popular?
4.9 What is the term used for an objective function of a goal program?
4.10 Once the objectives, deviational variables, and decision variables have been defined for a goal programming situation, what type of preference information must be defined by the DM to completely define the goal program?
4.11 In a goal program, if one deviational variable associated with a goal has a positive value in a solution, what must be the value associated with the other deviational variable for the goal?
4.12 A lexicographic goal program is typically solved by solving a sequence of minimization problems (*true* or *false*).
4.13 What are the three types of goals associated with a goal program?
4.14 If a particular goal in a goal program involves the minimization of the overachievement of a target value, which of the two deviational variables associated with this goal will have a weight of 0 in the achievement function?
4.15 During the earlier years of goal programming applications, which variant of goal programming was most popular?
4.16 What is the basic form of a goal constraint in a goal program?
4.17 In a lexicographic goal program, it is not possible to achieve a lower-level goal when a higher-level goal has not been achieved (*true* or *false*).
4.18 A standard approach for solving a lexicographic linear goal program involves solving a sequence of linear programs, one for each lexicographic category (*true* or *false*).
4.19 Briefly describe the "Big P" method for solving a lexicographic goal program.
4.20 A weighted goal program might be considered a special case of a lexicographic goal program with a single priority category (*true* or *false*).
4.21 What is lexicographic redundancy?

4.22 What are two basic reasons for the existence of lexicographic redundancy in a goal program?

4.23 What is a general rule for the maximum number of priority levels in a lexicographic goal program?

4.24 Define Pareto inefficiency in a goal program.

4.25 What are the two situations for which nonzero weights need to be determined for deviational variables in the achievement function of a goal program?

4.26 Why is it important to set target values (for goals in a goal program) that are relatively close in value to each other?

4.27 Of the two formal methods for determining weights for deviational variables discussed in the chapter, which one would typically represent the preference structure of the decision maker in a more accurate fashion?

Exercises

4.1 Consider a weighted goal program with three deviational variables: n_1, p_2, p_3 for three respective goals. Three single-objective optimization models, involving the optimization of each of the three deviational variables separately, have been solved. The minimum and maximum values for the three deviational variables from the three single-objective problems solved are given by

n_1: 0 and 10

p_2: 0 and 15

p_3: 0 and 5

The DM for this problem has expressed the following set of preference relations:

(0, 15, 5) is preferred to (10, 0, 5), and (10, 0, 5) is preferred to (10, 15, 0),

where each of the vectors refers to an outcome: (n_1, p_2, p_3). Suppose further that the DM has expressed the following sets of indifferences related to the outcomes:

1. Indifference between (2.1, 15, 15) and (10, 0, 15)
2. Indifference between (6.7, 15, 15) and (10, 15, 0)

Using the approach discussed in the chapter for determining the deviational variable weights, what should the achievement function be for the goal program?

4.2 Consider the lexicographic goal program associated with Example 4.4, for distribution center location and transportation planning. Solve modified versions of this goal program, with different orderings for the lexicographic goals:

1. Lex Min Z = $\{p_1, p_3, p_2\}$, subject to (4.46) through (4.54)
2. Lex Min Z = $\{p_2, p_1, p_3\}$, subject to (4.46) through (4.54)
3. Lex Min Z = $\{p_2, p_3, p_1\}$, subject to (4.46) through (4.54)
4. Lex Min Z = $\{p_3, p_1, p_2\}$, subject to (4.46) through (4.54)
5. Lex Min Z = $\{p_3, p_2, p_1\}$, subject to (4.46) through (4.54)

Discuss how the solutions found differ from one another.

4.3 Consider the lexicographic goal program associated with Example 4.4. Use "trial and error" to experiment with the goal for total cost to see if you can obtain a solution, which satisfies all three of the goals. Is it possible to achieve all three goals by experimenting with the goal for total cost only?

4.4 Consider again the lexicographic goal program associated with Example 4.4. Experiment with the goal for transportation cost to see if all three goals can be satisfied. Is it possible to achieve all three goals by experimenting with the goal for transportation cost only?

4.5 Consider Example 4.3, involving the selection of suppliers. Formulate and solve this problem as a lexicographic goal program in which the first priority goal is to achieve a cost of at most $160,000, the second priority goal is to purchase at most 100 units from Supplier 1, and the third priority goal is to purchase at least 100 units from Supplier 2.

4.6 Experiment with the goal of $160,000 set for cost in Exercise 4.5 (Example 4.3) to determine if all three goals could be met with a different cost goal.

5

A Brief Review of Probability Theory

5.1 Introduction

Decisions are often made in the presence of forecasts concerning uncertain quantities. How these quantities are represented, for example, through the use of particular probability distribution or density functions, will affect the outcome forecast for an alternative decision. For example, the production rate associated with a particular scheduling policy will depend upon the reliability of the machines in the production facility; machine reliability might be represented through the use of a density function, which in turn represents the times between machine breakdowns. Additional examples of these uncertain quantities/parameters would include the following:

1. Arrival times of customers, patients, and trucks to their appropriate facilities
2. Fractions of customers or patients belonging to various categories
3. Activity/task times associated with various activities of a process or of a project
4. Amount of sales associated with a new product
5. Vendor lead times
6. Number of competitors bidding on a contract

The primary approach used to handle uncertainty involves the use of concepts from probability theory. This chapter will review probability theory, especially with respect to how it relates to decision making in the presence of uncertainty. More specifically, we will be concerned with how to develop evaluation models, which reflect the uncertainty in the outcome associated with a decision.

Much, if not all, of the material in this chapter is covered in an introductory course in probability and statistics. Hence, the reader can skip this chapter if he or she is already familiar with the material.

5.2 Events/Experiments, Sample Space, Outcomes, and Partitions

An *event* or an *experiment* is a procedure or process with an outcome that cannot be known in advance. A *sample space* is the set of all possible *outcomes* associated with an event/experiment.

Note that an experiment can correspond to a purposeful activity associated with some human endeavor (e.g., a doctor performs surgery on a patient, a corporation decides to locate a new production facility in Denver, Colorado, or an analyst runs 30 replications of a simulation model of a production facility), or to some process for which the decision maker, or any human, has no direct influence (e.g., tomorrow's weather).

The *sample space* associated with any specific experiment can typically be defined in any of several different ways. For example, when a surgeon performs an operation on a patient, the sample space might be defined in any of the following three ways, among others:

SS_1 = {survival of longer than one day, survival of less than one day}
SS_2 = {number of days spent in the hospital after the surgery}
SS_3 = {quality-adjusted life years following the surgery}

If the experiment involves operating a production facility over the next 6 months, the sample space could be any of the following:

SS_1 = {percentage of production orders meeting a due date}
SS_2 = {mean inventory level over the 6 months}

A *partition* of the sample space is a collection of subsets from the sample space such that the intersection of any two subsets is the null set and the union of all subsets is the sample space.

Note that the way that the sample space is defined is often tied into either the attributes associated with a decision problem or the possible data, which one can collect from an experiment. For example, if a decision situation involves which inventory policy to choose for the next 6 months, the sample space might be

{expected holding cost, expected ordering cost, expected shortage},

where each of the three quantities is evaluated per day, over the 6-month period. If a decision situation involved a plan to schedule activities and allocate resources in a construction project, the sample space might be

{project duration, direct project cost}.

5.3 Probability

A *probability* is associated with an outcome; hence, we refer to the "probability of an outcome or a set of outcomes," such as the probability that a patient will survive at least 1 day following an operation, or the probability that at least 98% of the production orders over the next 6 months will be met by their due dates.

As such, probability might be thought of as a function that maps from the space of outcomes onto a closed interval of real number line: [0, 1]. Such a function, $P: A \rightarrow [0,1]$, a mapping from any outcome or subset of outcomes, A, contained in the sample space, S, onto the closed interval [0, 1], should satisfy the following axioms:

1. $0 < = P(A) < = 1$
2. $P(S) = 1$, where S is the entire sample space
3. If A_i, for $i = 1,2, \ldots, n$ are a set of pairwise, mutually exclusive sets of outcomes, then

$$P(A_1 U\, A_2 U \ldots U\, A_n) = P(A_1) + P(A_2) + \cdots + P(A_n)$$

(Note that two sets of outcomes, A_1 and A_2, are mutually exclusive if their intersection is the null set—i.e., there is no intersection.)

These three axioms are sufficient to derive the other properties of probability, including

4. $P(\Phi) = 0$, where Φ refers to the null space, the space of no outcomes
5. $P(A)$ = probability of not $A = 1 - P(A)$

5.4 Random Variables

A *random variable* is a function, which is a mapping from the outcome space to the space of real numbers. Note that a random variable is neither random nor variable. Typically, we think of the range of the function as being the random variable.

There are two basic types of random variables: *discrete random variables* and *continuous random variables*. A discrete random variable is one for which the values are integers or can be mapped as integers. For example, the number of customers arriving at a fast-food restaurant between 11 a.m. and 12 noon is a discrete random variable. The fraction of the items in a lot of 1000 parts, which are defective, is also a discrete random variable, since the possible values for this fraction can be enumerated as 0, 1/1000, 2/1000, 3/1000, …, 1. So another way to think about discrete random variables is that their possible values can be enumerated.

A continuous random variable is one for which the possible values are all real numbers within some interval. Typically, these random variables would represent time (e.g., the amount of time between arrivals of customers to a bank branch), volume, length (e.g., the length of a part), weight (the weight of the contents of a bag of potato chips), velocity, and so on.

Example 5.1: Sample Space, Random Variable, and Probabilities Associated with the Experiment of Throwing a Die

Suppose that we throw a dice three times and observe the sequence of numbers that emerge. The sample space might be defined as

$$SS = \{(1,1,1),\ (1,1,2),\ (1,1,3),\ldots,(1,1,6),\ (1,2,1),\ (1,2,2),\ldots(1,2,6),\ldots,(6,6,6)\}.$$

That is, the sample space is every combination of three consecutive integers ranging from 1 to 6. Each sequence of three integers is an outcome in the sample space. Now, define a random variable as a function, which maps from the outcome space to Y: the number of "ones," which occurs in the three rolls of the dice. Typically, we would say that a random variable would be "the number of ones that occur in three rolls of the dice," that is, the range of the mapping. Refer to the following table for an example:

Outcome	→	Y: Random Variable
(1, 1, 1)	→	3
(1, 2, 1)	→	2
(6, 6, 4)	→	0

Now, one can observe that there are $6 * 6 * 6 = 216$ possible combinations of three consecutive integers, or there are 216 mutually exclusive outcomes within the sample space. Each of these outcomes has an equal chance of occurrence. One could compute the probability associated with Y equaling a particular value:

$$P(Y = 3) = 1/216 = .00463.$$

Note that this probability could have been computed by observing that only 1 out of the 216 outcomes in the outcome space had three consecutive "ones." If one wanted to take a "more scientific approach," to the computation, it could be observed that

$$P(Y = 3) = P(\text{"one" on the first roll } \textit{and} \text{ "one" on the second roll } \textit{and} \text{ "one" on the third roll}),$$

$$= P(\text{"one" on the first roll}) * P(\text{"one" on the second roll}) * P(\text{"one" on the third roll})$$

$$= (1/6) * (1/6) * (1/6)$$

$$= .00463.$$

Example 5.2: Random Variable Associated with Waiting Time in a Hospital Emergency Department

Suppose that we conduct an experiment in which we observe the patients who enter the emergency department of a hospital from 5 to 7 p.m. over 10 consecutive days. In particular, we gather data on the waiting time (the amount of time from when the patient enters the emergency room until treatment begins) for each patient. An example of a random variable in this case would be the fraction of patients who must wait longer than 30 minutes for treatment.

There is a final note to be made here concerning terminology and notation. Some authors refer to a *chance event* as "something about which the decision maker is uncertain" (see Clemen and Reilly, 2001, p. 249). Thus, a chance event is a random variable as defined earlier. In addition, random variables (or chance events) are denoted as capital letters (e.g., A, B,..., X, Y, or Z), while values (or outcomes) associated with these random variables are denoted by small letters such as a, b, and c.

5.5 Conditional Probabilities, Forecasts, Joint Probabilities, Independence, Causality, and Conditional Independence

Let a_1 and b_1 be two respective outcomes; then $P(a_1 | b_1)$ denotes the *conditional probability of* a_1 *occurring given that* b_1 *has occurred or will occur.* Note that both a_1 and b_1 represent *specific outcomes* associated with random variables and not the random variables themselves.

Note also that b_1 does not have to *actually occur* prior to a_1. For example, one could consider the conditional probability of a weather forecast predicting more than 3 in. of rain given that there will be 3 in. of rain:

P(forecast is more than 3 inches of rain | there will be at least 3 inches of rain).

Forecasts are common in decision analysis; another word for forecast is *predictor.* Some examples of forecasts/predictors include medical tests, surveys, and lot sampling.

Now, the *joint probability of both a_1 and b_1 occurring* is denoted as $P(a_1,b_1)$ or $P(a_1 \text{ and } b_1)$.

Two random variables, X (with possible outcomes of $x_1, x_2, \ldots, x_m$) and Y (with possible outcomes of $y_1, y_2, \ldots, y_n$), are *independent* if and only if $P(x_i|y_j) = P(x_i)$ for all outcomes x_i and y_j. In other words, the outcome associated with Y has no bearing on the outcome associated with X.

If two events, X and Y, are independent, then we can say the following:

$$P(x_i) = P(x_i \mid y_j) = \frac{P(x_i, y_j)}{P(y_j)},$$

or

$$P(x_i, y_j) = P(x_i)P(y_j).$$

Consider the following two events, each with two outcomes:

1. University of Louisville's men's basketball team plays a game on a Saturday afternoon with outcomes of Louisville wins or Louisville loses.
2. University of Kentucky's men's basketball team plays a game on a Saturday afternoon with outcomes of Kentucky wins or Kentucky loses.

These two events can be considered as independent (as long as the teams are not playing each other). So we can say that

$$\text{P (Louisville wins and Kentucky wins)} = \text{P (Louisville wins)} \times \text{P (Kentucky wins)}.$$

Independent events are not to be confused with *mutually exclusive* outcomes. For example, as noted, Louisville winning a basketball game on Saturday is independent of Kentucky winning a game on Saturday, but Louisville winning and Kentucky winning are *not* mutually exclusive outcomes.

In Chapter 7, we will present a type of model used in decision analysis called an *influence diagram.* In an influence diagram, nodes are used to represent events in a decision situation, and an arc from one event node to another indicates probabilistic dependence.

A final note concerning probabilistic dependence concerns causality. If two outcomes are probabilistically dependent, this *does not necessarily imply a causal relationship.* For example, a weather forecast that predicts that a hurricane will hit New Orleans does not cause the hurricane to hit New Orleans. But these two outcomes (1) weather forecast predicting that a hurricane will hit New Orleans and (2) hurricane hitting New Orleans are *probabilistically dependent.*

Conditional independence can apply in a situation involving three events, say X, Y, and Z. In particular, one can say that the *events X and Y are conditionally independent, given event Z* if and only if the following holds:

$$P(X \mid Y, Z) = P(X \mid Z).$$

Example 5.3: Conditional Independence Regarding Two Different Types of Medical Tests

Consider three different sample spaces, each with two outcomes:

1. A particular child has lead poisoning (LP) or does not have lead poisoning (NLP).
2. An imperfect urine test for the child is negative for lead poisoning (NUT) or is positive for lead poisoning (PUT).
3. A perfect blood test for the child is negative for lead poisoning (NBT) or is positive for lead poisoning (PBT).

Now, considering the urine test, we have $P(NLP|NUT) < 1$ and $P(LP|PUT) < 1$, but both conditional probabilities are close to 1. Considering the perfect blood test, we have the conditional probabilities of $P(NLP|NBT) = 1$ and $P(LP|PBT) = 1$. The reader will recall from Chapter 2 that the conditional probabilities of $P(NLP|NUT)$ and $P(NLP|NBT)$ are called the specificities of the urine test and blood test, respectively; $P(LP|PUT)$ and $P(LP|PBT)$ are called the sensitivities of the urine test and blood test, respectively.

We also have

$$P(I \mid J \text{ and } K) = P(I \mid K) \text{ for } I = \{LP, NLP\}; J = \{NUT, PUT\};$$

$$\text{and } K = \{NBT, PBT\}.$$

In other words, we are interested in knowing whether or not the child has lead poisoning; if I and J are *conditionally independent* given K, then learning the outcome of J (the urine test outcome) given K (the blood test outcome) adds no new information regarding I (whether or not the child has lead poisoning) if the outcome of K (the blood test) is known.

Therefore, I (the outcome of whether or not the child has lead poisoning) is *conditionally independent* of J (the outcome of the urine test) given K (the outcome of the blood test).

5.6 Probability Functions

The function, p(x), is a *probability distribution function* of the discrete random variable, X, if for each possible outcome x:

1. $p(x) \geq 0$.
2. $\sum_{\text{all } x} p(x) = 1$.
3. $p(x) = P(X = x)$.

The function, F(x), called the "cumulative distribution function" of the discrete random variable, X, is given by $F(x) = \sum_{t \le x} p(t)$. Note that F(x) gives the probability that the random variable, X, will have a value less than or equal to x, $P(X \le x)$.

The function f(x) is a *probability density function* for the continuous random variable, X, defined over **R**, the space of real numbers, if

1. $f(x) \ge 0$ for $x \in \mathbf{R}$
2. $\int_{-\infty}^{\infty} f(x)dx = 1$
3. $P(a < X < b) = \int_{a}^{b} f(x)dx$

The *cumulative distribution function* (or sometimes just called "distribution function"), F(x), of a continuous random variable, X, is given by $F(x) = \int_{-\infty}^{x} f(t)dt$, where f is the probability density function for X. Note that F(x) gives the probability that the random variable, X, will have a value less than or equal to x, $P(X \le x)$. Also, note that $f(x) = dF(x)/dx$; that is, f(x) is the first derivative of F with respect to x.

A summary of the terminology and notation for these probability functions is given in Table 5.1.

There are two types of distribution/density functions: empirical and theoretical. Examples of theoretical distribution functions are binomial, hypergeometric, and Poisson for a discrete random variable; normal, uniform, triangular, exponential, beta; and so on for a continuous random variable. These theoretical distributions often correspond to some type of event or experiment.

Empirical distributions are typically formed to correspond to data that have been gathered from a process. Example 5.4 provides an illustration of an empirical distribution function.

TABLE 5.1

Summary of Terminology and Notations for Probability Functions

Type of Random Variable	Name and Notation for Function Type 1	Name and Notation for Function Type 2
Discrete: X	p(x) = P(X = x), probability distribution function	F(x), (cumulative) distribution function
Continuous: Y	f(y), probability density function	F(y), (cumulative) distribution function

TABLE 5.2

Number of Data Points Collected for the Various Numbers of Customers Entering a Fast-Food Restaurant between 10:30 and 10:45 a.m. on Monday through Thursday

Data Point Value Collected	Number of Data Points
0	2
1	3
2	5
3	4
4	8
5	5
6	12
7	3
8	10
9	9
10	2

Example 5.4: Empirical Probability Distribution Function for Customer Arrivals

Suppose that you have been hired as an analyst to determine the staffing policy for a fast-food restaurant. You have decided to develop a simulation model of the restaurant's operations in order to experiment with different staffing policies. One of the parameters for this model is the number of customers arriving between 10:30 and 10:45 a.m. to the restaurant. Through consultation with the restaurant's management, you have determined that any 10:30–10:45 a.m. time period from Monday through Thursday can be treated as the same with respect to the characteristics of customer arrivals. The restaurant has collected data for several consecutive 10:30–10:45 a.m. time periods for Monday through Thursday and has obtained the following results, as shown in Table 5.2.

Using the proportional values allows the formation of an empirical distribution function to represent the number of customers arriving between 10:30 and 10:45 a.m. on Monday through Thursday. For example, if X represents this random variable, then $P(X = 0) = 2/(2 + 3 + 5 + \cdots + 2) = 2/63 \approx .031476$. The entire empirical distribution function is shown in Table 5.3.

Note also that there is no distinction between open and closed intervals for continuous random variables; that is, $P(X = a) = 0$ and $P(X \leq a) = P(X < a)$ for any continuous random variable X, and any constant, a.

5.7 Parameters Associated with Probability Functions

Probability functions can be characterized by their parameters, including the mean (or expected value), variance, minimum, maximum, and quantiles.

TABLE 5.3

Empirical Distribution Function for the Number of Customers Entering a Fast-Food Restaurant between 10:30 and 10:45 a.m. on Monday through Thursday

x	P(X = x)
0	.031476
1	.047619
2	.079365
3	.063492
4	.126984
5	.079365
6	.190476
7	.047619
8	.158730
9	.142857
10	.031476

Often, an attribute for a decision situation is specified in terms of such a parameter; for example, instead of considering the entire distribution of waiting times for customers in a service system in order to evaluate the quality of customer service, the mean waiting time or the fraction of the number of customers waiting longer than some prespecified number of minutes might be used. For example, one of the performance measures typically used for 911 (emergency) call centers is the fraction of callers who wait more than 12 seconds to have their call answered.

The *expected value* (or *mean*) of a random variable is a weighted (by the associated probability) average of the values that the random variable can assume. The expected value of a random variable is also a measure of the central tendency of that random variable. As such, at least as a single measure associated with a distribution, this parameter can give a good representation for important performance measures such as mean waiting time and mean utilization. "Expected value" is actually a misnomer, since the expected value of a random variable could very easily be a value with 0 probability of occurring; this is certainly true for a continuous random variable, and very possibly true for a discrete random variable. For example, a random variable with a .5 probability of realizing a value of 0, and a .5 probability of realizing a value of 1 would have an expected value of .5.

The *variance* of a random variable can be thought of as a measure of the amount of spread for a distribution. The *standard deviation* for a random variable is the square root of the variance.

Often, the *minimum* and *maximum values* for a random variable are minus infinity and plus infinity, respectively, or zero and plus infinity, respectively. For example, normally distributed random variables have a

TABLE 5.4

Important Parameters of a Random Variable

Type of Random Variable	Expected Value	Variance	Quantiles
Discrete	$E(X) = \mu = \sum_x x\,p(x)$	$E[(X-\mu)^2]$	x_q, such that $F(x_q) = q$, for $q = .25, .5, .75$
Continuous	$E(X) = \mu = \int_{-\infty}^{+\infty} x\,f(x)dx$	$E[(X-\mu)^2]$	x_q, such that $F(x_q) = q$, for $q = .25, .5, .75$

range of $(-\infty, +\infty)$, and gamma distributed random variables have a range of $(0, +\infty)$. Other random variables, such as those that are uniformly distributed, have finite values for minimum and maximum.

There are several quantiles for a random variable; in particular, a q-quantile for a random variable is that number, x_q, such that $F(x_q) = q$, where F is the (cumulative) distribution function for the random variable, as denoted in Table 5.1. As examples, the median (or .5-quantile) is denoted as $x_{.5}$, and the octiles for a random variable are given by $x_{.125}$ and $x_{.875}$.

Table 5.4 gives a summary for the various parameters of a random variable. Additional parameters include location, scale, and shape parameters.

In addition to the parameters that are associated with a random variable, there are also parameters associated with *samples* of data, collected from a population. For example, if one has a sample, denoted as $X_1, X_2, \ldots, X_n$, where, for example, these values might denote waiting time for sequential customers, then the sample mean would be given by $(X_1 + X_2 + \cdots + X_n)/n$. The standard deviation, variance, quantiles, minimum, maximum, and mode could also be computed from the sample data.

5.8 Some Specific Theoretical Probability Distribution Functions

There are several specific theoretical distribution functions, for both discrete and continuous random variables, which correspond to real-world processes or experiments, and are therefore useful for modeling in decision analysis. Examples of such functions for discrete random variables include those associated with the binomial distribution, the hypergeometric distribution, and the Poisson distribution. Examples associated with continuous random variables include the normal distribution, the uniform distribution, the triangular distribution, the exponential distribution, and the beta distribution, among others. In this section, we will describe some of the characteristics of these distribution functions, as well as relationships between them. Table 5.8 provides a summary of some of this information.

The *binomial distribution* can be used to represent the number of items in a sample of size n with a particular characteristic, where there is a probability of p associated with each item having that characteristic (often termed as a *success*). As such, this distribution is often used to represent "sampling with replacement." More specifically, if one has a group of items for which some fraction of these items has a particular characteristic, and one randomly samples some subset of these items (while placing each item sampled back into the group after being sampled), then the number of items in the sample having the characteristic will be a random variable with a binomial distribution. As such, the binomial distribution has applications in acceptance sampling for quality control as well as in testing a subset of people from a larger population for a disease.

The *hypergeometric distribution*, on the other hand, can be used to represent "sampling without replacement." The distribution can be used to represent the number of items in the sample with the identified characteristic, except that after sampling, the item is not placed back in the group—hence, "sampling without replacement." As the reader can readily determine, the hypergeometric distribution is much more complicated in nature than the binomial distribution, since the probability that an item sampled has the relevant characteristic will *change* with each item sampled when the item sampled is not placed back in the group—this is not the case when sampling with replacement where the probability *does not change* with each item sampled.

When the number of items in the group is large relative to the sample size, then the binomial distribution represents a good approximation to the hypergeometric distribution. Hence, for a situation where, for example, a medical organization is setting up a screening test for a disease at a state fair and has an estimate of the percentage of people in the overall population with the disease, the number of people going through the screening test with the disease could be represented with a binomial distribution (even though a screened test subject obviously would have zero chance of being tested again).

To see the difference between the binomial and hypergeometric distributions more vividly, let's consider two situations. In the first situation, let's suppose that we have 10 items in the population and that 5 items are sampled from this population. Three items out of 10 (or 30%) have the characteristic of interest. The probability distributions for this situation are shown in Table 5.5. One can see that in this situation, a binomial approximation to the hypergeometric distribution would be poor. For example, the probabilities of the random variable having a value of 0 are .1681 for the binomial and .0833 for the hypergeometric. In addition, note that for this hypergeometric distribution $P(Y = y) = 0$ for $y = 4$ or 5, since when sampling without replacement the probability of achieving more successes in the sample than there are in the entire population must be 0.

Now, let's consider a second situation involving a comparison of the binomial and hypergeometric distributions. In this second situation, suppose that the population size is 100, the sample size is 5, and the number of characteristic items in the population is 30. The comparison of the values for the binomial

TABLE 5.5

Binomial Distribution Values versus Hypergeometric Distribution Values for a Situation with a Population Size of 10, Sample Size of 5, and the Number of Characteristic Items in the Population Is 3

Binomial distribution, X						
x	0	1	2	3	4	5
P(X = x)	.1681	.3601	.3087	.1323	.0283	.0024
Hypergeometric distribution, Y						
y	0	1	2	3	4	5
P(Y = y)	.0833	.4167	.4167	.0833	0	0

TABLE 5.6

Binomial Distribution Values versus Hypergeometric Distribution Values for a Situation with a Population Size of 100, Sample Size of 5, and the Number of Characteristic Items in the Population Is 30

Binomial distribution, X						
x	0	1	2	3	4	5
P(X = x)	.1681	.3601	.3087	.1323	.0283	.0024
Hypergeometric distribution, Y						
y	0	1	2	3	4	5
P(Y = y)	.1608	.3654	.3163	.1302	.0255	.0019

and hypergeometric distributions is given in Table 5.6. From viewing the probabilities given in Table 5.6, the binomial is a much better approximation of the hypergeometric distribution for this situation with a population size of 100 than the first situation with a population size of 10. Note also that P(Y = y) for y = 4 or 5 is nonzero in this situation, since the number of characteristic items in the population is greater than 5, that is, equal to 30.

A rule of thumb that is often used is that if the population size is at least 20 times the sample size, then the binomial will be a good approximation of the hypergeometric distribution.

The *Poisson distribution* has a wide range of applications. In particular, it is often used to represent the number of arrivals to some type of service system, such as a hospital emergency department, a bank, or a restaurant within some period of time. This is accomplished by representing the arrivals as a nonstationary Poisson arrival process (see Law, 2007, pp. 377–379) in which the number of arrivals for a particular period of time is distributed according to a Poisson random variable, but the average number of arrivals varies by period; for example, the average number of arrivals to a fast-food restaurant during the 7–8 a.m. time period would be much larger than the average number during the 10–11 a.m. period, but the random variable describing each could still be Poisson distributed.

In addition, it can be shown that if the number of events occurring during some period of time is Poisson distributed, then the time between events must be distributed according to an exponential distribution. In particular, if the number of events occurring within a period of time, say 1 hour, is distributed according to a Poisson random variable with a mean of λ, then the time between events is distributed according to an exponential random variable with a mean of $1/\lambda$ hour—for example, if λ is 5 (per hour), then the average time between events is 1/5 hour, or 12 minutes.

Finally, the Poisson distribution can be used as an approximation of the binomial distribution if n is large and p is small enough so that np is not "too large" (see Ross, 1998, p. 154). In this case, the approximation of the binomial distribution with the Poisson distribution will have a parameter value of $\lambda = np$.

As an example of a Poisson approximation to a binomial distribution, consider a situation where n = 100 and p = .1; then the Poisson approximation to the binomial distribution will have $\lambda = np = 10$. A way to think about this situation for the binomial distribution is that you are sampling 100 items, and there is a .1 probability that any one item sampled will have a certain characteristic; the number of items in the sample of 100 that have the characteristic is the value for the random variable. From Table 5.7, the probability that 10 items in the sample of 100 will have the characteristic is .132.

One way to think about this situation for the Poisson distribution is an arrival process in which on average 10 (=.1 * 100) items arrive on average per hour, but the actual number arriving is a Poisson distributed random variable. As mentioned previously, the time between arrivals will be exponentially distributed with a mean value of 6 minutes (=(60 minutes/hour)/(10 arrivals/hour)). One reason for the simplicity of the Poisson and exponential distributions is that they are each completely defined by a single parameter.

The density function values for this situation for various values of x for both the binomial distribution and the Poisson approximation are shown in Table 5.7.

The *normal distribution* is often used to represent the sum of a large number of independent quantities; such a representation is valid as a result of the central

TABLE 5.7

Poisson Distribution Approximation to the Binomial Distribution with n = 100 and p = .1

Binomial distributed random variable, X					
x	8	9	10	11	12
P(X = x)	.115	.130	.132	.120	.099
Poisson distributed random variable, Y					
Y	8	9	10	11	12
P(Y = y)	.113	.125	.125	.114	.095

limit theorem. A special case of this distribution, which turns out to be quite useful, is the standard normal distribution function, which is a normal distribution with a mean of 0 and a standard deviation of 1; this specific normal distribution is one for which lookup tables are typically given in the appendix for introductory books in statistics. In particular, if X is a normally distributed random variable with a mean of μ and a standard deviation of σ, then $Z = (X - \mu)/\sigma$ is a random variable with the standard normal distribution. This relationship between Z and X, in which $X = Z\sigma + \mu$, allows us to determine the probability that X lies within some range of values through the use of the standard normal tables. For example, if X is normally distributed with a mean of 20 and a standard deviation of 5, then the probability that X is less than 28 is given by

$$\begin{aligned} P(X < 28) &= P((Z\sigma + \mu) < 28) \\ &= P(Z < (28 - \mu)/\sigma) \\ &= P(Z < (28 - 20)/5) \\ &= P(Z < 1.6). \end{aligned}$$

From the standard normal tables, $P(Z < 1.6) = .9452$. Hence, $P(X < 28) = .9452$.

The *uniform distribution* can be used to represent an uncertain, but continuous, quantity about which little is known, other than its minimum and maximum possible values. As noted by Law (2007, p. 282), it is often used as a "first" representation of a quantity, prior to the collection of any data. As such, it might be used to represent a random variable that is not thought to be too important with respect to the decision under consideration.

The *triangular distribution*, as its name indicates, has a density function, which looks like a triangle. As such, it has three parameters: a minimum possible value (denoted as a), a maximum possible value (denoted as b), and a mode (denoted as m); the mode represents the y value at which the density function (f(y)) reaches its maximum. It is often used in situations where an "expert" might provide numbers that correspond to a minimum, maximum, and most likely values for a quantity, such as the minimum, maximum, and most likely values for the amount of time required to perform some task. One thing to keep in mind about this distribution function is that the mode (i.e., the most likely value) will not be the expected value unless the density function is symmetric.

The *exponential distribution* is often used as a representation of the time to failure for a machine or as the time between arrivals, for example, for customers arriving to a service system, as noted in the discussion of the Poisson distribution. It has some nice properties, which are especially useful in queuing theory when the service times and interarrival times are represented as exponentially distributed random variables. As mentioned earlier, another useful feature is the fact that this distribution can be completely described by a single parameter (its mean value) (Table 5.8).

TABLE 5.8

Characteristics of Specific Distribution Functions

Distribution	Type of Random Variable	Parameters	Mean	Variance
Binomial	Discrete	n = number of trials p = probability of success	np	np(1−p)
Hypergeometric	Discrete	N = population size k = no. of successes in population n = sample size	nk/N	$\frac{N-n}{N-1}\cdot n\cdot\frac{k}{N}\left(1-\frac{k}{N}\right)$
Poisson	Discrete	λ = average number of events	λ	λ
Normal	Continuous	μ = mean σ^2 = variance	μ	σ^2
Uniform	Continuous	a = minimum value b = maximum value	(a + b)/2	$(b-a)^2/12$
Triangular	Continuous	a = minimum value m = mode b = maximum value	(a + m + b)/3	$(a^2 + m^2 + b^2 - ab - am - bm)/18$
Exponential	Continuous	μ = mean	μ	μ^2
Beta	Continuous	α_1 α_2	$\alpha_1/(\alpha_1 + \alpha_2)$	$\alpha_1\alpha_2/(\alpha_1+\alpha_2)^2(\alpha_1+\alpha_2+1)$

In addition to its relationship with the Poisson distribution as discussed earlier, the exponential distribution has relationships to the gamma and Weibull distributions, as well as the m-Erlang distribution (see Law, 2007, p. 284).

5.9 Bayes' Theorem

Bayes' theorem allows one to compute a conditional probability as a function of other conditional and joint probabilities. One form of this theorem is given by

$$\Pr(B_k|A) = \frac{P(B_k, A)}{\left(\sum_{k=1}^{K} P(B_k, A)\right)},$$

where the $B_1, B_2, \ldots, B_K$ form a *partition* of outcomes over the sample space. In many cases, there are only two outcomes considered in the sample space (e.g., B_1 is a patient who has a particular disease and B_2 is a patient who does not have that particular disease).

Example 5.5: Bayes' Theorem Applied in a Disease-Testing Problem

Let's consider an example involving an application of Bayes' theorem. Suppose that you know that "about" 1,000 people out of 100,000 in a city have a certain disease. You have a medical test available with the following characteristics: *sensitivity* of .95 (remember that sensitivity in this case is the conditional probability of a person testing positive for the disease given that the person has the disease) and *specificity* of .90 (remember that specificity in this case is the conditional probability of a person testing negative for the disease given that the person does not have the disease). Suppose that 1000 people, randomly selected from the city, are tested for the disease and that we want to determine expected values for the following quantities:

1. The number of people testing positive for the disease
2. The number of people testing negative for the disease
3. The number of people testing positive who do have the disease
4. The number of people testing positive who do not have the disease
5. The number of people testing negative who do have the disease
6. The number of people testing negative who do not have the disease

First, define the various outcomes associated with a randomly selected person from the city being given the medical test:

D is the outcome that the selected person has the disease.
ND is the outcome that the selected person does not have the disease.
PT is the outcome that the selected person tests positive for the disease.
NT is the outcome that the selected person tests negative for the disease.

Now, let's give the probabilities that we know:

$$P(D) = \frac{1{,}000}{100{,}000} = .01, \ P(ND) = 1 - P(D) = .99,$$

$$P(PT \mid D) = .95, \ P(NT \mid ND) = .9.$$

Therefore, we have that $P(NT|D) = 1 - P(PT|D) = .05$ and $P(PT|ND) = 1 - P\{NT|ND) = .1$

Using Bayes' theorem, we want to derive probabilities like $P(D|PT)$. So we will let the partition $B_1, B_2, \ldots$ associated with Bayes' theorem be given as D, ND.

Therefore, using Bayes' theorem, we obtain

$$\begin{aligned}
P(D \mid PT) &= P(D,PT)/(P(D,PT) + P(ND,PT)) \\
&= P(PT,D)/(P(PT,D) + P(PT,ND)) \\
&= P(PT \mid D)\, P(D)/(P(PT \mid D)\, P(D) + P(PT \mid ND)\, P(ND)) \\
&= .95(.01)/(.95(.01) + .1(.99)) \\
&= .0876.
\end{aligned}$$

Also, we obtain

$$P(ND \mid PT) = 1 - P(D \mid PT) = 1 - .0876 = .9124,$$

and

$$\begin{aligned}
P(ND \mid NT) &= P(ND,NT)/(P(ND,NT) + P(D,NT)) \\
&= P(NT,ND)/(P(NT,ND) + P(NT,D)) \\
&= P(NT \mid ND)P(ND)/(P(NT \mid ND)P(ND) + P(NT \mid D)P(D)) \\
&= .1(.99)/(.1(.99) + .05(.01)) \\
&= .99497,
\end{aligned}$$

and

$$\begin{aligned}
P(D \mid NT) &= 1 - P(ND \mid NT) \\
&= 1 - .99497 \\
&= .00503.
\end{aligned}$$

Finally, we can obtain the probability associated with a randomly selected person getting a positive test result as follows:

$$\begin{aligned}
P(PT) &= P(D,PT)/P(D \mid PT) \\
&= P(PT,D)/P(D \mid PT) \\
&= P(PT \mid D)P(D)/P(D \mid PT) \\
&= .95(.01)/.0876 \\
&= .1084,
\end{aligned}$$

and the probability of a randomly selected person getting a negative test result:

$$P(NT) = 1 - P(PT)$$
$$= 1 - .1084$$
$$= .8916.$$

A summary of the various probabilities is shown in Table 5.9.

A "probability table" is often helpful to summarize the probabilities in a situation such as this. In this case, the probability table is given in Table 5.10.

Note that there is some round-off error in Table 5.10. Also, note that the marginal probabilities are formed by summing the joint probabilities, for example, P(PT) = P(PT,D) + P(PT,ND) = .1084, and that the conditional probabilities can be found by dividing the joint probabilities by the marginal probabilities, for example, P(PT|D) = P(PT,D)/P(D) = .0095/.01 = .95 and P(D|PT) = P(D,PT)/P(PT) = .0095/.1084 = .0876.

A perhaps surprising result, among others, is the relatively low value of the probability that a randomly selected person who tests positive for the disease actually has the disease (i.e., P(D|PT) = .0876). At first, this seems counterintuitive given that the sensitivity of the test (.95) seems to be fairly good. However, this relatively low value for P(D|PT) can be explained by the low value of P(D) (= .01) in the first place. P(D) is called a "prior probability," and P(D|PT) is called a "revised probability" (or a "posterior probability"), given the additional information of a positive test.

Now, to determine the quantities for which we were originally interested, we should realize that the testing process represents 1000 trials, which are

TABLE 5.9

Summary of Various Probabilities Found for the Disease-Testing Situation

Probability Found	P(D|PT)	P(ND|PT)	P(ND|NT)	P(D|NT)	P(PT)	P(NT)
Probability	.0876	.9124	.99497	.00503	.1084	.8916

TABLE 5.10

Probability Table for the Disease-Testing Situation

	D	ND	
PT	P(PT,D) = .0095	P(PT,ND) = .099	P(PT) = .1084
NT	P(NT,D) = .0005	P(NT,ND) = .891	P(NT) = .8916
	P(D) = .01	P(ND) = .99	1.

"almost independent." In fact, the expected values that we want to find are expected values associated with a hypergeometric random variable. In order to keep things relatively simple, we will use a binomial approximation to the hypergeometric distribution; this would usually be permissible in this case, since the number of people being tested (1,000) is relatively small compared to the number of people in the entire city (100,000). As noted previously, the expected value associated with a binomial random variable is just the probability of a "success" multiplied by the number in the sample. Hence, our expected values are given by

1. Expected value of the number of people testing positive = .1084(1000) = *108.4*
2. Expected value of the number of people testing negative = 1000 – 108.4 = *891.6*
3. Expected value of the number of people tested who will have the disease = 1000P(D) = 1000(.01) = *10*
4. Expected value of the number of people tested who will not have the disease = 1000P(ND) = 1000(.99) = *990*
5. Expected value of the number of people testing positive who do have the disease = 1000P(PT,D) = 1000P(PT|D)P(D) = 1000(.95)(.01) = 1000(.0095) = *9.5*
6. Expected value of the number of people testing positive who do not have the disease = 1000P(PT,ND) = 1000P(PT|ND)P(ND) = 1000(.1)(.99) = *99*
7. Expected value of the number of people testing negative who do have the disease = 1000P(NT,D) = 1000P(NT|D)P(D) = 1000(.05)(.01) = *.5*
8. Expected value of the number of people testing negative who do not have the disease = 1000P(NT,ND) = 1000P(NT|ND)P(ND) = 1000(.9)(.99) = 1000(.891) = *891*

So, by testing 1000 people at random, we would have on average 99 people who tested positive who do not have the disease and *only* 9.5 people who test positive who do have the disease. This is not a very good result, which is why one often gives a screening test prior to a more sophisticated test.

Material Review Questions

5.1 Define the following terms: event/experiment and sample space. Give an example of each, not contained in the book.

5.2 There is a unique definition for a sample space for any specific event/experiment (*true* or *false*).

5.3 What is a partition of a sample space? Give an example of a partition, not contained in the book.

5.4 A probability can be thought of as a mapping from the space of outcomes into the [0, 1] interval (*true* or *false*).

5.5 What is a random variable?

5.6 What are the two basic types of random variables?

5.7 Give two examples of discrete random variables.

5.8 Give two examples of continuous random variables.

5.9 Typically, several different random variables can be defined for any particular experiment (*true* or *false*).

5.10 For a conditional probability, denoted as $P(x_1 | y_1)$, the outcome denoted as y_1 must occur prior to the outcome of x_1 (*true* or *false*).

5.11 Independent events are the same as mutually exclusive events (*true* or *false*).

5.12 If two outcomes are probabilistically dependent, then one of the outcomes must have a causal effect on the other (*true* or *false*).

5.13 For a continuous random variable, X, the value of $P(X = 5)$ will be 0 (*true* or *false*).

5.14 What does an empirical distribution function typically correspond to?

5.15 What is a rule that one can use to determine whether or not a binomial distribution can be used in place of a hypergeometric distribution?

5.16 What distribution is typically used to model arrivals of customers to a service system?

5.17 If the number of "events" (such as traffic accidents) that occur during a particular unit of time (such as a day) is Poisson distributed, then what is the distribution for the time between these events?

5.18 When can a Poisson distribution be used as an approximation for a binomial distribution?

5.19 Since a Poisson distribution is associated with a discrete random variable, its mean must be an integer (*true* or *false*).

Exercises

5.1 On August 11, 2015, the home team won all 15 major league baseball games. This was the first time that the home teams had won all of their major league baseball games since (when the home team record was 12–0) May 23, 1914. Show why such a long stretch between home teams winning all of the games is not unusual.

More specifically, consider the situation where (1) the probability of the home team winning is .5 and (2) there are 100 days in a season where there are 15 games. What is the probability associated with there being one day out of 10,000 days (i.e., 100 seasons) where the home teams win all of the games? What about 2 days out of 10,000 days where the home teams win all of the games?

5.2 Suppose that there are three manufacturing facilities, each producing the same item. The three facilities have a defect rates of 2%, 8%, and 4%, respectively. In addition, the three facilities produce 60%, 10%, and 30%, respectively, of the total output of the three facilities. You choose one item at random from the output of all three facilities. What is the probability that the item chosen is defective?

5.3 The amount of time required for students to finish a final exam is normally distributed with a mean of 130 minutes and a standard deviation of 30 minutes. What is the probability that a randomly selected student will finish the exam in less than 150 minutes?

5.4 The number of arrivals per hour to a hospital emergency department is a Poisson random variable with a mean value of 15. An arrival to the department has just occurred. What is the probability that the next arrival will occur within 5 minutes? Within 10 minutes?

5.5 A high school has 1000 students. It is estimated that 10% of the students have a particular disease. Out of 40 randomly selected students, what would be the probability that 5 of these students have the disease? What would be the expected number of students out of 40 with the disease?

5.6 A parts supplier stipulates that a particular part has a 1% defect rate. If the 1% claim is true, what is the probability associated with finding 5 or more defective parts in a sampling of size 50 from a lot of size 1000?

5.7 The number of calls to an emergency call center in a major metropolitan area during a particular hour of the week is distributed according to a Poisson distribution with a mean value of 100. What is the probability associated with receiving between 97 and 100 calls, inclusive, during this hour?

6

Modeling Preferences over Risky/ Uncertain Outcomes

6.1 Introduction

Any decision will result in an outcome that is both multidimensional and uncertain or risky in nature. The basic reason for the uncertain or risky aspect of the outcome is that the decision will be made sometime in the future and the outcome will occur over time, after the decision is made (see Figure 6.1). The future is always uncertain.

Within the context of a decision analysis, this uncertainty manifests itself in uncertainty in the multiple attribute values associated with the decision's effects.

Even though there is uncertainty associated with the effects of any decision, we do not always model or represent this uncertainty, as noted in Section 3.1 of Chapter 3. But when it is important to represent this uncertainty, such as when the decision is only to be made once (and is therefore not repetitive in nature) and there is much variability in the outcome, then it is also usually important to represent the decision maker's (DM's) preferences over uncertain or risky outcomes.

In this chapter, we address the issue of how to model the preferences of the DM over uncertain or risky outcomes, not how to determine the uncertainty in the outcome, which results from the decision (see Figure 6.2). Modeling the uncertainty in the outcome as a function of the alternative chosen (represented by the mapping M1 in Figure 6.2) is discussed in subsequent chapters; this area of study, involving the mapping M1 in Figure 6.2, employs the general methodologies of simulation modeling, influence diagrams, decision trees, and others.

The typical way to represent uncertainty in an outcome is through the use of a probability distribution over the attribute values associated with the problem (either a simple marginal probability distribution for those situations with a single attribute or a joint probability distribution for those situations involving multiple attributes). However, there are other ways, such as the use of confidence intervals, a mean value coupled with a variance, or

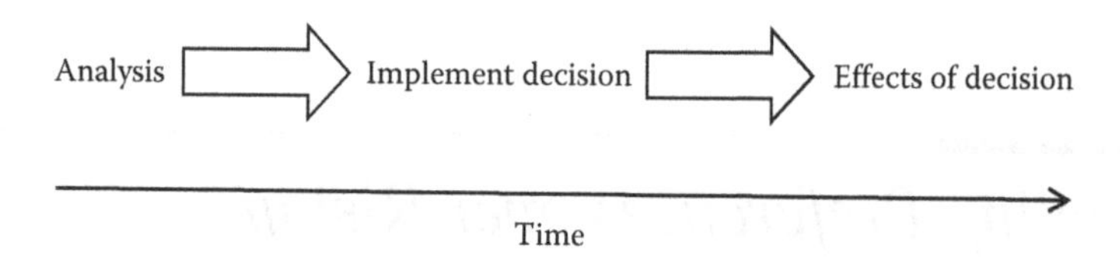

FIGURE 6.1
Effects of a decision occurring after the decision is made.

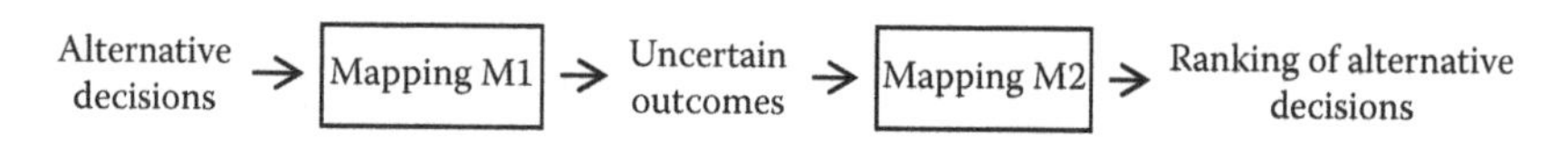

FIGURE 6.2
Two types of mappings involved in decision making under uncertainty or risk.

outcomes associated with alternative scenario combinations. Each of these other ways to represent the outcomes, however, stems from the original probability distributions associated with the attributes.

Thus far in this chapter, we have used the words *uncertainty* and *risk* in an interchangeable fashion. The convention, however, is to consider (1) situations involving uncertainty and (2) situations involving risk as separate in nature. In particular, *decision making under uncertainty* is defined as a situation in which there can be more than one outcome from any decision, but the respective probabilities of occurrence associated with those outcomes are unknown. *Decision making under risk* is defined as a situation in which the respective probabilities of occurrence associated with the outcomes are "known," meaning that we can at least *estimate* these probabilities.

Example 6.1 presents a simple hypothetical example involving a decision situation with a single attribute, but an uncertain outcome. Section 6.2 presents methods for decision making under uncertainty, as defined earlier.

Sections 6.3 and 6.4 provide material on single attribute and multiattribute utility (MAU) functions, respectively. These functions allow one to represent preferences over single dimensional and multidimensional *risky outcomes*, respectively, for a DM. In addition to defining these concepts, these sections provide the reader with material on how to assess these functions. Section 6.4 contains Examples 6.2 and 6.3, involving the use of MAU functions for acceptance sampling, and Example 6.4 involving the use of MAU functions for medical diagnosis.

Example 6.1: A Simple Project Investment Example Involving a Single Attribute

Suppose that we have two possible projects in which to invest. Each project can result in one of the three possible outcomes, depending on whether the economy is good, satisfactory, or poor. This "state of the economy" is typically called the "state of nature or scenario." Note

TABLE 6.1
Payoff Table for a Simple Example

State of Nature, Θ	Probability, P(Θ)	Project 1 Outcome	Project 2 Outcome
Good	.3	$10000	$9000
Satisfactory	.4	$5000	$5000
Poor	.3	$3000	$4000

that there is one thing that we can control (the *decision*, i.e., the project in which to invest) and one thing that we cannot control (the *state of nature*).

Let

P_{ij} = Net profit from alternative i (Project i) under state of nature j, for i = 1, 2 and j = 1, 2, 3.

Now, suppose that a model has been developed to give the value for P_{ij} in all cases and that probabilities associated with the various states of nature have been developed; these are typically illustrated through the use of a *payoff table* (see Table 6.1).

Note that Project 1 does *better* than Project 2 under the *good state of nature*, but *worse* than Project 2 under the *poor state of nature*. Hence, neither project *stochastically dominates* the other; that is, there is no value of i such that $P_{ij} \geq P_{kj}$ for all $k \neq i$ and $j = 1, 2, 3$ and $P_{ij} > P_{kj}$ for all $k \neq i$ and at least one $j = 1, 2, 3$.

6.2 Methods That Do Not Consider the Probabilities Associated with the States of Nature

Let's suppose for the time being that we do not know the probabilities associated with the states of nature. As discussed in Section 6.1, this situation is often called *decision making under uncertainty*, as opposed to *decision making under risk*, where the probabilities are known. Basically, in decision making under uncertainty, one assumes that there is not enough confidence in the situation to attach probabilities to any of the states of nature.

There are at least three potential approaches that one could use in decision making under uncertainty: the *maximin* approach, the *maximax* approach, and the *minimax regret* approach.

The *maximin approach* involves choosing the alternative that *maximizes the minimum gain* over all states of nature. For example, the minimum gain from Project 1 is $3000, and the minimum gain from Project 2 is $4000, so under this criterion, one would choose Project 2. This "rule" is often called the "pessimistic criterion," since it assumes that the worst scenario will occur. Note here that if the outcomes are measured in terms of losses rather than gains, one would *minimize the maximum loss* under this approach.

The *maximax approach* involves choosing the alternative that *maximizes the maximum gain* over all states of nature. For example, the maximum gain from Project 1 is \$10000 and the maximum gain from Project 2 is \$9000, so under this criterion, one would choose Project 1. This "rule" is often called the "optimistic criterion," since it assumes that the best scenario will occur. Note here that if the outcomes are measured in terms of losses rather than gains, one would *minimize the minimum loss* under this approach.

The *minimax regret approach* involves choosing the alternative that *minimizes the maximum regret*, where regret, denoted as R_{ij}, is defined for every alternative scenario combination as follows:

$$\begin{aligned} R_{ij} &= \text{Regret if scenario j occurs and alternative i is selected,} \\ &= \text{the difference in gain between the alternative with the best value} \\ &\quad \text{for gain and the value for gain associated with alternative i} \\ &\quad \text{if scenario j occurs,} \\ &= \underset{i'}{\text{Sup}}\{P_{i'j}\} - P_{ij}. \end{aligned}$$

"Sup" is short for "supremum" or the maximum value over a set of values. Hence, $\text{Sup}\{P_{i'j}\}$ over all i′ for this example would be the maximum value over all alternatives (projects) for a particular value of j.

For the example, R_{ij} is given in Table 6.2.

Note, for example, that R_{13} (= \$1000) is the regret that would occur if alternative 1 is chosen and state of nature 3 occurs; this is computed from the following fact:

$$\begin{aligned} R_{13} &= \text{Sup}\{P_{13}, P_{23}\} - P_{13} \\ &= \text{Sup}\{\$3000, \$4000\} - \$3000 \\ &= \$4000 - \$3000 \\ &= \$1000. \end{aligned}$$

TABLE 6.2

Regret, R_{ij}, for the Simple Example

i|j	1	2	3
1	\$0	\$0	\$1000
2	\$1000	\$0	\$0

Hence, if Project 2 had been chosen, one could have had a gain of $4000 under Project 2 instead of the $3000 under Project 1 ($1000 = $4000 – $3000).

Now, with Project 1, the maximum regret is given by Sup($0, $0, $1000) = $1000, while with Project 2, the maximum regret is given by Sup($1000, $0, $0) = $1000. Hence, there is a tie for the best alternative between Project 1 and Project 2 under the criterion of minimize the maximum regret (*minimax* regret).

Rarely will there be a decision-making situation for which nothing is known about the probabilities associated with the outcomes for the different alternatives. In the next section, we will discuss methodologies that do consider probabilities associated with the outcomes for the various alternatives.

6.3 Single Attribute Utility Functions

6.3.1 Using Expected Payoff as a Criterion

One approach that employs the probabilities associated with the outcomes of the alternatives involves the selection of the alternative, which maximizes the expected value of the payoff. In our simple example, the expected payoff for Project 1 is given by

$$\begin{aligned}\text{Expected payoff for Project 1} &= .3(\$10000) + .4(\$5000) + .3(\$3000)\\ &= \$5900,\end{aligned}$$

while the expected payoff for Project 2 is given by

$$\begin{aligned}\text{Expected payoff for Project 2} &= .3(\$9000) + .4(\$5000) + .3(\$4000)\\ &= \$5900.\end{aligned}$$

Hence, under the criterion of maximizing expected payoff, Project 1 and Project 2 would be equally preferred.

As noted in Chapter 5 on the review of probability, the expected value of a random variable might be thought of as a measure of the central tendency of that random variable; however, also as noted in that chapter, the term *expected value* is a misnomer, since the expected value may not actually occur. In our example, we note that it would be *impossible* for a gain of $5900 to occur for either alternative. This might be thought of as one of the drawbacks associated with the use of this criterion.

Another drawback associated with this criterion is that it does not consider the amount of risk associated with an outcome. For example, let's

TABLE 6.3
A Payoff Table Associated with a Lottery Decision

State of Nature, Θ	Probability, P(Θ)	Bet on Lottery	Do Not Bet on Lottery
Win lottery	.0002	\$9995	\$0
Do not win lottery	.9998	−\$5	\$0

suppose that one had the opportunity to purchase a lottery ticket for \$5 to win \$10000. Let's also suppose that the odds of winning the \$10000 is 1 in 5000, or 1/5000 = .0002. The payoff table associated with this situation is given in Table 6.3.

Note that in Table 6.3, the payoff associated with betting on the lottery, given that the lottery is won, is the \$10000 won minus the \$5 bet on the lottery, or \$9995; the payoff associated with betting on the lottery, given that the lottery is lost, is the \$0 won from the lottery, minus the \$5 bet, or −\$5.

The expected gain associated with each decision is given by

$$E\,(\text{bet on lottery}) = .0002(\$9995) + .9998(-\$5) = -\$3$$

and

$$E\,(\text{do not bet on lottery}) = .0002(\$0) + .9998(\$0) = \$0.$$

Even though the expected gain from betting on the lottery is less than the expected gain from not betting on the lottery, there are still many people who would undoubtedly choose the decision of betting on the lottery.

6.3.2 Accounting for Risk in a Decision Situation

Often, one wants to account for risk in making a decision. Using the expected value of a performance measure as the criterion is not the way to do this. For example, if one used expected value as a criterion, one would probably never purchase an insurance policy, purchase a warranty on an appliance, buy a lottery ticket (as discussed in the aforementioned example), place a bet on a horse race, get a physical exam, place a bet in a casino, or wear a seatbelt.

Note that there are two sides to many of these decision situations. For example, the company that sells the appliance with a warranty must have a situation in which their expected income will be positive through the offer of the warranty; this can only occur if the expected income from the purchase of the warranty on the consumer side is negative. From the appliance manufacturer's viewpoint, some of the consumers will actually gain by taking out a warranty; but the majority of consumers will lose money on the deal, resulting in an expected gain by the producer.

There are many different ways to account for the risk in a decision situation. One classical approach that arises from modern portfolio theory (Elton and

Gruber, 1997) is called "mean–variance optimization." The basic idea is that if one has a performance measure to be maximized, while minimizing the amount of risk associated with the value obtained, then one will want to maximize the expected value of the measure subject to a constraint on the maximum variance of the measure. In this case, risk is measured by the variance of the outcome measure. Looking at the simple example as given in Table 6.1, the mean and variance of the payoff associated with each project are given by

$$\text{Project 1 Mean} = \$5900,$$

$$\text{Project 2 Mean} = \$5900,$$

$$\text{Project 1 Variance} = .3(\$10000)^2 + .4(\$5000)^2 + .3(\$3000)^2 - (\$5900)^2$$
$$= \$7890000,$$

$$\text{Project 2 Variance} = .3(\$9000)^2 + .4(\$5000)^2 + .3(\$4000)^2 - (\$5900)^2$$
$$= \$4290000.$$

Hence, since the two projects have the same expected payoff, but Project 2 has the lower variance, Project 2 would be preferred. One of the complications associated with modern portfolio theory is that one is choosing some subset (i.e., a portfolio) of projects from a larger set, and therefore, the correlations between the projects must be considered in determining the mean and variance of the payoff.

Now in some cases, a DM might prefer more risk rather than less. This case would be an example of where the concept of a single attribute utility function becomes useful. A (single attribute) utility function is a mapping from X (a single dimensional attribute/outcome space) to a subset of the real number line (usually [0, 1]) such that the following applies:

1. A probability distribution over X is preferred by the DM to another probability distribution over X *if and only if* the expected utility associated with the first probability distribution is *greater than* the expected utility associated with the second probability distribution.
2. A DM is indifferent between two probability distributions over the outcome space *if and only if* their expected utilities are *equal.*

Note that the expected utility associated with a probabilistic outcome is just the expected value of a function (i.e., the utility function) of a random variable (where the random variable corresponds to the probabilistic outcome).

Note that the definition for the utility function given is operational in nature, in that it defines utility in terms of a test of its validity. Note also that a utility function is a *subjective* concept in that it is for a specific DM—that is, different DMs will typically have different utility functions for the same situation, depending upon how they evaluate risk.

6.3.3 Assessment of a Single Attribute Utility Function

Determining a utility function for a particular DM and situation is a relatively straightforward process, which we will illustrate for the simple example. The basic idea is to determine points on the function, denoted as u(x) (i.e., values for u(x) for various values of x), and then either fit a curve to the u(x) points or just use linear interpolation for the values of u for those points not assessed directly through the assessment. The process involves one or more sessions in which an analyst (or analysts) interviews a DM (or a group of DMs who answer as one).

The first step is to assign utility function values of 0 and 1 for the worst possible value and the best possible value, respectively, of X. Note that since u will range from 0 to 1, we are assessing what is called a "scaled utility function." In the case of the simple example, as illustrated through Table 6.1, we set initial values for the utility function as follows:

$$u(\$3000) = 0 \quad \text{and} \quad u(\$10000) = 1.$$

Note that oftentimes in the assessment process, an analyst may want to choose best and worst values that "bound" the actual best and worst values, in order to allow for new alternatives, which might arise. In this example though, we will just use the current best and worst values.

Now, the next step is to find what is called the "certainty equivalent" for the lottery associated with a 50% chance (or a .5 probability) of receiving $3000 and a 50% chance of receiving $10000. Note that the two values chosen for the lottery are values for which we already have utility function evaluations. The certainty equivalent for a lottery is defined as the certain value for an attribute for which the DM is indifferent between receiving the outcome associated with that value and the outcome associated with the lottery.

The 50–50 lottery is denoted by ⟨$3000, $10000⟩, and the certainty equivalent is typically found by having the DM "hone in" on its value; this is accomplished through the analyst having the DM rank pairs of outcomes, with one of these in each pair being the 50–50 lottery: ⟨$3000, $10000⟩. As an example, consider the sequence of questions and answers shown in Table 6.4.

TABLE 6.4

Honing In on the Certainty Equivalent for ⟨$3000, $10000⟩
Analyst: Do you prefer $4000 for certain or the 50–50 lottery <$3000, $10000>?
DM: I prefer the 50–50 lottery.
Analyst: Do you prefer $6000 for certain or the 50–50 lottery <$3000, $10000>?
DM: I prefer $6000 for certain.
Analyst: Do you prefer $5500 for certain or the 50–50 lottery <$3000, $10000>?
DM: I am indifferent between these two outcomes.

Note that after the DM answers the first question shown in Table 6.4, the analyst knows that the DM prefers the 50–50 lottery to receiving the certain outcome of \$4000; hence, the certainty equivalent for ⟨\$3000, \$10000⟩ must be more than \$4000. So, in the next question involving a pairwise ranking, the DM tries a value greater than \$4000, or in this case \$6000.

Following the answer to the last question shown in Table 6.4, one knows that the expected utility of the 50–50 lottery: ⟨\$3000, \$10000⟩ is equal to the expected utility of \$5500 for certain, or just 1.*u(\$5500). Hence, we have the following:

$$\$5500.\ \mathbf{I}\ \langle \$3000,\ \$10000\rangle$$

or

$$\begin{aligned} u(\$5500) &= .5u(\$3000) + .5u(\$10000) \\ &= .5(0) + .5(1) \\ &= .5. \end{aligned}$$

Note that we are using the same notation, **I**, to indicate indifference between two outcomes as we did in Chapter 3; except in this case, we are allowing probabilistic outcomes.

Now, we have three points on the utility function curve. As a next step, the analyst could obtain the certainty equivalent for either ⟨\$3000, \$5500⟩ or for ⟨\$5500, \$10000⟩. Let's suppose that he or she obtains the certainty equivalent for ⟨\$3000, \$5500⟩, again through a series of questions, which allows the DM to hone in on the true value. Suppose the analyst finds that

$$\$4000\ \mathbf{I}\ \langle \$3000, \$5500\rangle$$

or

$$\begin{aligned} u(\$4000) &= .5u(\$3000) + .5u(\$5500) \\ &= .5(0) + .5(.5) \\ &= .25. \end{aligned}$$

Finally, we find the certainty equivalent for ⟨\$5500, \$10000⟩. Let's suppose that this turns out to be \$7400; hence, we have u(\$7400) = .5u(\$5500) + .5u(\$10000) = .5(.5) + .5(1) = .75. So, at this point, we have five points on the utility function curve, as seen in Table 6.5.

A graph associated with this utility function is shown in Figure 6.3.

TABLE 6.5
Five Points on the Utility Function Curve

x($)	u(x)
3000	0
4000	0.25
5500	0.5
7400	0.75
10000	1

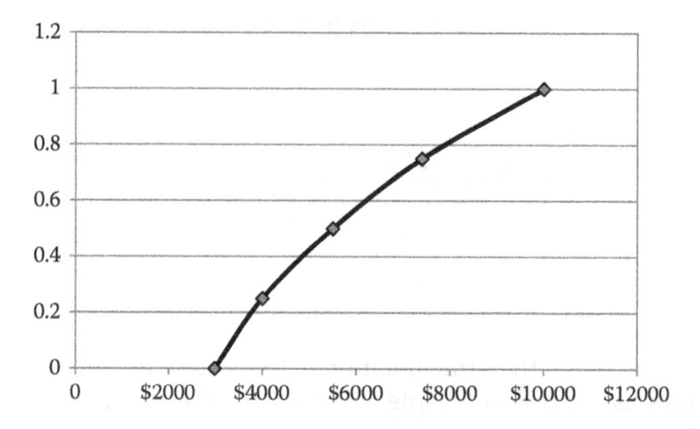

FIGURE 6.3
Graph corresponding to the utility function of Table 6.5.

Additional points on the curve could be found by, for example, finding the certainty equivalent for ⟨$3000, $4000⟩, which would give the x value for u(x) = .125.

One important aspect of utility function assessment is the use of consistency checks. For example, the analyst should check for this assessment: the certainty equivalent of ⟨$3000, $10000⟩, which was found to be $5500, is the same as the certainty equivalent for ⟨$4000, $7400⟩; if this turned out to not be the case, then the analyst would want to have the DM rethink some of his or her answers.

Now, with respect to using the utility function to evaluate alternative outcomes for the simple example, as mentioned earlier, one could either fit a curve to the points found or use linear interpolation. Let's use linear interpolation to evaluate those points, which have not been directly evaluated in the assessment as shown in the following:

$$\begin{aligned} u(\$5000) &= u(\$4000) + \left(\frac{(\$5000 - \$4000)}{(\$5500 - \$4000)} \right) (u(\$5500) - u(\$4000)) \\ &= .25 + .66(.5 - .25) \\ &= .415. \end{aligned}$$

Now, we have already assessed u($4000) = .25, and in a fashion similar to the aforementioned, we obtain u($9000) = .9038.

The expected utility for each project would then be computed as follows:

- Eu (Project 1) = .3u($10000) + .4u($5000) + .3u($3000) = .3(1) + .4(.415) + .3(0) = *.466.*
- Eu (Project 2) = .3u($9000) + .4u($5000) + .3u($4000) = .3(.9038) + .4(.415) + .3(.25) = *.512.*

So, under the expected utility criterion, this DM would prefer Project 2 to Project 1, since Project 2 gives the larger expected utility. Keep in mind that this utility function, and therefore the ranking of projects, is specific to this DM.

In summary, the basic steps for forming a single attribute, scaled utility function are

1. Determine the best (x^b) and worst (x^w) possible values for the attribute over all alternatives.
2. Set $u(x^b) = 1$ and $u(x^w) = 0$.
3. Determine the certainty equivalent, $x^{.5}$, for the 50–50 lottery, $\langle x^w, x^b \rangle$. Set $u(x^{.5}) = .5$.
4. Determine the additional certainty equivalents: $x^{.25}, x^{.75}, x^{.125}, x^{.375}, x^{.625}$, $x^{.875}$, and so on, as desired. Set $u(x^{.25}) = .25, u(x^{.75}) = .75$, and so on.
5. Perform consistency checks, and reevaluate, if needed.

6.3.4 Risk Attitudes

An important characteristic of a DM is his or her *risk attitude,* which can be quantified for his or her utility function. For example, the DM might be inclined to take risks, or the DM might be inclined to avoid risks over a certain range of the outcome as defined by the attribute under consideration.

In order to quantify a DM's risk attitude, we need to quantify the concept of a *risk premium* (*RP*). An RP for a probabilistic outcome over a single attribute is defined as the expected value of the probabilistic outcome minus the certainty equivalent for that probabilistic outcome. More specifically, we have Definition 6.1.

Definition 6.1: An *RP* for a probabilistic outcome over a single attribute (defined by a random variable y) is defined as RP(y) = EV(y) − CE(y), where EV(y) is the expected value of y and CE(y) is the certainty equivalent for y for the relevant DM.

As an example, consider the RP for our DM for Example 6.1. For the 50–50 lottery, ⟨$3000, $10000⟩, we have

$$\begin{aligned} RP(\langle \$3000, \$10000 \rangle) &= EV(\langle \$3000, \$10000 \rangle) - CE(\langle \$3000, \$10000 \rangle) \\ &= .5(\$3000) + .5(\$10000) - \$5500 \\ &= \$1000. \end{aligned}$$

With the RP for a DM, we can define various risk attitudes, as shown in Definition 6.2.

Definition 6.2: A DM is

1. *Risk averse* if the RP is positive
2. *Risk neutral* if the RP is 0
3. *Risk prone* if the RP is negative

We also note that these risk attitudes correspond to specific types of utility functions. For example, a risk-averse (risk-neutral, risk-prone) DM will have a concave (linear, convex) utility function.

In the Example 6.1, since the DM's utility function is concave (i.e., all RPs computed over the domain of [$3000, $10000] are positive), the DM is risk averse. Note that a DM could be risk averse over some domains, risk neutral over others, and risk prone over still others. For example, it may very well be that a typical individual of reasonable wealth would be risk averse over a range of $0–$1 million of income but risk prone over a range of $10 million–$20 million of income.

6.3.5 Caveats in Interpreting Utility

There are several things to remember in the interpretation of utility. First, utility theory is a *normative*, not a *descriptive* theory. That is, utility theory is used to describe how people should make decisions, not how they actually do make decisions; in some sense, one could say that is a good thing since otherwise there would be no need for people to study or use the theory.

Second, utility functions, as was mentioned earlier, are *subjective* in nature. That is, two different reasonable people could, and probably would, have different utility functions for a given situation, depending on their attitude toward risk (in the case of a single attribute) and/or the way they would trade off between different attributes of the situation (in the case of multiple attributes).

Third, *utility differences do not convey strength of preference*. That is, utility should be thought of as an ordinal measure, not an interval measure. For example, if the expected utilities of three specific outcomes A, B, and C are, respectively, .9, .89, and .2, then one cannot a priori say that the difference in preference between A and B is much less than the difference in preference between B and C. Hence, the expected utility values can only be used to rank the outcomes (and their corresponding alternatives).

6.3.6 Alternative Approaches in the Assessment of Single Attribute Utility Functions

In the assessment procedure presented earlier, the DM was asked to provide certainty equivalents for risky outcomes (lotteries). The DM's response allowed the formation of an equation with one unknown, which could then be solved to give a point on the DM's utility function curve.

In a more general sense, the DM was presented with two outcomes involving four parameters (O_1, O_2, O_3, and p):

1. *Outcome 1*: O_1, a certain value for the attribute
2. *Outcome 2*: A value of O_2 with probability p, or a value of O_3 with probability (1 – p), where $O_2 > O_3$

In the assessment procedure discussed earlier, the values for O_2, O_3, and p were set; in particular, the value for p was set at .5, and hence, the DM was asked to provide the certainty equivalent for $\langle O_2, O_3 \rangle$. This approach is called the "certainty equivalence response mode." The other response modes for the assessment process depend upon which of the four parameters that the DM is to provide a value for, given values for the other three parameters, so that he or she is indifferent between the two outcomes. Hence, the four response modes are as follows:

1. *Certainty equivalence*, in which the values for O_2, O_3, and p are set and the DM is asked to provide the value for O_1
2. *Probability equivalence*, in which the values for O_1, O_2, and O_3 are set and the DM is asked to provide the value for p
3. *Gain equivalence*, in which the values for O_1, O_3, and p are set and the DM is asked to provide the value for O_2
4. *Loss equivalence*, in which the values for O_1, O_2, and p are set and the DM is asked to provide the value for O_3

As an example, consider the probability equivalence response mode for Example 6.1. The analyst might ask a series of questions with corresponding answers as follows:

Analyst: Suppose that you were to choose between the following outcomes. Which would you prefer?

Outcome 1: \$6500 for certain

Outcome 2: Receiving \$10000 with the probability of 2/3 or receiving \$3000 with a probability of 1/3

DM: I would prefer the second outcome (Outcome 2).

At this point, the analyst knows that the probability equivalence value (i.e., the probability of receiving \$10000 in the second outcome) must be set at a value less than 2/3. So, the analyst might ask a second question as follows:

Analyst: OK. Suppose that you were to choose between the following outcomes. Which would you prefer?

Outcome 1: $6500 for certain

Outcome 2: Receiving $10000 with a probability of .6 or receiving $3000 with a probability of .4

DM: I think that I would be indifferent between these two outcomes.

At this point, the analyst knows that the expected utilities of these two outcomes must be equal. Hence, the following equations hold

$$\text{Eu(Outcome 1)} = \text{Eu(Outcome 2)}$$

or

$$u(\$6500) = .6u(\$10000) + .4u(\$3000)$$

or

$$u(\$6500) = .4.$$

Note that in most cases, the preferred mode for an assessment is the certainty equivalence response mode; one reason for this is that people often have trouble in providing answers to questions regarding preferences in terms of probabilities.

6.3.7 Some Standard Functional Forms for a Single Attribute Utility Function

Once the "points" of a single attribute utility function have been assessed as described, there are at least two approaches for specifying a functional format. (Throughout this chapter, we will assume that the utility function is scaled such that a functional value of 1(0) is associated with the best [worst] attribute value.) We will also assume, at least initially, that this utility function is increasing as X increases in value—that is, more of an attribute is desired. Later in this section, we will discuss decreasing utility functions.

One approach is to just assume that the function is linear in between the assessed points. While this approach would obviously be an approximation, it may be sufficient for the purpose of the study being undertaken. In addition, the larger the number of points assessed, the more accurate this approach of linear interpolation will be in terms of actually representing the preferences of the DM.

Let's suppose that we want to use this approach of having a linear utility function "in between" the assessed points. In some rare cases, this function may be linear over the entire range of the attribute. In this case, the function could just be represented as (6.1):

$$u(x) = c_1 x + c_2, \tag{6.1}$$

where c_1 and c_2 are appropriate constants. As mentioned previously, this type of utility function represents a risk-neutral DM.

In most cases, however, a "piecewise linear" function would be a more accurate representation of the DM's preference structure than a linear function. Let's suppose that the assessed points for the utility function are denoted by $(x_1, u(x_1)), (x_2, u(x_2)), \ldots, (x_n, u(x_n))$, where n is the number of assessed points and the x_i, which are increasing as i increases from 1 to n, represent various attribute values and $u(x_i)$ is the assessed value of utility for x_i. Let's also suppose that x_1 represents the "worst value" for the attribute and that x_n represents the "best value" for the attribute. Therefore, for our scaled, increasing utility function, we would have $u(x_1) = 0$ and $u(x_n) = 1$. A piecewise linear function connecting the assessed points would be given by

$$u(x) = a_i x + b_i \quad \text{for } x \in (x_i, x_{i+1}), \tag{6.2}$$

where

$$a_i = (u(x_{i+1}) - u(x_i))/(x_{i+1} - x_i)$$
$$b_i = u(x_i) - x_i(u(x_{i+1}) - u(x_i))/(x_{i+1} - x_i), \quad \text{for } i = 1,2,\ldots, n-1$$

As an example, let's suppose that we were assessing a utility function over an attribute, X, where the worst value, x^w, was 0 and the best value, x^b, was 40. Suppose also that the utility function values for 10 points ($n = 10$) were assessed, including $u(0) = 0$ and $u(40) = 1$. Given two assessed points, $u(x_3) = u(15) = .42$ and $u(x_4) = u(20) = .54$, the value for $u(17)$ could be interpolated as $u(17) = a_3x + b_3$, where $a_3 = (.54 - .42)/(20 - 15) = .024$ and $b_3 = .42 - 15(.54 - .42)/(20 - 15) = .06$. Hence, $u(17) = .024(17) + .06 = .468$.

A second category of approaches for representing a utility function would be to use a particular functional form, which is a best fit for the set of assessed points. There are several possibilities, which could be tried. For example, a general *exponential utility function* appears as

$$u(x) = c_1 - c_2 \exp(-c_3(x - c_4)), \tag{6.3}$$

where $c_1 \geq 0$, $c_2 > 0$, $c_3 > 0$, and c_4 (unrestricted) are all constants. The function given by (6.3) is concave, thereby representing a risk-averse DM. In addition, smaller values of c_3 make the graph of u(x) "flatter," thereby representing an increasingly less risk-averse DM as c_3 decreases in value, given that c_1 and c_2 remain the same.

The constant, c_4, represents a shift parameter for the function; that is, the graph for an exponential utility function of type (6.3) with $c_4 = 10$ will appear the same as the graph with $c_4 = 0$ (with c_1, c_2, and c_3 having the same values

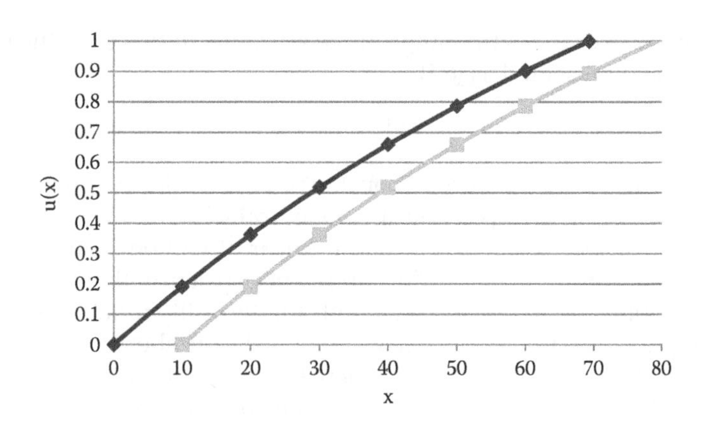

FIGURE 6.4
Two single attribute utility functions, $2 - 2e^{-.01x}$, and its shifted (by 10 units of X) counterpart, $2 - 2e^{-.01(x-10)}$.

in each case), except that the first graph will be shifted to the right by 10 units of the attribute represented by X. If c_4 is assigned a negative value, the graph will be shifted to the left. See Figure 6.4 for an illustration of this shift.

The reader will note that in (6.3), $u(c_4) = c_1 - c_2$ since $\exp(0) = 1$. Also, if the user of this function wants to have $u(x^w) = 0$ and $u(x^b) = 1$, where x^w and x^b are the worst and best possible values of x, respectively, then c_1, c_2, c_3, and c_4 must have values such that (6.4) and (6.5) are satisfied:

$$x^w = \frac{-\ln(c_1/c_2)}{c_3} + c_4, \tag{6.4}$$

$$x^b = \frac{-\ln((c_1 - 1)/c_2)}{c_3} + c_4. \tag{6.5}$$

Since $\ln(1) = 0$, in order to have $x^w = c_4$ (the shift parameter value), one must have $c_1 = c_2$. In addition, since $\ln(x)$ is undefined for $x \le 0$, c_1 must be greater than 1 in order for x^b to achieve a finite value in (6.5). Various values for x^w and x^b for different values of c_1, c_2, and c_3 (which satisfy (6.4) and (6.5)) are shown in Table 6.6, where c_4 is set to 0.

Additional functional forms for a single attribute utility function are given as

$$u(x) = c_1 + c_2 \log_{10}(c_3(x - c_4)), \tag{6.6}$$

$$u(x) = c_1 + c_2(x - c_4)^{c_3}, \tag{6.7}$$

$$u(x) = c_1 + c_2 x + c_3 \exp(c_5(x - c_4)), \tag{6.8}$$

TABLE 6.6

Resulting Values for x^w and x^b for Various Values of c_1, c_2, and c_3 in Equation 6.3 (Where x^w and x^b Are Defined as $u(x^w) = 0$ and $u(x^b) = 1$)

c_1	c_2	c_3	c_4	Resulting Value for x^w	Resulting Value for x^b
1	1	.01	0	0	Infinity
2	2	.01	0	0	69.3
3	3	.01	0	0	40.5
4	4	.01	0	0	28.8
1	1	.05	0	0	Infinity
2	2	.05	0	0	13.9
3	3	.05	0	0	8.2
4	4	.05	0	0	5.75
2	3	.01	0	40.5	109.9
2	4	.01	0	69.3	138.6
2	5	.01	0	91.6	160.9
2	6	.01	0	109.8	179.1

where c_1, c_2, c_3, c_4, and c_5 are all constants, set so that the resulting function fits the set of points assessed by the analyst for the DM. In each case for the functions of (6.3), (6.6), (6.7), and (6.8), c_4 represents the shift parameter discussed earlier.

The utility function shown in Equation 6.6 is concave and hence corresponds to a risk-averse DM just as a function of type (6.3). The utility function of type (6.7), however, is convex and hence would correspond to a risk-seeking DM. As noted by Clemen and Reilly (2001), (6.8) is called a linear plus exponential utility function.

Now, in many cases, one would want a utility function, $u(x)$, that decreases in value as x increases. This would be the case, for example, if X represents an attribute such as cost, fraction defective, or percentage of dissatisfied customers. In these situations, one could use 1 minus one of the respective utility functions given in (6.3), (6.6), (6.7), or (6.8); for example, the negative of (6.3) is given by

$$u(x) = 1 - c_1 + c_2 \exp(-c_3(x - c_4)), \tag{6.9}$$

where, as before, c_1, c_2, c_3, and c_4 are all constant values. Note that (6.9) represents a convex, as opposed to a concave function as in (6.3) and (6.6). Hence, a utility function such as (6.9) would represent a risk-seeking DM. Any time that a scaled, concave utility function is replaced by 1 minus that function, the resulting function will be convex; similarly, any time a scaled convex utility function is replaced by 1 minus that function, the resulting function will be concave.

TABLE 6.7

Listing of Single Attribute Utility Functions and Relevant Comments (Values for x^w and x^b Are Given Such That $u(x^w) = 0$ and $u(x^b) = 1$)

Utility Function	Functional Form	Comments
Linear	$u(x) = c_1 + c_2x$	Corresponds to a risk-neutral DM
Piecewise linear	$u(x) = a_ix + b_i$ for $x \in (x_i, x_{i+1})$ where a_i and b_i are as shown just below Equation 6.2.	Increased accuracy with increased number of assessed points
Exponential	$u(x) = c_1 - c_2 \exp(-c_3(x - c_4))$	Concave function (risk-averse DM) $x^w = -\ln(c_1/c_2)/c_3 + c_4$, $x^b = -\ln((c_1 - 1)/c_2)/c_3 + c_4$
Logarithmic	$u(x) = c_1 + c_2 \log_{10}(c_3(x - c_4))$	Concave function (risk-averse DM) $x^w = 10^{-c_1/c_2}/c_3 + c_4$ $x^b = 10^{(1-c_1)/c_2}/c_3 + c_4$
Polynomial	$u(x) = c_1 + c_2(x - c_4)^{c_3}$	Convex function (risk-seeking DM) $x^w = (-c_1/c_2)^{1/c_3} + c_4$ $x^b = ((1 - c_1)/c_2)^{1/c_3} + c_4$
Linear plus exponential	$u(x) = c_1 + c_2x + c_3 \exp(c_5(x - c_4))$	
Decreasing utility function	$u(x) = 1 - u'(x)$ where $u'(x)$ is an increasing function	$u(x)$ is concave if $u'(x)$ is convex and vice versa

A listing of various single attribute utility functions along with some of their respective characteristics is shown in Table 6.7. Except for the last entry, only increasing utility functions are shown in Table 6.7. As noted earlier, decreasing functions can be formed by subtracting the functions given in the table from the number 1.

6.3.8 Axioms Associated with Utility Theory and the Allais Paradox

Von Neumann and Morgenstern (1947) provided seven axioms, which assured that the maximization of expected utility is an appropriate criterion to use in a risky decision situation. These seven axioms are also discussed in Luce and Raiffa (1957), Bunn (1984, pp. 53–56), Clemen and Reilly (2001, pp. 572–575), and Goodwin and Wright (2009, pp. 131–133). As noted by Bell and Farquhar (1986) and as described in Fishburn (1970), von Neumann and Morgenstern's initial seven axioms have been refined by various researchers and reduced to three axioms, as described in the following.

Axiom 1 (rationality): For any pair of probabilistic outcomes, denoted as O_1 and O_2, either O_1 is preferred to O_2, O_2 is preferred to O_1, or the DM is indifferent between O_1 and O_2.

Axiom 2 (independence): Consider three probabilistic outcomes: O_1, O_2, and O_3 and a probability, p. If O_1 is preferred to O_2, then the compound lottery $pO_1 + (1 - p)O_3$ is preferred to the compound lottery $pO_2 + (1 - p)O_3$ for all outcomes O_3 and for all $p \in [0, 1]$.

Axiom 3 (continuity): Consider three deterministic outcomes: D_1, D_2, and D_3, which are ranked by the DM as D_1 is preferred to D_2 and D_2 is preferred to D_3. There exists some probability p such that the DM will be indifferent between D_2 and the probabilistic outcome: D_1 with probability p and D_3 with probability $(1 - p)$.

Bell and Farquhar (1986) note that DMs often do not behave in a way that satisfies the von Neumann–Morgenstern axioms, especially as this behavior relates to the independence axiom (Axiom 2). Even when it is pointed out to individuals that their decisions do not satisfy the von Neumann–Morgenstern axioms, these individuals often will not alter these decisions. This concept of decision behavior violating axioms of utility theory is embodied in the Allais Paradox (see Allais, 1953).

The Allais Paradox involves a hypothetical situation in which a DM is presented with four outcomes:

1. O_1: Receive \$1 million for certain (i.e., with a probability of 1)
2. O_2: Receive \$5 million with a probability of 0.1, \$1 million with a probability of 0.89, or \$0 with a probability of 0.01
3. O_3: Receive \$5 million with a probability of 0.1 or \$0 with a probability of 0.9
4. O_4: Receive \$1 million with a probability of 0.11 or \$0 with a probability of 0.89

The DM is then asked to rank the following pairs of outcomes in decreasing order of preference: O_1 versus O_2 and O_3 versus O_4.

Now, various studies have been conducted, which show that most people prefer O_1 to O_2 *and* O_3 to O_4. However, this pair of rankings under the criterion of maximization of expected utility is not possible for any utility function via the following argument. Let our attribute, X, be payoff in *millions* of dollars, and let u be the DM's utility function over this payoff. Suppose we scale the utility function so that $u(0) = 0$ and $u(5) = 1$. Then the expected utilities for outcomes O_1 and O_2, denoted as $Eu(O_1)$ and $Eu(O_2)$, are given by

$$Eu(O_1) = u(1)$$

and

$$Eu(O_2) = .1 + .89u(1).$$

Therefore, O_1 is preferred to O_2 if and only if u(1) > .1 + .89u(1) or equivalently if .11u(1) > .1.

Now, similarly,

$$Eu(O_3) = .1 \quad \text{and} \quad Eu(O_4) = .11u(1).$$

Therefore, O_3 is preferred to O_4 if and only if .1 > .11u(1), which contradicts the previous inequality. Therefore, O_1 preferred to O_2 *and* O_3 preferred to O_4 are inconsistent choices (as are O_2 preferred to O_1 *and* O_4 preferred to O_3). The only coherent choices are O_1 preferred to O_2 *and* O_4 preferred to O_3 *or* O_2 preferred to O_1 *and* O_3 preferred to O_4

Besides indicating that people do not necessarily make choices according to expected utility theory, the Allais Paradox suggests that when assessing a DM's utility function, the analyst should perform many consistency checks as well as much sensitivity analysis.

6.4 Multiattribute Utility Functions

6.4.1 Introduction

MAU functions are used to represent preferences over risky outcomes with two or more attributes.

Examples of applications involving MAU functions from the literature are shown in Table 6.8.

So, what is an MAU function? Basically, an MAU function is a mapping from an attribute space with two or more attributes into the space of real numbers; usually, the utility function will be scaled so that it is mapped into the [0, 1] closed interval. Such a function has the characteristic that a joint probability distribution over the outcome space is preferred by the relevant DM to another joint probability distribution over the outcome space (i.e., the space of attributes) if and only if the expected utility of the first joint probability distribution is greater than the expected utility of the second joint probability distribution. Example 6.2 is used to illustrate the use of an MAU function.

Example 6.2: A Simple Acceptance Sampling Problem

Let's consider a simple hypothetical problem to illustrate a decision situation in which an MAU function might be used. Suppose we have a quality control problem involving a lot of two items. Suppose from past experience with the supplier for these items that there is a 50–50 chance that one item in the lot will be defective. That is, either (1) one of the items in the lot will be defective (with a probability of .5) or (2) none of the items in the lot will be defective (also with a probability of .5). We have two alternatives to investigate:

TABLE 6.8

Applications of Multiattribute Utility Functions from the Literature

Application	References
Considered attributes related to nonproliferation, operational effectiveness, environment, safety, and health with respect to evaluation of alternatives for the disposition of excess waste plutonium from excess nuclear weapons	Butler et al. (2005)
Considered attributes related to software capabilities, hardware capabilities, and vendor performance in the selection of a GIS	Ozernoy et al. (1981)
Considered attributes related to fishing stock size, long-term and short-term economic benefits, social acceptability, and opportunities for learning in evaluating alternatives for the opening day of fishing season	McDaniels (1995)
Considered attributes related to cost, malfunction cost, failure rate, and proportion of each failure mode related to aspects of circuit design and packaging technology	Ronen and Pliskin (1981)
Employed utility functions in developing a methodology for the allocation of resources to competing risk-reducing activities	Bodily (1980)
Considered attributes related to ecological and environmental effects, toxicology, pollution control, process and operational data analysis, and industrial hygiene and occupational health to develop a portfolio of environmental and health research programs for a commercial-scale synthetic fuels facility	Peerenboom et al. (1989)
Considered attributes related to nonproliferation, operational effectiveness, and environment, safety, and health in the evaluation of alternatives related to selection of a technology for the disposition of surplus weapons-grade plutonium	Dyer et al. (1998)
Considered attributes related to economics, management, environment, socioeconomics, health/safety, public attitudes, and feasibility with respect to alternative generation technologies for Baltimore Gas and Electric Company	Keeney et al. (1986)
Considered attributes related to health status and life years for choosing between alternative treatments for coronary artery disease and chronic kidney disease	Pliskin et al. (1980)
Illustrative applications involving air pollution control, preference trade-offs among instructional programs, fire department operations, structuring of corporate preferences, evaluating computer systems, and siting and licensing of nuclear power facilities	Keeney and Raiffa (1993, Chapter 7)

Alternative A1: Let the lot proceed with no inspection.

Alternative A2: Randomly sample one item from the lot. If it is defective, rework that item and let the lot proceed. If it is not defective, let the lot proceed with no further inspection. The cost for inspection of a part is $20, while the cost for reworking a part is $50.

There are two attributes to consider in making the decision:

X_1: The total cost associated with the alternative

X_2: The quality of the outgoing lot as measured by the fraction of defective items in this outgoing lot

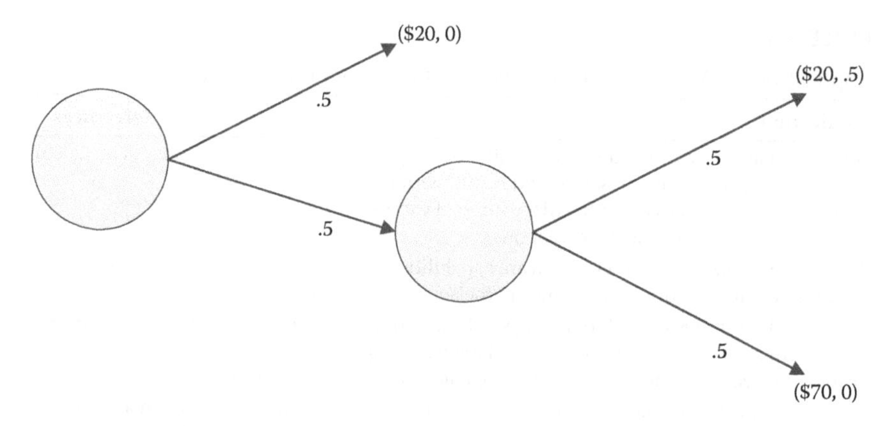

FIGURE 6.5
Illustration of the outcome associated with alternative A2 of Example 6.2.

X_1 (total cost) includes two types of cost—the inspection cost and the reworking cost.

Now, the choice of either alternative will lead to an uncertain outcome, as measured by the two attribute values. In other words, we have a situation where an MAU function might be useful.

If alternative A1 is chosen, the cost will be 0 ($X_1 = 0$), while the fraction of defectives in the outgoing lot (X_2) will be .5 (i.e., one defective item out of the two items in the lot) with probability .5 and 0 with probability .5.

If alternative A2 is chosen, the computation of the probabilistic outcome is a little more, but not much more, complicated. In particular, if there are 0 defectives in the incoming lot (probability of .5), there will obviously be 0 defectives in the sample, and the lot will proceed with 0 defectives with a cost (for the one inspected part) of $20. If there is one defective in the incoming lot (probability of .5), there will be a .5 probability of sampling this defective and reworking it, leading to an outgoing lot with 0 fraction defective and a cost of $70 (the $20 cost of the sampling plus the $50 cost for the reworking); there will be a .5 probability of not discovering the defective item in the sample and then the lot will proceed with .5 fraction defective and the cost will be $20 (the cost of sampling 1 part). The outcome associated with alternative A2 is illustrated in Figure 6.5.

In Figure 6.5, the first set of branches portrays the state of nature of the incoming lot (zero defective items with a probability of .5 or one defective item with a probability of .5). The second set of branches portrays the probability associated with the one defective item being sampled (.5), or not being sampled (.5). The outcomes associated with this situation are given at the "ends" of the branches as (X_1, X_2). Note that the two events (a lot with zero or one defective item, and the event of sampling or not sampling the defective item given that there is a defective item in the lot) are independent.

In summary, the outcomes associated with the alternatives A1 and A2 are shown in Tables 6.9 and 6.10, respectively.

TABLE 6.9

Outcome Associated with Alternative A1 of Example 6.2

Outcome (X_1, X_2)	Probability
($0, 0)	0.5
($0, .5)	0.5

TABLE 6.10

Outcome Associated with Alternative A2 of Example 6.2

Outcome (X_1, X_2)	Probability
($20, 0)	0.5
($20, .5)	0.25
($70, 0)	0.25

Now, it is not obvious as to which alternative should be chosen. For example, alternative A1 will certainly have a lower cost ($0) as compared to A2, which will have a cost of either $20 (with a probability of .75) or a cost of $70 (with a probability of .25). However, alternative A2 will have a higher probability of having 0 defectives in the outgoing lot (.75 probability versus .5 probability for alternative A2).

Although this example is relatively simple, it does illustrate the basic aspects associated with applications involving MAU functions:

1. A set of multiple alternatives from which one is to be selected
2. Multiple attributes
3. Probabilistic outcomes over the attribute values associated with one or more of the alternatives

In this case, the probabilistic outcomes could be computed through the use of a very simple analytic model. In other, more complex situations, decision trees, influence diagrams, general simulation models, and/or expert judgments may be needed.

Now, in order to employ expected utility as a criterion, we need to discuss how to compute the expected utility for an alternative decision, given (1) a joint probability distribution corresponding to the alternative and (2) the DM's MAU function. The expected utility of a probabilistic outcome (sometimes called a "lottery") with multiple attributes is computed as follows. Suppose that you have p attributes and that the joint probability distribution function representing the outcome for alternative i is represented as $f_i(x_1, x_2,\ldots, x_p)$ for $i = 1,\ldots, n$ where n is the number of alternatives.

The expected utility associated with the ith alternative is given by

$$E_i(u(x_1,x_2,\ldots,x_p)) = \sum_1 \sum_2 \sum_3 \cdots \sum_p u(x_1,x_2,\ldots,x_p),\ f_i(x_1,x_2,\ldots,x_p),\ \text{if the } x_1,\ x_2,\ldots,x_p$$

are discrete,

or by

$$E_i(u(x_1,x_2,\ldots,x_p))$$
$$= \int\int\ldots\int u(x_1,x_2,\ldots,x_p) f_i(x_1,x_2,\ldots,x_p) dx_1 dx_2 \ldots dx_p \quad \text{if the } x_1,x_2,\ldots,x_p$$
are continuous.

Note that this computation is just the *expected value* of a function (i.e., *the utility function*) of *several random variables*: $X_1, X_2, \ldots, X_p$.

Let's consider how this expected utility computation would work for our simple example involving acceptance sampling, Example 6.2. Suppose that we have assessed our DM's utility function as follows (the topic of utility function assessment will be discussed in greater detail later in this chapter):

$$u(x_1,x_2) = w_1u_1(x_1) + w_2u_2(x_2) + ww_1w_2u_1(x_1)u_2(x_2),$$

where

the functions u_1 and u_2 are called the individual attribute utility functions

w_1, w_2, and w are the scaling constants

The function u is a multiplicative type of utility function. This type of utility function will be discussed in more detail later.

Let's also suppose that we have $w = 3$, $w_1 = .2$, $w_2 = .5$. Determining these values for the scaling constants would normally be a part of the assessment process; in addition, there is a specific functional relationship between w, w_1, and w_2, so once w_1 and w_2 are determined through the assessment process, the value for w is automatically determined.

The best and worst values for cost are given as $x_1^b = \$0$ and $x_1^w = \$70$. The best and worst values for fraction defective are given as $x_2^b = 0$ and $x_2^w = .5$. Hence, we want to have $u_1(\$0) = 1$, $u_1(\$70) = 0$, $u_2(0) = 1$, and $u_2(.5) = 0$; therefore, both of our individual attribute utility functions will be decreasing functions.

Now, let's also suppose that the individual attribute utility functions are given as decreasing exponential utility function for the cost attribute and as a decreasing linear utility function for the fraction defective attribute, as shown:

$$u_1(x_1) = 1 - .01x_1^{1.083} \quad \text{and} \quad u_2(x_2) = 1 - 2x_2.$$

The reader will note that for each attribute, the utility function value for the best (worst) possible value for the attribute is 1(0), at least in the approximation; that is,

$$u_1(\$0) = 1, \quad u_1(\$70) = 0, \quad u_2(0) = 1, \quad u_2(.5) = 0.$$

Now, evaluating the utility function values for the various outcomes given in Tables 6.9 and 6.10, we obtain

$$u(\$0,0) = .2(1) + .5(1) + 3(.2)(.5)(1)(1) = 1,$$

$$u(\$0,.5) = .2(1) + .5(0) + 3(.2)(.5)(1)(0) = .2,$$

$$u(\$20,0) = .2(.74) + .5(1) + 3(.2)(.5)(.74)(1) = .87,$$

$$u(\$20,.5) = .2(.74) + .5(0) + 3(.2)(.5)(.74)(0) = .148,$$

$$u(\$70,0) = .2(0) + .5(1) + 3(.2)(.5)(0)(1) = .5.$$

Therefore, the expected utility associated with alternative 1 is given by

$$.5u(\$0,0) + .5u(\$0,.5) = .5(1.) + .5(.2) = .6,$$

and the expected utility associated with alternative 2 is given by

$$.5u(\$20,0) + .25u(\$20,.5) + .25u(\$70,0) = .5(.87) + .25(.148) + .25(.5) = .597.$$

Hence, under the criterion of maximization of expected utility, alternative 1 would be preferred to alternative 2. Of course given the closeness of the expected utility values between the two alternatives, a slight change in the utility function could easily result in a change in the ranking of these alternatives.

6.4.2 Independence Conditions and the Form of the Multiattribute Utility Function

Just as in the case of a multiattribute value function, we can assume a specific functional form for an MAU function if the DM's preferences satisfy certain independence conditions. The first of these is utility independence, as defined in Definition 6.3.

Definition 6.3: A set of attributes X is *utility independent* (UI) of its complementary set X′ if the conditional preference structure over lotteries on X given values for the attributes in X′ does not depend on the values for those attributes.

It is virtually impossible to "prove" utility independence for any particular situation since the condition must hold over the entire range of the attributes. Instead, the typical approach is to provide evidence that utility independence holds. This can be accomplished, for example, for the two-attribute case, by showing that the certainty equivalent for a 50–50 lottery over one attribute is not dependent on the value for the other attribute.

Let's consider a situation similar to that portrayed in Example 6.2, involving two attributes: X_1 (cost associated with inspection and reworking) and X_2 (fraction of defective items in the outgoing lot). Let's also suppose, in this more general situation, that the best and worst values associated with X_1 are \$0 and \$1000, respectively, and that the best and worst values associated with

X_2 are 0 and .3, respectively. In order to test if X_1 is utility independent (UI) of X_2, the analyst might ask the DM the following sequence of questions:

Analyst: Suppose that $X_2 = .1$ (i.e., there is 10% defective items in the outgoing lot), what would be your certainty equivalent for the 50–50 lottery over X_1 given by ⟨\$0, \$500⟩?

DM: I would be indifferent between a 50–50 lottery for the cost of ⟨\$0, \$500⟩ and a certain cost of \$280.

Analyst: OK. If X_2 had a value of .01, would your certainty equivalent for the 50–50 lottery of ⟨\$0, \$500⟩ change from \$280?

DM: No, it would stay at the value of \$280.

At this point, the analyst might be able to assume that preferences with respect to lotteries over X_1 do not depend upon the value for X_2 and that therefore X_1 is UI of X_2. Note that X_1 being UI of X_2 does not necessarily mean that X_2 is UI of X_1. The analyst would need to ask similar questions with X_1 and X_2 reversed in order to show this.

Now, if one wants to show that X is UI of X′ where there are two or more attributes in the set X, then the lotteries on the set X must correspond to joint probability distribution functions; these questions posed to the DM would be obviously much more difficult to answer.

As shown by Keeney and Raiffa (1993, pp. 293–294), if one has a set of attributes: $X_1, X_2, \ldots, X_p$ for a decision situation and X_i is UI of its complement for $i = 1, \ldots, p$, then the MAU function will have a multilinear form; the multilinear form is the most general form used with any regularity for an MAU function, as given by Definition 6.4.

Definition 6.4: The *multilinear form* for a utility function is given by

$$u(x) = \sum_{i=1}^{p} w_i u_i(x_i) + \sum_{i=1}^{p} \sum_{j>i} w_{ij} u_i(x_i) u_j(x_j)$$

$$+ \sum_{i=1}^{p} \sum_{j>i} \sum_{m>j>i} w_{ijm} u_i(x_i) u_j(x_j) u_m(x_m) + \cdots$$

$$+ w_{12\ldots p} u_1(x_1) u_2(x_2) \ldots u_p(x_p)$$

where

u_i is a single attribute utility function over X_i scaled from 0 to 1

w_i, also scaled from 0 to 1, is the scaling constant for attribute i

w_{ijm} are scaling constants, which measure the impact of the interaction among attributes i, j, and m on preferences

As an example, for the case with three attributes: X_1, X_2, and X_3, in order for X_i to be UI of X_j for all $i \neq j$, one must have X_1 UI of X_2, X_1 UI of X_3, X_2 UI of X_1, X_2 UI of X_3, X_3 UI of X_1, and X_3 UI of X_2, or six individual utility independence conditions. The corresponding multilinear utility function for this situation will have the following form:

$$u\,(x_1,x_2,x_3) = w_1u_1(x_1) + w_2u_2(x_2) + w_3u_3(x_3) + w_{12}u_1(x_1)u_2(x_2)$$
$$+ w_{13}u_1(x_1)u_3(x_3) + w_{23}u_2(x_2)u_3(x_3) + w_{123}u_1(x_1)u_2(x_2)u_3(x_3).$$

In general, when one has p attributes, showing that the corresponding MAU function is multilinear means showing that $p * (p-1)$ individual utility independence conditions hold.

A special case of the multilinear utility function is the multiplicative utility function. An MAU function will have this form if the corresponding set of attributes is mutually utility independent (MUI), as defined in Definition 6.5.

Definition 6.5: A set of attributes X is *mutually utility independent* (MUI) if every subset $X' \in X$ is utility independent of its complement.

As an example, a set of three attributes X_1, X_2, X_3 is MUI if and only if six individual utility independence conditions are satisfied:

1. X_1 is UI of X_2,X_3
2. X_2 is UI of X_1,X_3
3. X_3 is UI of X_1,X_2
4. X_1,X_2 is UI of X_3
5. X_1, X_3 is UI of X_2
6. X_2,X_3 is UI of X_1

In general, if one has p attributes, the number of individual utility independence conditions that must be satisfied for mutual utility independence is the sum of the number of combinations of p things taken i at a time, summed from $i = 1$ to $i = p - 1$, or $\sum_{i=1}^{p-1} p!/i!(p-i)!$.

As mentioned earlier, if the set of attributes $X = (X_1,X_2,\ldots,X_p)$ is MUI, then one has a multiplicative form for the MAU function, as given by Definition 6.6.

Definition 6.6: The *multiplicative form* for an MAU function with p attributes is given by $\prod_{i=1}^{p}((1 + ww_iu_i(x_i)) - 1)/w)$ where the $u_i(x_i)$ are the individual attribute utility functions scaled to have a minimum value of 0 and a maximum value of 1 and the w, w_1, w_2,..., w_p are the weights (constants) associated with the individual attribute utility functions.

Another, less compact but probably more understandable, form for the *multiplicative* MAU function is given by (6.10) through (6.14):

$$u(x_1,x_2,\ldots,x_p) = \sum_{i=1}^{p} w_i u_i(x_i) + w \sum_{i=1}^{p} \sum_{j>i} w_i w_j u_i(x_i) u_j(x_j)$$

$$+ w^2 \sum_{i=1}^{p} \sum_{j>i} \sum_{m>j} w_i w_j w_m u_i(x_i)\, u_j(x_j)\, u_m(x_m) + \cdots$$

$$+ w^{p-1} w_1 w_2 \ldots w_p u_1(x_1) u_2(x_2) \ldots u_p(x_p), \quad (6.10)$$

where

The utility function, u, is scaled from 0 to 1, (6.11)

The individual attribute utility functions (the u_i) are scaled from 0 to 1, (6.12)

w_i is equal to u with all attribute values except x_i at their worst possible values and x_i at its best possible value for i = 1,..., p, (6.13)

w is a constant such that $1 + w = \prod_{i=1}^{p} (1 + w w_i)$. (6.14)

As mentioned earlier, w is found by solving (6.14) for w once the individual w_i are found. Bunn (1984, p. 95) notes that if the sum of the weights w_i is greater than 1 (i.e., $\sum_{i=1}^{p} w_i > 1$), then w must be between −1 and 0:

$$-1 < w < 0.$$

If the sum of the weights w_i is less than 1 (i.e., $\sum_{i=1}^{p} w_i < 1$), then w must be greater than 0:

$$w > 0.$$

(6.14) can be rewritten as

$$w = \prod_{i=1}^{p} (1 + w w_i) - 1. \quad (6.15)$$

(6.15) can be solved in an iterative fashion by choosing a reasonable value for w, substituting its value into the right-hand side of (6.15), and then computing a new value based on the right-hand side of (6.15). This is repeated until convergence is achieved.

For example, let's suppose that we have a multiplicative utility function with three attributes and that the following values have been determined for the weights:

$$w_1 = .2, \quad w_2 = .7, \quad \text{and} \quad w_3 = .4.$$

Since the weights do not sum to 1, we know that the utility function is not additive and therefore $w \neq 0$; and since the sum is larger than 1, we know that w must be between –1 and 0. Let's begin our iterations by setting $w = -.4$. Substituting this value and the assessed values of $w_1 = .2$, $w_2 = .7$, and $w_3 = .4$ into the right-hand side of (6.15), we obtain

$$w = (1 + (-.4)(.2))(1 + (-.4)(.7))(1 + (-.4)(.4)) - 1 = -.4436.$$

Substituting $w = -.4436$ into the right-hand side of (6.15) and continuing, we achieve the following sequence of values for w:

$$-.4436, -.483, -.5177, \ldots, -.6466, -.6467.$$

Hence, we could reasonably assume a value of –.6467 for w.

If there are only two attributes ($p = 2$), (6.14) is relatively easy to solve for w in terms of w_1 and w_2, as shown in the following. (6.14) can be rewritten as

$$1 + w = (1 + ww_1)(1 + ww_2),$$

or

$$1 + w = 1 + ww_1 + ww_2 + w^2 w_1 w_2,$$

or

$$w = w(w_1 + w_2 + ww_1 w_2),$$

or assuming that w is not 0 and dividing both sides by w, one obtains

$$1 = w_1 + w_2 + ww_1 w_2,$$

and solving for w,

$$w = \frac{(1 - w_1 - w_2)}{w_1 w_2}. \tag{6.16}$$

Substituting this for w into the multiplicative form for the MAU function for p = 2, the two-attribute case yields

$$u(x_1,x_2) = w_1u_1(x_1) + w_2u_2(x_2) + (1 - w_1 - w_2)u_1(x_1)u_2(x_2). \quad (6.17)$$

For the case of three attributes, a multiplicative MAU function would appear as

$$\begin{aligned} u(x_1,x_2,x_3) = {} & w_1u_1(x_1) + w_2u_2(x_2) + w_3u_3(x_3) + ww_1w_2u_1(x_1)u_2(x_2) \\ & + ww_1w_3u_1(x_1)u_3(x_3) + ww_2w_3u_2(x_2)u_3(x_3) \\ & + w^2w_1w_2w_3u_1(x_1)u_2(x_2)u_3(x_3). \end{aligned} \quad (6.18)$$

The corresponding multilinear utility function for three attributes is given by

$$\begin{aligned} u(x_1,x_2,x_3) = {} & w_1u_1(x_1) + w_2u_2(x_2) + w_3u_3(x_3) + w_{12}u_1(x_1)u_2(x_2) \\ & + w_{13}u_1(x_1)u_3(x_3) + w_{23}u_2(x_2)u_3(x_3) + w_{123}u_1(x_1)u_2(x_2)u_3(x_3). \end{aligned} \quad (6.19)$$

Note the differences between the multilinear and multiplicative MAU functions as indicated in (6.18) and (6.19). The general multilinear utility function has many more scaling constants to assess than the multiplicative utility function; for example, with three attributes, the general multilinear utility function has seven scaling constants (w_1, w_2, w_3, w_{12}, w_{13}, w_{23}, and w_{123}), while the multiplicative utility function will have four scaling constants (w_1, w_2, w_3, and w) to assess. The difference in the number of scaling constants between the multiplicative and the general multilinear utility function increases as the number of attributes increases. As noted by Butler et al. (2001), for a multiplicative utility function, "the strength of interactions among all criteria is the same."

Testing for MUI when one has three or more attributes can be a prohibitive task, but Keeney and Raiffa (1993, p. 292) state an *equivalent condition* for MUI:

If $X = (X_1, X_2, \ldots, X_p)$ denotes a set of p attributes, then the following conditions are equivalent:

1. Attributes X are MUI
2. X_1 is UI of $(X_2, \ldots, X_p)$ and (X_1, X_i) is preferentially independent (PI) of its complement for $i = 2, 3, \ldots, p$ where $p \geq 3$

The equivalence condition indicates that in order to show MUI for a set of attributes, instead of performing $2^p - 2$ UI tests, one only has to perform 1 UI test and n − 1 PI tests. In addition to greatly reducing the number of tests

TABLE 6.11

Results of the Equivalence Theorem for Various Numbers of Attributes (p) in the Decision Problem

Value of p	Number of UI Tests Required to Show MUI	Number of UI and PI Tests Required as a Result of the Equivalence Theorem
3	6	1 UI test and 2 PI tests
4	14	1 UI test and 3 PI tests
5	30	1 UI test and 4 PI tests
6	62	1 UI test and 5 PI tests
7	126	1 UI test and 6 PI tests

required, the questions associated with a preferential independence test are typically much easier for a DM to answer than the questions associated with a utility independence test. This difference in the number of tests required to show MUI is shown in Table 6.11.

Now, it may be difficult for the reader to conceive of a situation for which an attribute is *not* utility independent of another attribute. However, Keeney and Raiffa (1993, p. 232) provide an example of such a situation: a farmer's preferences with respect to the amount of sunshine received (denoted as an attribute X_1) *would not be utility independent* of the amount of rain received (denoted as an attribute X_2). In particular, a farmer might prefer a 50–50 lottery of 40% days of sunshine and 60% days of sunshine to a certainty of 98% days of sunshine if the rainfall is only 2 inches over a 3-month period, but that same farmer might prefer the certainty of 98% days of sunshine to the 50–50 lottery if the rain is 25 inches over a 3-month period.

The additive form for an MAU function is given by (6.20) through (6.22):

$$u(x_1, x_2, \ldots, x_p) = \sum_{i=1}^{p} w_i u_i(x_i), \tag{6.20}$$

where

$$\sum_{i=1}^{p} w_i = 1, \tag{6.21}$$

$$w_i > 0 \quad \text{for } i = 1, \ldots, p. \tag{6.22}$$

The multiplicative form for an MAU function reduces to the special case of the additive form when $\sum_{i=1}^{p} w_i = 1$ or equivalently when w = 0 in (6.10) through (6.14).

The DM's utility function will have the additive form of (6.20) through (6.22) if and only if additive independence (AI) holds, as given by Definition 6.7.

Definition 6.7: AI occurs if preferences over lotteries on {X} depend only on the marginal probability distributions of the x_i and not on the overall joint probability distribution over the {X}.

AI is a very restrictive condition and therefore rarely holds. For example, in the case of Example 6.2, AI would be satisfied only if the DM were indifferent between the following two outcomes:

$$\text{Outcome 1: } \langle(\$70, 0), (\$0, 0.5)\rangle, \quad \text{Outcome 2: } \langle(\$0, 0), (\$70, 0.5)\rangle.$$

Note that in our notation, Outcome 1 corresponds to a .5 probability of achieving a \$70 cost with 0 defectives in the outgoing lot and a .5 probability of achieving a \$0 cost with 0.5 fraction defective in the outgoing lot, while Outcome 2 corresponds to a .5 probability of achieving a \$0 cost with 0 defectives in the outgoing lot and a .5 probability of achieving a \$70 cost with 0.5 fraction defective in the outgoing lot. Note that the marginal probabilities of achieving particular attribute values are the same for each outcome. Note also that the fact that the DM is indifferent between these two outcomes is a necessary but not a sufficient condition, since the condition must be satisfied for any pair of outcomes for which the marginal distributions are the same. In reality though, if one is able to show that the condition holds for one or two pairs of such outcomes, that is typically considered enough to indicate AI.

The steps associated with assessing an MAU function are as follows:

1. Determine the worst and best possible values for each respective attribute over all alternatives.
2. Assess each of the scaled individual attribute utility functions, as shown in Section 6.3.3.
3. Test for the various independence conditions described earlier in order to show that a particular form for the MAU function is appropriate. Start with the less restrictive conditions (e.g., those leading to a multilinear or multiplicative utility function) and work toward the more restrictive conditions (those leading to an additive utility function).
4. Determine the scaling constants for the MAU function.

We have not discussed the fourth step in the process outlined earlier, but typically, the first activity in determining the scaling constants $w_1, w_2, \ldots, w_p$ for an MAU utility function is to determine their order in terms of decreasing value. This can be done in a relatively simple fashion by asking the DM to rank order, from most to least preferred, the p outcomes associated with the ith attribute having its best value, and all other attributes at their worst values for i = 1,…, p:

$$\left(x_1^b, x_2^w, x_3^w, \ldots, x_p^w\right), \left(x_1^w, x_2^b, x_3^w, \ldots, x_p^w\right), \ldots, \left(x_1^w, x_2^w, x_3^w, \ldots, x_p^b\right).$$

The reader will note from the definitions given for these weights that the utility for the ith outcome is equal to w_i; hence, in ranking these p outcomes, the DM is equivalently ordering the weights from largest to smallest.

This activity gives the DM some experience in thinking about these outcomes, but we still need the actual weights. As the reader might surmise, determining the actual weights involves having the DM provide respective values for probabilities or attribute values such that he or she will be indifferent between sets of two outcomes, leading to a series of equations that can be solved for the scaling constants. The number of equations needed is equal to the number of scaling constants for the MAU function. For example, a multiplicative MAU function with three attributes has four scaling constants (w_1, w_2, w_3, and w); since w is already expressed as a function of w_1, w_2, and w_3 in (6.15), we need three additional equations in order to determine the values for w_1, w_2, and w_3.

Keeney and Raiffa (1993, p. 303) note that two types of questions can be asked to form these scaling constants, one type involving probabilistic outcomes and the other type involving deterministic outcomes, as shown in the following:

1. What is the value for the probability, q, such that you (the DM) are indifferent between the outcomes?

   O_1: Probability q of achieving the best outcome $(x_1^b, x_2^b, \ldots, x_p^b)$ and probability (1 − q) of achieving the worst outcome $(x_1^w, x_2^w, \ldots, x_p^w)$.

   O_2: The deterministic outcome of all attribute values at their worst levels except for attribute i, which is at its best level (i.e., outcome $(x_1^w, x_2^w, \ldots, x_i^b, \ldots, x_p^w)$).

2. What is the value for attribute i, denoted as x_i, and a value for attribute j, denoted as x_j, such that you are indifferent between the following two outcomes:

   O_1: $(x_1^w, x_2^w, \ldots, x_i, \ldots, x_j^w, \ldots, x_p^w)$,

   O_2: $(x_1^w, x_2^w, \ldots, x_i^w, \ldots, x_j, \ldots, x_p^w)$,

where all attributes except for i and j are set at their worst levels: $x_1^w, x_2^w, \ldots, x_{i-1}^w, x_{i+1}^w, \ldots, x_{j-1}^w, x_{j+1}^w$, for O_1 and O_2, respectively.

Note that the fact that both the overall utility function and the individual attribute utility functions are scaled from 0 to 1 makes the equations (resulting from setting the expected utilities associated with the aforementioned outcomes equal to each other) easy to solve. Specifically, many of the terms from the resulting equations will "zero out." For example, setting the expected utilities for the two outcomes associated with the first question equal to each other yields the following equation:

$$qu\left(x_1^b, x_2^b, \ldots, x_p^b\right) + (1-q)u\left(x_1^w, x_2^w, \ldots, x_p^w\right) = u\left(x_1^w, x_2^w, \ldots, x_i^b, \ldots, x_p^w\right).$$

The left-hand side of this equation reduces to q, and the right-hand side reduces to w_i; hence, we have $w_i = q$.

The answer to the second question results in the following equation:

$$w_i u_i(x_i) = w_j u_j(x_j).$$

Since we already have the values for $u_i(x_i)$ and $u_j(x_j)$, we can determine w_i as a function of w_j.

This fourth step of the process will be illustrated in the next section through the use of an example.

As was noted with the assessment of a multiattribute value function, even if the form selected for the MAU function does not exactly match the preference structure of the DM, the function can still be very useful for decision-making purposes.

Example 6.3: Assessing and Using a Multiattribute Utility Function in Acceptance Sampling

Consider a problem involving a lot of 100 purchased parts from a vendor by a manufacturer. There are three alternatives that the manufacturer is considering regarding possible inspection of the lot:

A_1: There is no inspection, and the parts proceed to assembly.

A_2: There is inspection of the entire lot with subsequent reworking of any defective items found.

A_3: A single sample acceptance sampling plan, in which 10 items would be randomly sampled from the lot. If two or fewer of the sampled items are found to be defective, then only the defective items found in the sample will be reworked, and the lot will proceed to assembly. If more than two items are found defective in the sample, then those defective items will be reworked, and the rest of the lot will be inspected, along with the reworking of any defective items being found in the remainder of the lot.

The reader will note that the third alternative has n = 10 and a = 2 as parameters for the single sample acceptance sampling plan.

Prior experience with this vendor indicates that about 10% of any lot will be defective.

The cost for inspection per unit is $3, and the cost of rework per unit is $15.

In order to evaluate the three alternatives described, the quality manager would like to assess his or her MAU function over two attributes:

X_1: The cost associated with inspection and reworking of the lot
X_2: The fraction of defective items in the lot once one of the three alternatives has been implemented

As noted earlier, the first step in assessing the MAU function is to determine the worst and best attribute values over all alternatives for both attributes. Remember that all that is required is that the worst and best values be bounded by the values actually chosen.

The first alternative, A_1, would give a value of $0 for X_1, obviously the best possible value for this attribute, along with a value of .1 for X_2, since the previous experience indicates 10% defective. The second alternative, A_2, would lead to a value for X_1 of $450 ($300 for the inspection cost plus $150 for the rework cost), and a value for X_2 of 0 (the best possible value for this attribute).

The third alternative, A_3, involves a more difficult calculation, since the outcome for (X_1, X_2) will be probabilistic in nature. However, the worst and best possible values for X_1 and X_2 for A_3 will not be worse or better than those found thus far from examining alternatives A_1 and A_2.

Hence, the worst and best possible values for X_1 will be $450 and $0, respectively, while the worst and best possible values for X_2 will be .1 and 0, respectively. Let's bound these values for this situation and use values of $x_1^w = \$500$, $x_1^b = \$0$, $x_2^w = .15$, and $x_2^b = 0$, thereby possibly allowing the utility function to be used for other, similar, situations as long as these bounds still hold. Table 6.12 shows the best and worst values associated with each attribute.

Since we are assessing a scaled utility function, the initial utility function values are given by

$$u_1(\$500) = 0, \quad u_1(\$0) = 1, \quad u_2(.15) = 0, \quad \text{and} \quad u_2(0) = 1.$$

Now, using the certainty equivalence approach discussed earlier, we obtain the utility function values shown in Tables 6.13 and 6.14.

Tables 6.13 and 6.14 correspond to the individual utility function graphs shown in Figures 6.6 and 6.7.

TABLE 6.12

Best and Worst Values for Each Attribute

Attribute	x_i^b	x_i^w
X_1, cost	$0	$500
X_2, fraction defective	0.	0.15

TABLE 6.13

Individual Attribute Utility Function Values for X_1, Cost

x_1($)	$u_1(x_1)$
500	0.
480	0.125
440	0.25
400	0.375
330	0.5
269	0.625
210	0.75
120	0.875
0	1.

TABLE 6.14

Individual Attribute Utility Function Values for X_2, Fraction Defective in Outgoing Lot

x_2	$u_2(x_2)$
0.15	0
0.142	0.125
0.13	0.25
0.118	0.375
0.104	0.5
0.084	0.625
0.064	0.75
0.034	0.875
0.	1.

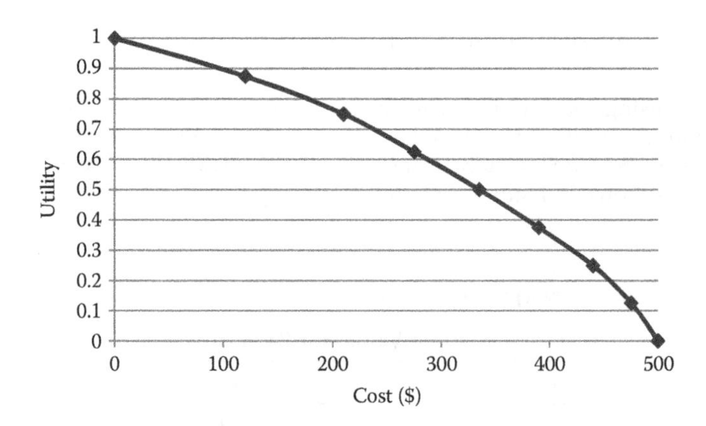

FIGURE 6.6
Individual attribute utility function for cost.

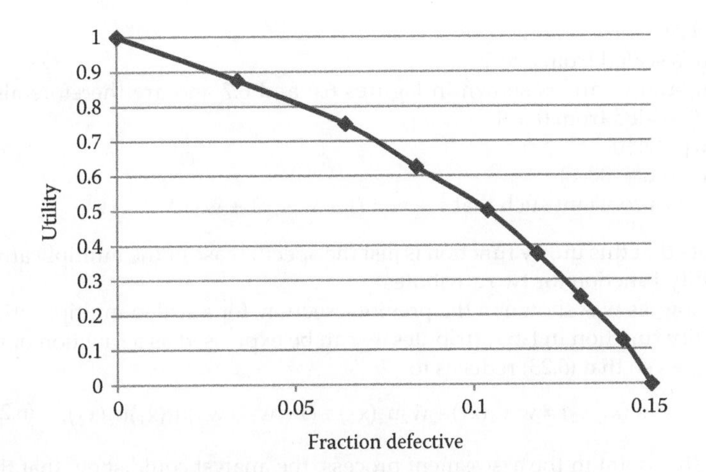

FIGURE 6.7
Individual attribute utility function for fraction defective.

Now, the next step in the assessment process is to verify the existence of any independence conditions. A first step would be to show that X_1 (cost) is UI of X_2 (fraction defective). This could be accomplished through the following set of questions and answers:

Analyst: Suppose that you have a lot with a fraction defective of .01. What would be your certainty equivalent for a 50–50 lottery of a cost of \$50 and \$450 for inspection and reworking for this lot?

DM: My certainty equivalent would be about \$300.

Analyst: Now suppose you had a lot with a fraction defective of .14, what would be your certainty equivalent for the 50–50 lottery of \$50 and \$450 change?

DM: No, it would have the same value of about \$300.

At this point, the analyst might reasonably assume that for this DM, cost is UI of fraction defective—that is, that preferences over lotteries on X_1 do not depend on the value of X_2. The question and answer session shown assumes the DM is familiar with the concept of certainty equivalence. If that were not the case, the analyst would have to be more basic in her questioning, for example, by "honing in" on the value for the certainty equivalent.

Once X_1 has been shown to be UI of X_2, the next step would be to show whether or not X_2 is UI of X_1 for this DM by asking questions similar to the aforementioned ones, but with X_1 and X_2 reversed. Let's suppose that X_2 is also UI of X_1. Hence, at this point in the assessment process, we can assume that (X_1, X_2) is MUI and the utility function has the multiplicative form shown in

$$u\,(x_1, x_2) = w_1u_1(x_1) + w_2u_2(x_2) + ww_1w_2u_1(x_1)u_2(x_2), \tag{6.23}$$

where

u is scaled from 0 to 1

u_1 and u_2 are as shown in Figures 6.6 and 6.7 and are therefore also scaled from 0 to 1

$w_1 = u(\$0, .15)$

$w_2 = u(\$500, 0)$

w is a constant such that $1 + w = (1 + ww_1)(1 + ww_2)$

Note that this utility function is just the specific case of the multiplicative utility function for two attributes.

Now, as was shown in the previous section, for a scaled multiplicative utility function in two attributes, w can be expressed as a function of w_1 and w_2 so that (6.23) reduces to

$$u(x_1, x_2) = w_1 u_1(x_1) + w_2 u_2(x_2) + (1 - w_1 - w_2) u_1(x_1) u_2(x_2). \quad (6.24)$$

At this point in the assessment process, the analyst could show that the utility function is additive by asking the DM questions concerning AI. For example, if the DM said that he or she was indifferent to the following two outcomes, then this would be evidence that the DM's utility function is additive:

O_1: (\$0, 0) with a probability of .5 and (\$500, .15) with a probability of .5
O_2: (\$500, 0) with a probability of .5 and (\$0, .15) with a probability of .5

Note that the marginal distributions for the outcomes associated with individual attribute values are the same—that is, for both O_1 and O_2, there is a .5 probability of achieving the best value for both cost and fraction defective and a .5 probability of achieving the worst value for both.

Of course, another way to show that the utility function is additive would be to continue the assessment process for the scaling constants, w_1 and w_2, and show that they sum to 1, in which case the third term in (6.24) would become 0.

Let's suppose that the DM has answered that he or she is not indifferent to the two outcomes O_1 and O_2. Continuing the assessment process, the next step would be to determine the scaling constants, w_1 and w_2. This step begins by ordering the scaling constants from largest to smallest in value. In this case, this can be accomplished by asking the DM which of the following two outcomes is preferred:

(\$0, .15) (an outcome with a cost of \$0 and a fraction defective of .15)
(\$500, 0) (an outcome with a cost of \$500 and a fraction defective of 0)

Let's suppose that the DM answers that he or she prefers the outcome of \$500 in cost with 0 fraction defective. Then, we know that

$$(\$500, 0)\ P(\$0, .15) \rightarrow u(\$500, 0) > u(\$0, .15) \rightarrow w_2 > w_1.$$

Next, the values for w_1 and w_2 are determined. Suppose that the following question and answer session ensues between the analyst and the DM.

Analyst: Which of the following two outcomes would you prefer?

O_1: A probability of .5 of the best possible outcome: (\$0., 0.) and a probability of .5 of the worst possible outcome: (\$500, .15)

O_2: The deterministic outcome: ($0, .15)
DM: I would prefer O_1.

At this point, the analyst knows that in order to decrease the attractiveness of outcome O_1 (to achieve indifference between O_1 and O_2), the probability associated with ($500, .15) would have to be increased. The analyst tries the probability of .6:

Analyst: OK, let's suppose that we increase the probability associated with ($500, .15) in O_1 to .6 and thereby decrease the probability associated with ($0, 0) to .4 in O_1. Now which outcome do you prefer?
DM: I would be indifferent to these two outcomes.

Since the expected utilities of the revised O_1 and O_2 are equal, we have the following:

$$Eu(O_1) = Eu(O_2)$$

or

$$.4u(\$0.,0) + .6u(\$500,.15) = u(\$0.,.15)$$

or

$$.4 = w_1.$$

At this point, the analyst asks a question of the second type in order to determine a relationship between w_1 and w_2—that is, determine values for x_1 and x_2 such that the DM is indifferent to the following outcomes: (x_1, x_2^w) and (x_1^w, x_2):

Analyst: Suppose that you had an outcome with a cost of $100 and a fraction defective of .15 compared to an outcome with a cost of $500 and a fraction defective of .05, which outcome would you prefer?
DM: I would prefer the second outcome—the one with a cost of $500 and a fraction defective of .05.

Since ($500, .05) is preferred to ($100, .15) by this DM, there must be a value for X_2, fraction defective, that is greater than .05 but less than .15, such that the DM will be indifferent between the two outcomes: ($500, x_2) and ($100, .15). Suppose that the analyst tries a value of .08 and therefore finds that the DM is indifferent between the outcomes of ($500, .08) and ($100, .15). Therefore, we can set the utility function values equal for these two outcomes, as shown in the following:

$$u(\$500,.08) = u(\$100,.15),$$

or since we have a multiplicative utility function,

$$w_1u_1(\$500) + w_2u_2(.08) + (1 - w_1 - w_2)u_1(\$500)u_2(.08)$$
$$= w_1u_1(\$100) + w_2u_2(.15) + (1 - w_1 - w_2)u_1(\$100)u_2(.15).$$

Since we have the individual attribute utility functions for u_1 and u_2, we can substitute the values for $u_1(\$500)$, $u_2(.08)$, $u_1(\$100)$, and $u_2(.15)$, namely, 0, .65, .896, and 0, respectively. Hence, several of the terms in the equation earlier are "zeroed out" and the equation becomes

$$.65w_2 = .896w_1 \quad \text{or} \quad w_2 = 1.378w_1.$$

Since we have already shown that $w_1 = .4$, we have $w_2 = .55$. Therefore, the final utility function is given by

$$u(x_1, x_2) = .4u_1(x_1) + .55u_2(x_2) + .05u_1(x_1)u_2(x_2),$$

where u_1 and u_2 are as shown in Figures 6.6 and 6.7, respectively.

Now we can compute the expected utilities for the respective alternatives discussed earlier. If the alternative A_1, no inspection, is chosen, then X_1, cost, will have a value of \$0, and fraction defective will be .1. Hence, the expected utility for this alternative is given by

$$\begin{aligned} Eu(A_1) &= .4u_1(\$0) + .55u_2(.1) + .05u_1(\$0)u_2(.1) \\ &= .4 + .55(.525) + .05(1)(.525) \\ &= .715. \end{aligned}$$

If alternative A_2, complete inspection, is selected, then X_1, cost, will be \$3 * 100 (the cost for inspecting the entire lot) + \$15 * 10 (the cost for repair of defective items), and X_1, fraction defective, will be 0, since any defective items will be repaired. Hence, the expected utility for this alternative will be

$$\begin{aligned} Eu(A_2) &= .4u_1(\$450) + .55u_2(0) + .05u_1(\$450)u_2(0) \\ &= .4(.21875) + .55(1) + .05(.21875)(1) \\ &= .648. \end{aligned}$$

Finally, if alternative A_3, the single sampling plan with $n = 10$ and $a = 2$, is chosen, the probability of having 0, 1, or 2 defective items in the sample is given by a hypergeometric distribution function; in particular, these probabilities are shown in Table 6.15.

TABLE 6.15

Probability Associated with Having x Defective Items in a Sample of 10 Items from a Lot of 100 Items in Which 10 Are Defective

x	**Pr(X = x)**
0	0.33
1	0.4
2	0.2

TABLE 6.16

Probability Distribution Over X_1, X_2, u_1, u_2, and u Associated with the Sampling Plan n = 10 and a = 2

X_1 Value	X_2 Value	$u_1(x_1)$	$u_2(x_2)$	$u(x_1, x_2)$	Probability
$30.	0.1	0.969	0.525	0.70175	0.33
$45.	0.09	0.953	0.5875	0.732325	0.4
$60.	0.08	0.9375	0.65	0.763	0.2
$450.	0.	0.21875	1.	0.6484	0.07

Adding up the probabilities associated with having 0, 1, or 2 defectives in the sample, we find that the probability of accepting the lot is .93 and the probability of *rejecting* the lot is .07. The probability distribution over X_1, X_2, u_1, u_2, and u is given in Table 6.16.

Hence, the expected utility associated with alternative A_3 is given by

$$Eu(A_3) = .33(.70175) + .4(.732325) + .2(.763) + .07(.6484)$$
$$= .722488$$

Alternative A_3, the use of an acceptance sampling plan with n = 10 and a = 2, is the one that maximizes expected utility.

Example 6.4: Assessing a Multiattribute Utility Function for a Medical Diagnosis Decision

In order to illustrate the computation of the scaling constants for an MAU function with more than two attributes, consider a situation in which the manager of a county health department has funds available to administer tests to detect lead poisoning in the children, numbering approximately 100000, of the county. There are two types of tests:

1. A urine test with a sensitivity of .9, a specificity of .85, and a cost of $10
2. A blood test with both a sensitivity and a specificity of 1, but a cost of $100

Recall from Chapter 2, that the sensitivity of a medical test is the conditional probability that the test will be positive given that the disease is present, and the specificity of a medical test is the conditional probability that the test is negative given that the disease is not present. Hence, the closer the values of sensitivity and specificity are to 1, the better is the test.

A prior estimate is that 10% of the children have lead poisoning. One of the two strategies is being considered:

1. Administer blood tests only.
2. Administer the urine test to a number of children, and for those children who test positive on the urine test, administer blood tests.

Note that these two strategies do not completely define the alternatives; the main thing missing is the number of each type of test to administer for each strategy. So, let's at least define some variables for each strategy in order to more completely define them:

A_1: Administer a_1 blood tests only.
A_2: Administer a_2 urine tests, followed by a blood test for each child who tests positive on the urine test.

Now, upon giving the problem some thought, one might develop the following attributes for this situation:

X_1: The cost for administering the tests for lead poisoning in thousands of dollars
X_2: The number of children definitely identified as having lead poisoning
X_3: The number of children incorrectly identified as having lead poisoning through the urine test (but later correctly identified as not having lead poisoning through the blood test)
X_4: The number of children who are incorrectly identified as *not* having lead poisoning through the urine test (but actually do have lead poisoning)

Note that the children who are counted in the "X_4 category" depart the process under the assumption that they *do not have lead poisoning* even though they do.

Let's place approximate bounds on the amount of money to be spent; a minimum of \$50000 can be spent and a maximum of \$100000 can be spent.

Now, if the first strategy of administering only the "perfect" blood test to a randomly selected children set of a_1 children from the population was chosen, then since the test is perfect, the number of children incorrectly identified as having lead poisoning when they do not, or as not having lead poisoning when they do will both be 0 (i.e., $X_3 = 0$ and $X_4 = 0$). The number of children correctly identified as having lead poisoning would be a hypergeometric random variable, with the exact distribution dependent on the value for a_1. In summary therefore, given a specific number of blood tests to administer, a_1, this strategy would result in the values for $X_1 = 100a_1/1000$, $X_3 = 0$, $X_4 = 0$, and X_2 being a hypergeometric random variable. See the summary of results for this strategy in Table 6.17.

TABLE 6.17

Summary of Results for the Attribute Values If Strategy A_1 (a_1 Blood Tests Administered)

Attribute	Characteristics of Values
X_1	$100a_1/1000$
X_2	Hypergeometric random variable with parameter values of $N = 100000$, $K = 10000$, and $n = a_1$
X_3	0
X_4	0

The parameter values associated with the hypergeometric distribution in Table 6.17 refer to the total population (N = 100000), the number of "success states" in the population (K = 10000), and the number of "draws" from the population ($n = a_1$). Note that a "success" for a hypergeometric random variable can be defined in any way; in this case, success is defined as a positive test.

Because of the large numbers involved, we could approximate the hypergeometric random variable for X_2 with a binomial random variable with parameter values of $n = a_1$ (number of "draws") and p = .1 (probability of "success").

Now, the second strategy of administering the urine test to a random sample of a_2 children followed by the administration of the blood test to those who test positive on the imperfect urine test involves some more difficult calculations. In particular, the number of children who are given the urine test fall into four separate, mutually exclusive, categories:

1. Number of children who have lead poisoning who test positive on the urine test
2. Number of children who have lead poisoning who test negative on the urine test
3. Number of children who do not have lead poisoning who test positive on the urine test
4. Number of children who do not have lead poisoning who test negative on the urine test

These numbers in each category will all be random variables. The children in categories 1 and 3 will all be correctly identified through the use of the blood test; but since these numbers are uncertain, the cost associated with the blood tests given will also be a random variable. Hence, in summary, all four of the attribute values will be random variables if this strategy is followed. See the summary of results for this strategy in Table 6.18.

Let's place some bounds on the attribute values as a prelude to discussing how to determine the scaling constants for an MAU function of these attributes. These bounds will be determined through what might be called "back of the envelope" calculations.

Suppose we consider only the first type of alternative, administer blood tests only. If we decide to spend \$50000, we will be able to give 500 blood tests; about 50 of the children tested will have lead poisoning; since the blood test is "perfect," we can expect a value for X_2 (number

TABLE 6.18

Summary of Results for the Attribute Values If Strategy A_2 (a_2 Urine Tests Followed by Blood Tests, for Those Who Test Positive on the Urine Test)

Attribute	Characteristics of Values
X_1	$(10a_2 + 100 * rv)/1000$
X_2	rv
X_3	rv
X_4	rv

of children correctly identified as having lead poisoning) of about 50. If we decide to spend $100000, by the same reasoning, we can expect X_2 to have a value of about 100. Both X_3 and X_4 will have values of 0 whether we spend $50000 or $100000 on only the blood tests.

Now, suppose we decide to follow the second policy: urine tests, followed by blood tests for those who test positive on the urine test. If we decide to spend about $50000 total, then we should give about 1500 urine tests at a cost of $15000. Of the 1500 children given urine tests, about 150 children will have lead poisoning, and about 1350 children will not have lead poisoning. Of the 150 children that do have lead poisoning, about 135 will test positive on the urine test (and thereby be positively identified on the follow-up blood test) and about 15 will test negative; therefore, X_4 will have a value of about 15. Of the 1350 children who do not have lead poisoning, about $(1-.85)*1350 \approx 202$ will test positive for the urine test, so the value for X_3 under this policy will be about 202. So, about 135 + 202 = 337 children will be given the follow-up blood test; of these, about 135 will test positive and about 202 will test negative. Hence, under the policy of urine tests followed by blood tests for those who test positive on the urine test, we will have approximately the following values for the attributes:

$$X_1 = \frac{(1500 * \$10 + 337 * \$100)}{1000} = 48.7 \text{ thousand dollars } (\$48700),$$

$$X_2 = 135, X_3 = 202, \text{ and } X_4 = 15.$$

If we decide to spend about $100000 under this policy, we will achieve the following approximate values for the attributes:

$$X_1 = 97.5 \text{ thousand dollars } (\$97500), \quad X_2 = 270, \quad X_3 = 405,$$

$$\text{and} \quad X_4 = 30.$$

The results for these "back of the envelope" calculations are shown in Table 6.19. Note that the values for X_1 indicate that we are spending a

TABLE 6.19

Approximate Attribute Values Associated with Various Testing Policies

Policy	Approximate X_1 Value	Approximate X_2 Value	Approximate X_3 Value	Approximate X_4 Value
1. Blood tests only, 50 K to spend	50.	50	0	0
2. Blood tests only, 100 K to spend	100.	100	0	0
3. Urine tests with follow-up blood tests, 50 K to spend	48.7	135	202	15
4. Urine tests with follow-up blood tests, 100 K to spend	97.5	270	405	30

TABLE 6.20

Best and Worst Values for Attributes in the Medical Testing Problem, where X_1 is given in thousands of dollars

Attribute	x_i^b, Best Value for Attribute	x_i^w, Worst Value for Attribute
X_1	40	120
X_2	300	40
X_3	0	500
X_4	0	50

TABLE 6.21

Assessed Values for Individual Attribute Utility Functions, where X_1 is given in thousands of dollars

x_1	$u_1(x_1)$	x_2	$u_2(x_2)$	x_3	$u_3(x_3)$	x_4	$u_4(x_4)$
120	0	40	0	500	0	50	0
117	0.125	62	0.125	464	0.125	48	0.125
112	0.25	86	0.25	420	0.25	45	0.25
106	0.375	111	0.375	372	0.375	41	0.375
98	0.5	138	0.5	308	0.5	36	0.5
88	0.625	166	0.625	242	0.625	29	0.625
75	0.75	197	0.75	175	0.75	21	0.75
60	0.875	241	0.875	95	0.875	12	0.875
40	1	300	1	0	1	0	1

little less than what we have available from our budget, but since these values are uncertain, this expected underspending is desirable.

The first thing that the reader might notice from Table 6.19 is the effectiveness of a screening test (the urine test) followed by a better test (the blood test). For example, we would expect to spend about half as much money with the third policy as we would with the second policy ($48700 versus $100000), while still identifying more children with lead poisoning (135 versus 100). However, this would be at the expense of increased uncertainty with respect to the attribute values, as well as nonzero positive values for X_3 and X_4.

The values given in Table 6.19 suggest bounds on the attribute values, as shown in Table 6.20. Note that we are setting the bounds a little outside of the values suggested in Table 6.19 in order to account for both the uncertainty associated with the attribute values and the investigation of other policies.

Now, let's suppose that we have assessed points on the single attribute value functions for X_1, X_2, X_3, and X_4, as shown in Table 6.21.

Also, suppose that (X_1,X_2,X_3,X_4) are MUI so that the utility function has the multiplicative form and can therefore be written as

$$u(x_1,x_2,x_3,x_4) = \sum_{i=1}^{4} w_i u_i(x_i) + w \sum_{i=1}^{3} \sum_{j>i} w_i w_j u_i(x_i) u_j(x_j)$$

$$+ w^2 \sum_{i=1}^{2} \sum_{j>i} \sum_{m>j} w_i w_j w_m u_i(x_i) u_j(x_j) u_m(x_m) + \cdots$$

$$+ w^3 w_1 w_2 w_3 w_4 u_1(x_1) u_2(x_2)\, u_3(x_3) u_4(x_4).$$

Now, the only task left in the assessment is the determination of values for w_1, w_2, w_3, w_4, and w. First, we would determine the order of w_1, w_2, w_3, and w_4, from the largest to the smallest by asking the manager to rank the following outcomes in order of preference:

(40, 40, 500, 50), (120, 300, 500, 50), (120, 40, 0, 50), and (120, 40, 500, 0).

Each of the outcomes listed above corresponds to having one attribute at its best value and the other attributes at their worst values. Suppose that the manager ranks these outcomes as first, second, fourth, and third, respectively, which means that we would have $w_1 > w_2 > w_4 > w_3$.

Now, suppose that the manager indicates indifference between the following sets of two outcomes each, corresponding to answers associated with the first type of question listed earlier to determine scaling constants:

1. O_1: Probability of .6 of achieving the best possible outcome (40, 300, 0, 0) and probability of .4 of achieving the worst possible outcome (120, 40, 500, 50)
   O_2: The outcome of attributes X_2, X_3, and X_4 at their worst possible values and attribute X_1 at its best possible level (40, 40, 500, 50)
2. O_1: Probability of .25 of achieving the best possible outcome (40, 300, 0, 0) and probability of .75 of achieving the worst possible outcome (120, 40, 500, 50)
   O_2: The outcome of attributes X_1, X_3, and X_4 at their worst possible values and attribute X_2 at its best possible level (120, 300, 500, 50)
3. O_1: Probability of .21 of achieving the best possible outcome (40, 300, 0, 0) and probability of .79 of achieving the worst possible outcome (120, 40, 500, 50)
   O_2: The outcome of attributes X_1, X_2, and X_3 at their worst possible values and attribute X_4 at its best possible level (120, 40, 500, 0)
4. O_1: Probability of .15 of achieving the best possible outcome (40, 300, 0, 0) and probability of .85 of achieving the worst possible outcome (120, 40, 500, 50)
   O_2: The outcome of attributes X_1, X_2, and X_4 at their worst possible values and attribute X_3 at its best possible level (120, 40, 0, 50)

These four sets of indifferences lead to the following values for the scaling constants:

$$w_1 = .6, \quad w_2 = .25, \quad w_3 = .15, \quad w_4 = .21.$$

TABLE 6.22

Utility Function Evaluation for Lead Poison Testing Policies

Policy	$u_1(x_1)$	$u_2(x_2)$	$u_3(x_3)$	$u_4(x_4)$	$u(x)$
1. Blood tests only, 50 K to spend	0.9375	0.0568	1.	1.	0.826
2. Blood tests only, 100 K to spend	0.46875	0.32	1.	1.	0.639
3. Urine tests with follow-up blood tests, 50 K to spend	0.946	0.486	0.7	0.833	0.845
4. Urine tests with follow-up blood tests, 100 K to spend	0.506	0.935	0.292	0.622	0.636

The value for w can be determined by solving equation (6.15) using the iterative method presented earlier. For this case, the equation is given by

$$w = (1+.6w)(1+.25w)(1+.15w)(1+.21w) - 1.$$

Since $\sum_{i=1}^{4} w_i > 1$, we know that $-1 < w < 0$. Starting the iterations with $w = -.7$, we obtain $w \approx -.467$, which gives us the final utility function.

Applying this utility function to the policies shown in Table 6.19 and treating the outcomes associated with these policies as deterministic (an admitted, but reasonable approximation), we would attain the evaluation shown in Table 6.22. As indicated by the results in the table, Policy 3 is preferred in this case.

Material Review Questions

6.1 Provide at least two examples of decision situations where it would be important to model the uncertainties in the outcomes associated with the decision.

6.2 Describe the various ways in which the uncertainty in the outcome associated with a decision can be described.

6.3 What is the difference between "decision making under uncertainty" and "decision making under risk"?

6.4 Define what is meant by a "state of nature."

6.5 Which word is sometimes used in place of "state of nature?"

6.6 Name the three approaches discussed in the chapter for decision making under uncertainty.

6.7 The rule, maximization of the minimum gain, is associated with what type of a criterion?

6.8 The rule, maximization of the maximum gain, is associated with what type of criterion?

6.9 Why is the term, expected value, considered to be a misnomer?

6.10 What are two of the drawbacks associated with using the expected value of an attribute as a criterion in a situation involving decision making under risk?

6.11 Provide two examples of alternatives that would never be chosen if one used the criterion of optimizing the expected value of a performance measure.

6.12 Briefly define "mean–variance optimization."

6.13 Two different rational DMs could typically be expected to have different utility functions for the same situation (*true* or *false*).

6.14 When a single attribute utility function has a value of 0 for the worst value of the relevant attribute and a value of 1 for the best value of the attribute, what type of utility function is this?

6.15 What is a "certainty equivalent" for a lottery?

6.16 What is a "risk premium" for a lottery?

6.17 Provide definitions for a

a. Risk-averse DM
b. Risk-neutral DM
c. Risk-prone DM

6.18 What type of utility function would correspond to a, b, and c, respectively, in material review exercise 6.17?

6.19 Utility theory is (normative, descriptive) in nature (choose one).

6.20 Utility should be thought of as an (ordinal, interval) measure (choose one).

6.21 What are the four response modes associated with the assessment of a utility function.

6.22 For an "exponential utility function": $u(x) = c_1 - c_2\exp(-c_3(x - c_4))$, a smaller value for c_3 represents what type of DM?

6.23 How would one transform a utility function, which is increasing with increasing attribute value, to a corresponding utility function, which is decreasing in increasing attribute value?

6.24 What does the Allais Paradox indicate about the behavior of DMs?

6.25 Suppose that you have two attributes, X_1 and X_2, for a decision situation. More of each attribute is desired. The attribute X_1 ranges from 0 to 50 in value, and the attribute X_2 ranges from 0 to 100 in value. Give an example of the type of question you would ask of a DM to show that X_1 is utility independent of X_2.

6.26 If an attribute X_1 is utility independent of X_2 for a DM, this necessarily means that X_2 is utility independent of X_1 for the DM (*true* or *false*).

6.27 Write down a multilinear utility function for a situation involving four attributes. Do not use any summation signs in writing the function.

6.28 For a situation involving four attributes, how many different individual utility independence conditions would need to hold in order for the corresponding utility function to be multilinear? What are these individual utility independence conditions?

6.29 All multiplicative MAU functions are also multilinear, but not all multilinear MAU functions are multiplicative (This is the same thing as saying that a multiplicative utility function is a special case of a multilinear utility function.) (*true* or *false*).

6.30 How many scaling constants need to be assessed for a multilinear utility function with four attributes?

6.31 How many scaling constants need to be assessed for a multiplicative utility function with four attributes?

6.32 Suppose that you had a decision situation involving two attributes: X_1 and X_2, where more of each attribute is preferred to less. Suppose also that both attributes ranged in value from 0 to 10. Give an example of the type of question and corresponding answer that would indicate that the utility function for this situation would be additive.

Exercises

6.1 Steve has decided to accept a lottery with a 40% chance of winning $1000 and a 60% chance of winning $400 rather than winning $600 for certain.

a. What can we infer about Steve's utility function for income over the range of $400–$1000?
b. Suppose Steve is indifferent between receiving $620 and receiving the results of the lottery. What is his RP for the lottery?

6.2 If the attributes for a problem are mutually utility independent, then what form will the utility function have?

6.3 Which condition would you expect to be met *less* frequently in terms of appropriately describing a DM's preference structure:

a. Mutual utility independence
b. AI

6.4 Suppose that you are in charge of airport security at Chicago O'Hare Airport. You must make a decision on which of three different types of machines to install at the airport for detecting guns/weapons/bombs in the passenger baggage. The machines vary in cost and reliability (e.g., ability to detect guns/weapons/bombs). Explain, in general, how you would use the techniques presented in this book (and other appropriate techniques) to make your decision. (Keep in mind that the machines can make both false-positive and false-negative diagnoses.)

6.5 Describe the main characteristic of the situation where one would use an MAU function instead of a multiattribute value function.

6.6 Suppose that a DM is risk seeking (i.e., risk prone) over a particular attribute: X_1, which ranges from 0 to 100. (More of the attribute is preferred to less.) What does this information indicate about the utility function value at $X_1 = 50$, given that the utility function is scaled from 0 to 1, with u(0) = 0 and u(100) = 1.

6.7 Suppose that you have a situation involving a multiplicative utility function over two attributes. You have assessed the value for two scaling constants: $w_1 = .2$, $w_2 = .4$. What is the value for w?

6.8 A DM's preference structure is such that the condition of AI is met. Given a situation involving two attributes and the following two outcomes, which of the following statement will be true for the DM (choose one):

A: .5 probability of (10, 40) and a .5 probability of (60, 30)
B: .5 probability of (60, 40) and a .5 probability of (10, 30)

a. The DM prefers A to B.
b. The DM is indifferent between A and B.
c. The DM prefers B to A.

6.9 Suppose that you want to show that a DM's utility function will be multiplicative in a situation involving four attributes. Given that you want to prove only one utility independence condition, how many preferential independence conditions would you have to demonstrate?

6.10 Suppose that you have assessed the following information concerning a scaled, additive utility function over two attributes:

x	0	20	60	100
$u_x(x)$	0	0.3	0.7	1
y	0	4	7	10
$u_y(y)$	0	0.5	0.8	1

The DM is indifferent between the following outcomes:

A: (0, 7) with probability .5 and (20, 0) with probability .5

B: (60, 0) for certain

What are the correct values for the scaling constants: w_1 and w_2?

6.11 Suppose that you want to show that an attribute, X_1, is utility independent of another attribute, X_2, for a particular DM, in a problem situation involving two attributes. Give an example of the types of questions you would ask to show this.

6.12 Application of utility function: NCAA Final Four Problem.
Suppose that you have an opportunity to bet on the team that will win the NCAA Division I Basketball Championship. The four teams in the Final Four are Louisville, Purdue, Kentucky, and Indiana. You can make one of four bets:

1. Bet \$100 on Louisville to win \$400.
2. Bet \$100 on Purdue to win \$450.

3. Bet $100 on Kentucky to win $500.
4. Bet $100 on Indiana to win $600.

(Note that, for example, if you bet the $100 on Indiana and Indiana wins, you will have a net income of 600 – 100 = $500.)

Suppose also that you have assessed the probabilities associated with one team beating another as follows:

		Would Beat			
		Louisville	Purdue	Kentucky	Indiana
Probability that	Louisville		.55	.6	.7
	Purdue			.52	.72
	Kentucky				.58
	Indiana				

Suppose that the initial pairings have Louisville playing Purdue and Kentucky playing Indiana.

a. Compute the probability associated with each team winning the tournament. Based on this only, rank the four bets.
b. Now assess your utility function over the appropriate range of income. Based on the criterion of maximizing expected utility, rank the four bets. Are there any changes from your initial ranking?

6.13 A trucking company is evaluating two alternative designs for its transportation network (which implies locations for its hubs and terminals). The company is mainly concerned with two attributes:

X_1: Expected number of deadhead miles driven in a typical year (in thousands)

X_2: Expected percentage of deliveries made on time during a typical year

There are other attributes for which the company is concerned (e.g., construction costs, workforce availability, desirability of locations chosen), but each design gives the same values for these other attributes, under each scenario evaluated.

A simulation model has been developed to evaluate each design under each of three scenarios (high demand [HD], medium demand [MD], and low demand [LD]). The following payoff table has been developed through the use of the simulation model. In this table the first number represents the value for X_1 and the second number represents the value for X_2.

	HD	MD	LD
Design 1	(25, 75)	(22, 85)	(15, 98)
Design 2	(30, 82)	(27, 90)	(25, 99)
Probability	.1	.6	.3

The utility function of the company's president has been assessed. The function has been shown to be multiplicative, with scaling constants of $w_1 = .1$, $w_2 = .5$. The individual attribute utility function values for the scaled utility functions are given as follows:

x_1	15	22	25	27	30	
$u_1(x_1)$	1	0.8	0.6	0.4	0	
x_2	75	82	85	90	98	99
$u_2(x_2)$	0	0.4	0.5	0.7	0.9	1

Compute the value for the scaling constant, w, the expected utilities for each design, and rank the two designs according to their expected utilities.

6.14 A simulation model has been developed to aid in the lunchtime (11 a.m.–1 p.m.) staffing decisions at a fast-food restaurant. Since the restaurant is located in a tourist area, one of the difficulties associated with the staffing decision is the uncertainty associated with the number of tour buses that will arrive during lunch. These bus arrivals are in addition to pedestrian and car arrivals, which are relatively stable. Historical data have indicated that the following probability distribution represents the uncertainty in the number of buses arriving.

Number of Buses	Probability Associated with This Number
0	0.1
1	0.42
2	0.3
3	0.18

Three different staffing policies are available for implementation, denoted as: low, moderate, and high. The staffing costs associated with these three levels can be easily computed. The simulation model is used to estimate the fraction of customers who wait more than 8 minutes, as a function of the number of buses arriving and the staffing level. These results from the simulation model, along with the cost associated with the three staffing policies, are shown in the following.

Staffing Policy	Fraction Waiting Longer Than 8 Minutes with 0 Buses Arriving	Fraction Waiting Longer Than 8 Minutes with 1 Bus Arriving	Fraction Waiting Longer Than 8 Minutes with 2 Buses Arriving	Fraction Waiting Longer Than 8 Minutes with 3 Buses Arriving
Low	.07	.12	.19	.28
Medium	.05	.08	.16	.21
High	.02	.05	.11	.14

Staffing Policy	Daily Cost
Low	$400
Medium	$600
High	$800

A multiplicative utility function has been assessed for the restaurant manager over two attributes:

X_1: Fraction of customers waiting longer than 8 minutes
X_2: Staffing cost

The scaling constants for this function are $w_1 = .6$ and $w_2 = .3$. Assessed values for the individual attribute utility functions are given by the following:

x_1	$u_1(x_1)$
0.3	0
0.27	0.25
0.21	0.5
0.12	0.75
0	1

x_2	$u_2(x_2)$
800	0
730	0.25
650	0.5
540	0.75
400	1

Evaluate and rank the three staffing policies using expected utility as a criterion. Use linear interpolation for any points not directly assessed for the attribute values.

7

Modeling Methodologies for Generating Probabilistic Outcomes: Decision Trees and Influence Diagrams

7.1 Introduction

In this chapter, we present methodologies, namely, decision trees and influence diagrams, which allow one to map from decisions/alternatives to a probability distribution over an outcome space. The decisions involved can be, and often are, sequential in nature. One reason why these methodologies are usually presented together is that a decision tree has a corresponding influence diagram and vice versa. In fact, the software packages decision programming language (DPL™) and Precision Tree™ will create a decision tree that is equivalent to the influence diagram constructed by the user.

As noted by Gass and Assad (2005, pp. 131–132), Magee (1964) introduced decision trees and Raiffa described concepts associated with decision trees in some detail in his book (Raiffa, 1968). Influence diagrams originated from work performed at the Stanford Research Institute (Miller et al., 1976) during the 1970s. In effect, influence diagrams represent a decision situation in a more compact form than decision trees, which tend to become unwieldy for many real decision situations. Also, as noted by Shachter (1986), an influence diagram represents a more natural way to represent a decision situation to a decision maker.

Decision trees and influence diagrams have been applied to decision making in such diverse areas as portfolio management (Skaf, 1999), commercialization of a new drug (Stonebraker, 2002), and scheduling of the refueling for a nuclear power plant (Dunning et al., 2001). Table 7.1 provides a listing of some of the applications involving decision trees/influence diagrams from the literature.

Section 7.2 provides a discussion of decision trees, including basic concepts, evaluation, calculation of probabilities using Bayes' theorem, qualities and values associated with perfect and imperfect predictors, strategies and risk profiles, and cumulative risk profiles and dominance. Section 7.3

TABLE 7.1

Selected Applications from the Literature Involving Decision Trees/Influence Diagrams

Application	References
Strategic planning for a video game software company	Matheson and Matheson (1999)
Strategic insights for pharmaceutical companies	Bodily and Allen (1999)
Soccer player performance rating	McHale et al. (2012)
Portfolio management in the oil and gas industry	Skaf (1999)
Evaluation of the commercial process for a new blood clot busting drug	Stonebraker (2002)
Allocation of risks within an RFP for the Department of Energy	Keisler et al. (2004)
Choosing an alternative for hazardous waste remediation for the Department of Energy	Toland et al. (1998)
Scheduling the refueling for a nuclear power plant	Dunning et al. (2001)

provides a discussion of influence diagrams, including basic concepts and their correspondence to decision trees.

Section 7.4 presents sensitivity analyses of decision trees and influence diagrams, while Section 7.5 discusses the use of expected utility as a performance measure within decision trees and influence diagrams. Section 7.6 discusses procedures for analyzing decision trees and influence diagrams involving outcome nodes/chance event nodes, which represent continuous random variables or at least random variables, with a large number of outcomes. Finally, an example of a bidding decision situation for a construction firm is presented at the end of the chapter.

7.2 Decision Trees

7.2.1 Basic Concepts

A decision tree represents a timed sequence of decisions and outcomes. It is composed of nodes and branches, with no "cycles" (i.e., one cannot trace a path that starts from any node and returns to that same node by following the branches in the diagram).

There are two types of nodes in a decision tree:

1. Decision nodes
2. Outcome nodes

Branches that emanate from a decision node represent a set of comprehensively exhaustive, mutually exclusive decisions at a point in time. Branches that emanate from an outcome node represent a set of comprehensively exhaustive, mutually exclusive outcomes; the outcomes from any outcome node represent a *partition* of the sample space.

Associated with each outcome emanating from an outcome node is a probability. This probability may be conditional in nature; that is, the probability can be dependent upon the alternative decisions and outcomes that occurred previously, as represented by the branches leading to that outcome node in the decision tree.

A cost or income can be associated with each branch (decision or outcome) in the tree; thus, a complete sequence of branches from the tree will correspond to an overall cost or income associated with the corresponding sequence of decisions and outcomes.

The basic objective associated with the analysis of a decision tree is the selection of the initial decision that optimizes expected monetary value (EMV) for the entire time frame represented by the decision tree. In determining this initial optimal decision, one needs to determine the optimal sequence of decisions that follow, *given current information.* Once an initial decision is made, outcomes may turn out to be different than expected. In selecting this initial decision, one must also select an optimal sequence of decisions that would follow this initial decision, but these following decisions may not (and probably will not) be implemented at the outset (i.e., one might wait and see which of the uncertain outcomes actually occurs).

7.2.2 Evaluation of a Decision Tree

As an example, suppose that an organization must decide whether or not to develop, produce, market, and sell a new product. Prior to this initial decision, the organization can hire a firm to survey the market to determine whether or not the product will be a success. Hence, there is a sequence of two decisions to make:

1. Hire the firm or not
2. Develop, produce, market, and sell the product or not (as a *single* decision)

There are *two, mutually exclusive, outcomes* associated with each decision—the market survey will indicate that the market will be good or bad, and the market for the product will be either good or bad.

The following costs and incomes are associated with this problem:

1. Income from product if market is good = \$500,000, if market is bad = −\$100,000
2. Cost of Market Survey = \$60,000

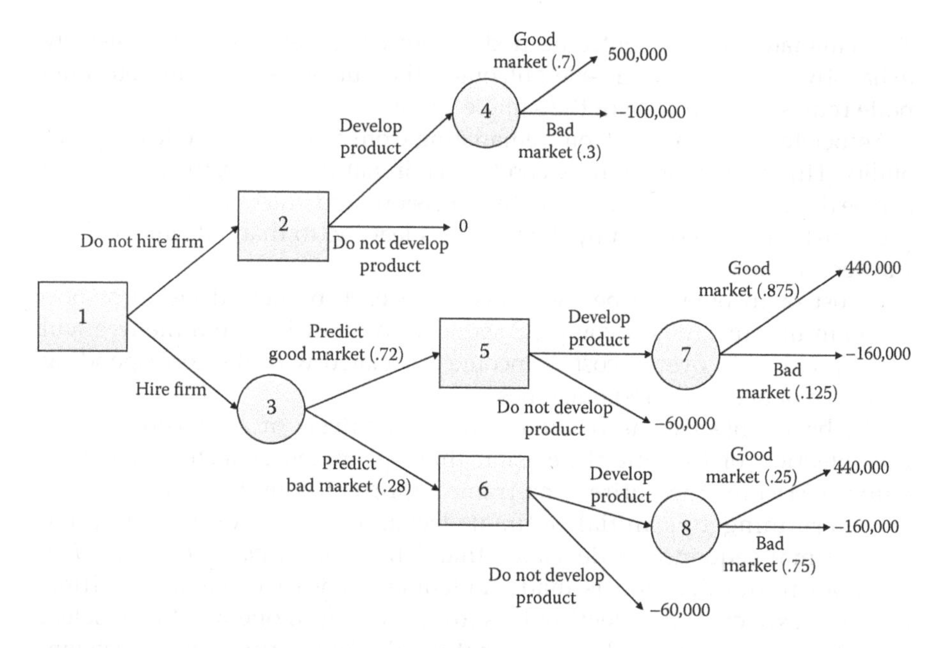

FIGURE 7.1
Decision tree for the example problem.

The decision tree for this simple, hypothetical, example is shown in Figure 7.1. Note that the decision nodes are denoted by squares while the outcome nodes are denoted by circles.

Note that in the decision tree of Figure 7.1, there is only *one branch* entering each node, although typically multiple branches will exit a node. Note also that outcome nodes preceding a decision node represent uncertainties that will be determined immediately before the decision is made.

Decision trees are somewhat arbitrary for any particular situation since typically a decision situation may be represented by one or multiple decisions. In the simple hypothetical example, one might decide to develop a prototype product prior to developing the product for sale; but to keep our example simple, we are not considering this prototype product decision. Often, just the process of developing the decision tree is useful—that is, one can gain value without actually analyzing the tree.

The numbers following the outcomes are probabilities—for example, the probability that the survey will predict a good market is given as .72. These probabilities are either *marginal probabilities* or *conditional probabilities*—for example, the outcome branch "good market," following Node 7, has a probability of *.875, conditioned* on the event that the market survey *predicts a good market*. This is denoted as P(Good Market|Market Survey Predicts Good Market) = .875. In decision analysis terminology, the market survey is called an *imperfect predictor* (or *imperfect information*). The computation of the conditional probabilities in the tree often requires the application of Bayes' theorem.

The numbers at the end of each "path" of branches are the values (e.g., costs or incomes) associated with all of the decisions and outcomes associated with that path/sequence of branches; for example, the value associated with the path of 1-3-5-7-good market is the income associated with developing the product for a good market minus the cost associated with the market survey = \$500,000 − \$60,000 = *\$440,000*. Typically, with decision tree software, the user will specify a cost or income associated with each branch of a tree; the software will then compute a net value for any sequence of branches in the tree. For example, the user would specify the cost of the survey as \$60,000 and the profit associated with having a "good market" as \$500,000, and then the software would compute the \$440,000 value associated with the sequence of 1-3-5-7-good market.

Remember that evaluation of a decision tree is used to determine the *initial decision* to make. The basic idea in evaluation of a decision tree is to work from right to left (i.e., work backward in time) with the decision tree, replacing each outcome node by its "certainty equivalent" (CE), and then at each decision node, choose the decision that optimizes (in this case, maximizes) the CE, thereby "blocking off" the dominated decisions. In many cases, the EMV is treated as the CE.

Starting at Node 4 of the decision tree of Figure 7.1, we compute the EMV for this node by computing the expected value associated with the probability distribution for that node:

$$\text{EMV (Node 4)} = .7(500{,}000) + .3(-100{,}000) = 320{,}000.$$

Node 4 and the branches that emanate from it can now be replaced with the EMV for that node, as shown in Figure 7.2.

Now, at Node 2, the EMV of *develop product* is 320,000, while the EMV of *do not develop product* is 0. Hence, the optimal decision at Node 2 is *develop product* with an EMV of 320,000.

As with Node 4, the EMVs for Nodes 7 and 8 are given by computing the respective expected values associated with the probability distributions for these nodes:

$$\text{EMV (Node 7)} = .875(440{,}000) + .125(-160{,}000) = 365{,}000,$$

$$\text{EMV (Node 8)} = .25(440{,}000) + .75(-160{,}000) = -10{,}000.$$

Now, Nodes 7 and 8, along with their emanating branches, can be replaced with their EMVs, as shown in Figure 7.3.

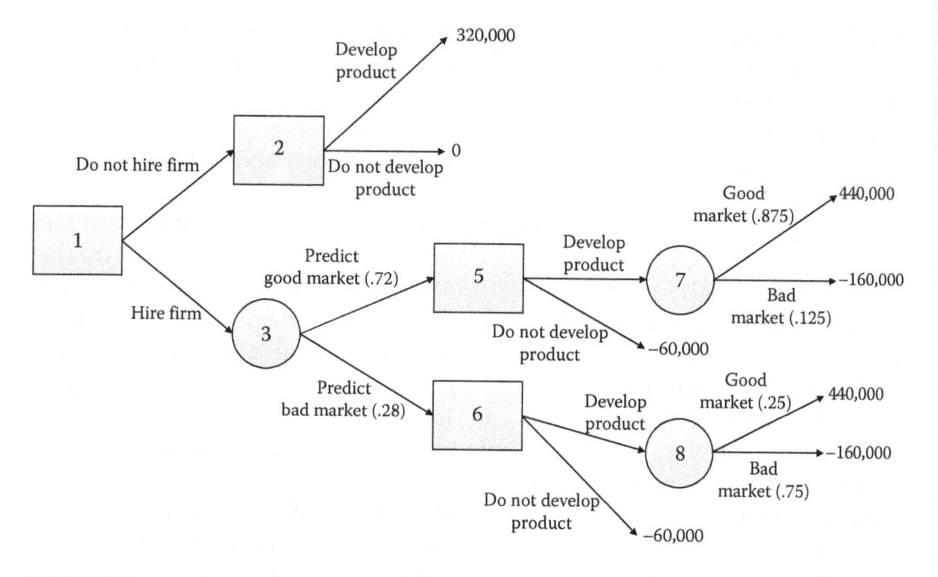

FIGURE 7.2
Decision tree for the example problem, after computation of EMV for Node 4.

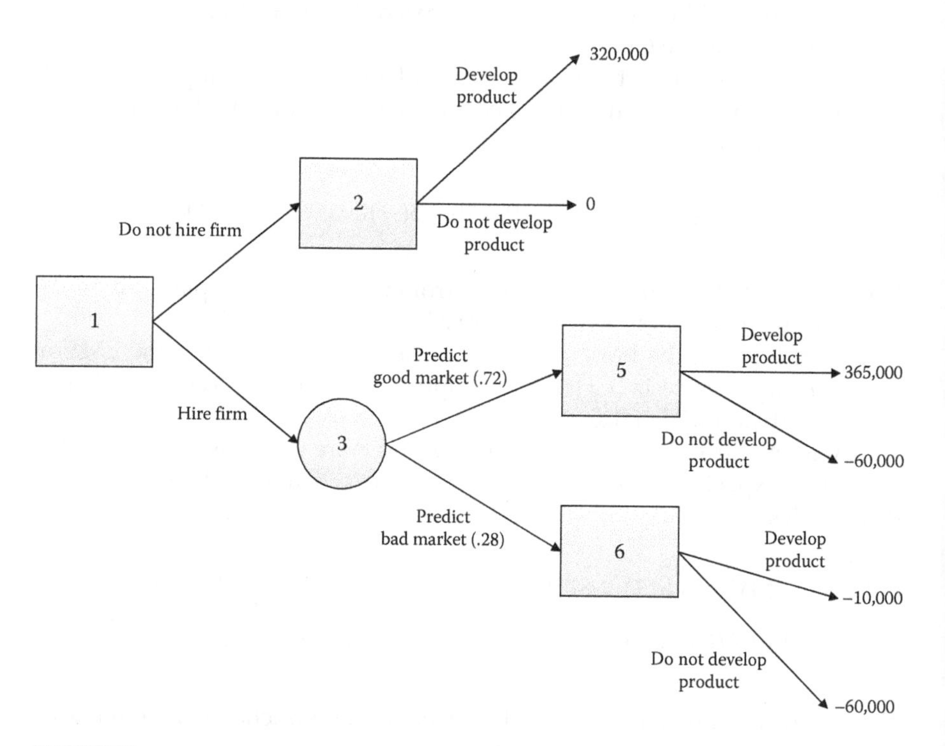

FIGURE 7.3
Decision tree for the example problem, after computation of EMVs for Nodes 4, 7, and 8.

At Decision Node 2 in Figure 7.3, the EMV of develop product is 320,000, while the EMV of do not develop product is 0. Hence, the optimal decision at Node 2 is develop product with an EMV of 320,000.

Similarly, at Node 5, the EMV of develop product is 365,000, while the EMV of do not develop product is –60,000. Hence, the optimal decision at Node 5 is develop product with an EMV of 365,000. At Node 6, the EMV of develop product is –10,000, while the EMV of do not develop product is –60,000. Hence, the optimal decision at Node 6 is develop product with an EMV of –10,000.

Now, the EMV for Outcome Node 3 is given as follows:

$$\text{EMV (Node 3)} = .72(365{,}000) + .28(-10{,}000) = 260{,}000.$$

Finally, at Decision Node 1, the EMV of do not hire firm is 320,000 while the EMV of hire firm is 260,000. Hence, the optimal decision at Node 1 is do not hire firm with an EMV of 320,000.

In summary, the main output of the analysis then is the initial optimal decision that is

Do not hire the firm (which will result in an EMV of $320,000 if optimal decisions are made following this decision).

7.2.3 Calculation of Probabilities in Decision Trees Using Bayes' Theorem

Typically, one needs to derive several of the probabilities associated with the outcomes in a decision tree from probabilities associated with events. The events associated with the decision tree in the simple example could be denoted as follows:

GM, good market; BM, bad market; PGM, predict a good market; PBM, predict a bad market

Suppose that initial probabilities given for the simple example were

$$P(GM) = .7,\ P(BM) = .3,\ P(PGM|GM) = .9,\ \text{and } P(PBM|BM) = .7.$$

These input probabilities could have been obtained in a subjective fashion from experts associated with the situation. From these initial probabilities, the following probabilities can be easily derived:

$$P(PBM \mid GM) = 1 - P(PGM \mid GM) = 1 - .9 = .1$$

and

$$P(PGM \mid BM) = 1 - P(PBM \mid BM) = 1 - .7 = .3.$$

Now, we need to derive P(GM|PGM) and P(BM|PBM) using Bayes' theorem:

$$
\begin{aligned}
P(GM|PGM) &= P(GM, PGM)/(P(GM, PGM) + P(BM, PGM)) \\
&= P(PGM, GM)/(P(PGM, GM) + P(PGM, BM)) \\
&= P(PGM|GM)P(GM)/(P(PGM|GM)P(GM) \\
&\quad + P(PGM|BM)P(BM)) \\
&= .9(.7)/(.9(.7) + .3(.3)) = .63/(.63 + .09) = .63/.72 = \mathit{.875}.
\end{aligned}
$$

From this, one can compute that

$$P(BM|PGM) = 1 - P(GM|PGM) = 1 - .875 = \mathit{.125}.$$

Similarly, one can derive using Bayes' theorem, P(BM|PBM) = *.75*, and therefore that

$$P(GM|PBM) = 1 - P(BM|PBM) = 1 - .75 = \mathit{.25}.$$

Finally, one can also obtain the marginal probabilities, P(PGM) and P(PBM), as follows:

$$
\begin{aligned}
P(PGM) &= P(GM, PGM)/P(GM|PGM) \\
&= P(PGM, GM)/P(GM|PGM) \\
&= P(PGM|GM)P(GM)/P(GM|PGM) \\
&= .9(.7)/.875 = .63/.875 \\
&= \mathit{.72}
\end{aligned}
$$

and

$$P(PBM) = 1 - P(PGM) = 1 - .72 = \mathit{.28}.$$

7.2.4 Quality and Value of a Predictor in a Decision Tree

As mentioned earlier, the market survey in our simple example is, in the general terminology for a decision tree, called a *predictor*. One almost always has the opportunity to gather information to improve a prediction associated with a decision tree. In addition to a market survey, such things as medical tests, samples, weather forecasts, expert assessments, and polls represent predictors for a decision situation. Sometimes, the value associated with obtaining additional information from a predictor must be traded off against the cost associated with obtaining the information and delaying a decision.

Within the context of a decision tree model, the quality of a predictor (sometimes called the quality of information) is indicated by the conditional probabilities associated with a predictor—the closer the conditional probabilities are to 1, the better the quality of a predictor. In our example, the closer that the conditional probabilities, P(PGM|GM) and P(PBM|BM), are to 1, the better the predictor (the market survey) is. In fact, a market survey for which

$$P(PGM|GM) = P(PBM|BM) = 1$$

would be called a *perfect predictor*. A market survey for which either or both of the conditional probabilities are less than 1 would be called an *imperfect predictor*. Of course, in the real world, there is no such thing as a perfect predictor; however, the concept is useful as seen by the following.

In many situations, one would want to know the value of a perfect predictor. One would never pay as much, or more for a predictor than this value for a perfect predictor. This is where the *expected value of a perfect predictor* (EVOPP) comes into play. (Sometimes, this is called the *expected value of perfect information* [EVPI].) The EVOPP is computed as follows:

$$EVOPP = EVWPP - EVWAI,$$

where

EVOPP is the expected value of a perfect predictor
EVWPP is the expected value with a perfect predictor
EVWAI is the expected value with available information

Let's consider the situation for the simple example discussed earlier, where the decisions are only to *develop the product* or *do not develop the product*. The decision tree for this situation (i.e., the situation with available information only) is shown in Figure 7.4.

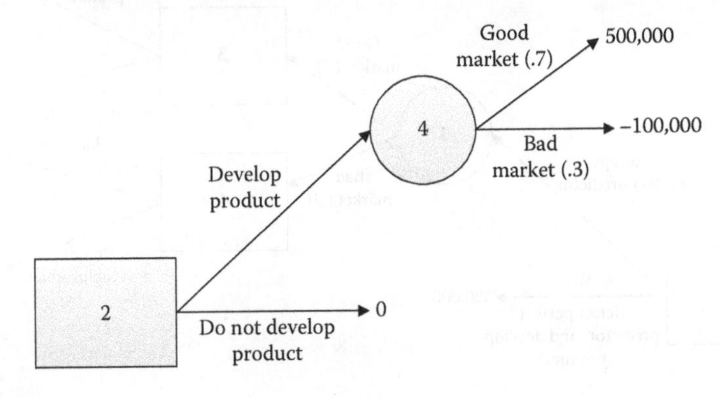

FIGURE 7.4
The decision tree for the example with available information.

For this decision tree, the optimal decision is *Develop Product*, with an *EMV of $320,000* (since the EMV of "Develop Product" = $.7 * 500{,}000 + .3 * (-100{,}000) = \$320{,}000$). Therefore, we say that

$$\text{Expected value with available information} = \$320{,}000$$

or

$$\text{EVWAI} = \$320{,}000.$$

Now, suppose that we have a perfect predictor available (e.g., a market survey that will indicate with certainty whether the market will be good or bad). In this situation, the decision tree will appear as in Figure 7.5.

Note that in the decision tree of Figure 7.5, the probabilities associated with a good market and a bad market, respectively, are the same as earlier; but this time, because *we have a perfect predictor*, we assume that we know this *prior to* making the decision of whether or not to develop the product.

In solving the decision tree of Figure 7.5, we have that at Decision Node 3, the optimal decision is *develop product*, with an EMV = $500,000, and at Decision Node 4, the optimal decision is *do not develop product*, with an EMV = 0. The EMV at Outcome Node 2 is .7(500,000) + .3(0) = $350,000.

At Node 1, the EMV of accept perfect predictor is $350,000, while the EMV of reject perfect predictor and develop product is $320,000. Hence, the optimal decision at Node 1 is accept perfect predictor with an EMV of $350,000. Hence, the EVWPP is $350,000, and

$$\text{EVOPP} = \text{EVWPP} - \text{EVWAI} = 350{,}000 - 320{,}000 = \$30{,}000.$$

Note that EVOPP is an *upper bound* on the amount that you would pay for the use of a predictor. Of course, the better the predictor, the more you would pay for it, *but you would never pay more than the EVOPP.*

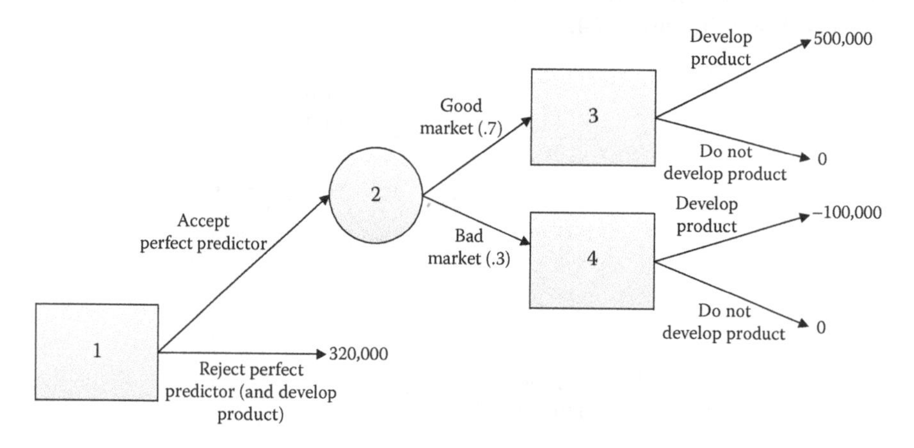

FIGURE 7.5
The decision tree for the example with perfect information.

Of course, all predictors are imperfect. So the question is how much should you pay for an imperfect predictor:

$$\text{EVOIP} = \text{EVWIP} - \text{EVWAI},$$

where
EVOIP is the expected value of an imperfect predictor
EVWIP is the expected value with an imperfect predictor
EVWAI is the expected value with available information

We saw earlier how to compute the value for EVWAI. EVWIP would also be evaluated with a decision tree, provided that the appropriate probabilities are known.

Let's consider the situation presented earlier with the simple hypothetical example and suppose that we consider the firm's survey as an imperfect predictor. In other words, we want to determine the *EVOIP*, where the predictor is the firm's survey. Note that in the calculation, *we do not consider the cost of the firm's survey,* since we just want to determine its value. Then this information could be used in negotiations with the firm.

Consider the decision tree shown in Figure 7.6. It is the same tree as shown in Figure 7.1 for the simple hypothetical example, except now, we do not consider the cost for the imperfect predictor (i.e., we just let the cost be 0).

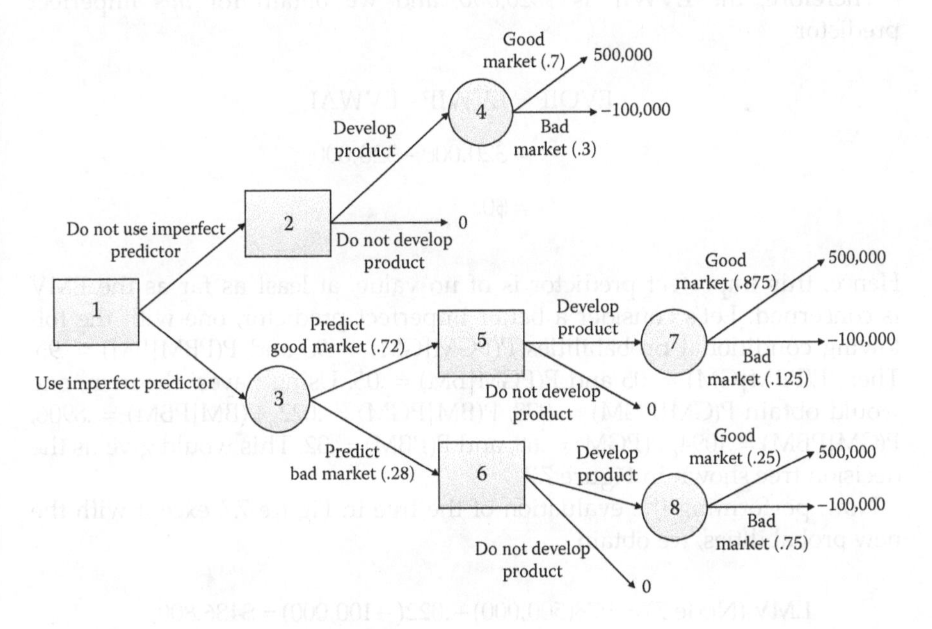

FIGURE 7.6
The decision tree for the example with an imperfect predictor.

In addition, we will replace the phrases "hire firm" and "do not hire firm" with the phrases "use imperfect predictor" and "do not use imperfect predictor" in order to consider a more general situation. Note also that the \$60,000 cost for "hiring the firm" has been reset to 0.

Now, performing the evaluation of this tree, we obtain

$$\text{EMV (Node 7)} = .875(500{,}000) + .125(-100{,}000) = \$425{,}000,$$

$$\text{EMV (Node 8)} = .25(500{,}000) + .75(-100{,}000) = \$50{,}000.$$

At Decision Node 5, we compare the decision of "develop product" with an EMV = 425,000 to the decision of "do not develop product" with an EMV of 0. So the optimal decision at Node 5 is develop product, with an EMV of 425,000. At Decision Node 6, we compare the decision of "develop product" with an EMV = 50,000 to the decision of "do not develop product" with an EMV of 0. So the optimal decision at Node 6 is develop product, with an EMV of 50,000.

Performing the appropriate evaluation for Node 3, we obtain

$$\text{EMV (Node 3)} = .72(425{,}000) + .28(50{,}000) = \$320{,}000.$$

So, at Node 1, either decision is optimal, with an EMV of \$320,000.

Therefore, the EVWIP is *\$320,000,* and we obtain for *this* imperfect predictor:

$$\begin{aligned}\text{EVOIP} &= \text{EVWIP} - \text{EVWAI}\\ &= 320{,}000 - 320{,}000\\ &= \$0.\end{aligned}$$

Hence, this imperfect predictor is of no value, at least as far as the EMV is concerned. Let's consider a better imperfect predictor, one with the following conditional probabilities P(PGM|GM) = .95 and P(PBM|BM) = .95. Then, P(PBM|GM) = .05 and P(PGM|BM) = .05. Using Bayes' theorem, one would obtain P(GM|PGM) = .978, P(BM|PGM) = .022, P(BM|PBM) = .8906, P(GM|PBM) = .1094, P(PGM) = .68, and P(PBM) = .32. This would give us the decision tree shown in Figure 7.7.

Now, performing the evaluation of the tree in Figure 7.7, except with the new probabilities, we obtain

$$\text{EMV (Node 7)} = .978(500{,}000) + .022(-100{,}000) = \$486{,}800,$$

$$\text{EMV (Node 8)} = .1094(500{,}000) + .8906(-100{,}000) = -\$34{,}360.$$

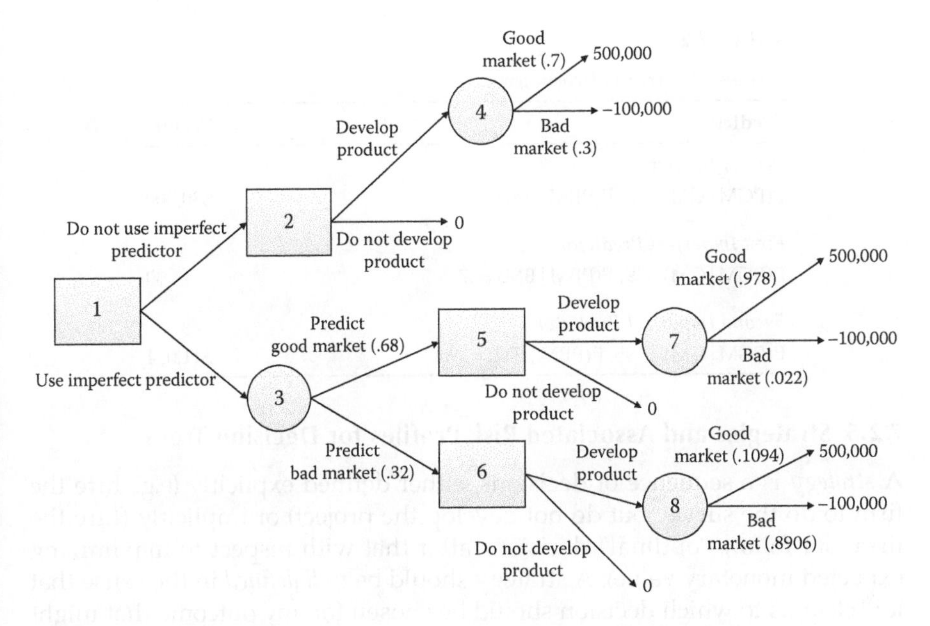

FIGURE 7.7
The decision tree for the example with the second imperfect predictor.

At Node 5, we compare the decision of "develop product" with an EMV = 486,800 to the decision of "do not develop product" with an EMV of 0. So the optimal decision at node 5 is *"develop product," with an EMV of $486,800.*

At Node 6, we compare the decision of "develop product" with an EMV = –34,360 to the decision of "do not develop product" with an EMV of 0. So, the optimal decision at Node 6 is develop product, with an EMV of 0.

Now, EMV (Node 3) = .68(486,800) + .32(0) = 331,024.

So, at Node 1, the optimal decision is hire firm, with an EMV of *$331,024.*

Therefore, the EVWIP = is *$331,024*, and

$$\begin{aligned} \text{EVOIP} &= \text{EVWIP} - \text{EVWAI} \\ &= 331{,}024 - 320{,}000 \\ &= \$11{,}024. \end{aligned}$$

Note that this is a strictly positive number, but still less than $30,000 that is the value of a perfect predictor.

A summary of the values associated for the various predictors for this example is given in Table 7.2. One thing to keep in mind with these various values is that if the cost for a particular predictor is less than its value, then it is worthwhile to use that predictor. For example, if the cost for the second imperfect predictor were $5,000, then it would be worthwhile to use that predictor since $5,000 < $11,024.

TABLE 7.2
Values of Various Predictors

Predictor	Value
Perfect Predictor	
P(PGM | GM) = 1., P(PBM | BM) = 1.	$30,000
First Imperfect Predictor	
P(PGM | GM) = .9, P(PBM | BM) = .7	$0
Second Imperfect Predictor	
P(PGM | GM) = .95, P(PBM | BM) = .95	$11,024

7.2.5 Strategies and Associated Risk Profiles for Decision Trees

A *strategy* is a sequence of decisions, either defined explicitly (e.g., hire the firm to do the survey but do not develop the project) or implicitly (hire the firm, but follow "optimal" decisions after that with respect to maximizing expected monetary value). A strategy should be *well defined* in the sense that it is clear as to which decision should be chosen for any outcome that might be encountered through the use of the strategy. So, in order to have a well-defined strategy, decisions must sometimes be stated in a conditional fashion, that is, dependent upon which outcome occurs.

For example, the strategy given by

Hire firm–develop product (if the survey predicts a good market),

is not a well-defined strategy since it does not indicate what should be done if the survey predicts a bad market. However, the strategy given by

Hire firm–develop product (if the survey predicts a good market),
Do not develop product (if the survey predicts a bad market), is a well-defined strategy since it prescribes what should be done under any outcome that occurs for any decision defined within the strategy.

A *risk profile* associated with a strategy is the probability distribution over the payoff associated with that strategy. A risk profile for a strategy gives you *more information* than the expected monetary value, since it gives the decision maker the various possible payoffs and associated probabilities for those payoffs for the strategy.

The risk profile associated with the strategy *do not hire firm–develop product* for the hypothetical example is shown in Table 7.3.

Note that the strategy *do not hire firm–develop product* is equivalent to the strategy *do not hire firm–follow optimal decisions after with respect to maximization of EMV*, since, as shown earlier, the optimal decision at Node 2 for Figure 7.1 is *develop product*.

In developing the risk profile for a particular strategy, we sum the probabilities associated with the various ways that a payoff can be achieved in

TABLE 7.3

Risk Profile for the Strategy *Do Not Hire Firm–Develop Product* for the Hypothetical Example

Net Income (Payoff), $	Probability
500,000	0.7
–100,000	0.3

order to obtain the probability of that payoff. In addition, we choose a specific branch for each decision node associated with the strategy, and then multiply the probabilities associated with sequential outcome nodes in the strategy.

For example, let's compute the risk profile for the strategy of *hire firm–follow optimal decisions after with respect to maximization of EMV.* Consider the probability of achieving a net income of $440,000 with this strategy. This $440,000 could be achieved through either of two sequences of outcomes and decisions from Node 3 in Figure 7.1:

1. *Predict good market–develop product–good market*
2. *Predict bad market–develop product–good market*

Note that from both Node 5 and Node 6 in Figure 7.1, the decision chosen was *develop product,* since this was the decision in each case that was dictated by the strategy. Now the probability of achieving the payoff of $440,000 through the first sequence is given by the probability that the firm's survey predicts a good market, multiplied by the probability that a good market results (given that the firm predicts a good market):

$$\text{P}(\textit{Predict Good Market} - \textit{Develop Product} - \textit{Good Market})$$

$$= \text{P}(\textit{Predict Good Market}) * \text{P}(\textit{Good Market given Predict Good Market})$$

$$= .72 * .875 = .63.$$

Similarly,

$$\text{P}(\textit{Predict Bad Market} - \textit{Develop Product} - \textit{Good Market})$$

$$= \text{P}(\textit{Predict Bad Market}) * \text{P}(\textit{Good Market given Predict Bad Market})$$

$$= .28 * .25 = .07.$$

Hence, for this strategy,

$$\text{P}(\$440{,}000) = \text{P}(\textit{Predict Good Market} - \textit{Develop Product} - \textit{Good Market})$$

$$+ \text{P}(\textit{Predict Bad Market} - \textit{Develop Product} - \textit{Good Market})$$

$$= .63 + .07 = .70.$$

Similarly, a payoff of −$160,000 could be achieved through either of two sequences of outcomes and decisions from Node 3:

1. *Predict good market–develop product–bad market*
2. *predict bad market–develop product–bad market*

Therefore,

$$\begin{aligned} P(-\$160{,}000) &= P(\textit{Predict Good Market} - \textit{Develop Product} - \textit{Bad Market}) \\ &\quad + P(\textit{Predict Bad Market} - \textit{Develop Product} - \textit{Bad Market}) \\ &= .72 * .125 + .28 * .75 = .09 + .21 = .30. \end{aligned}$$

Note that the probability of receiving a net income of −$60,000 from this strategy is 0, since the strategy does not dictate the decision of *do not develop product* at either of Node 5 or 6, and the only way that this outcome of −$60,000 can be achieved from Node 5 or 6 is by taking the decision of *do not develop product*.

In summary, the risk profile associated with the strategy for *hire firm–follow optimal decisions after with respect to maximization of EMV* is shown in Table 7.4.

In all, there are six possible strategies associated with the decision tree of Figure 7.1. These strategies (denoted as strategies S1 through S6), associated risk profiles, and EMVs are shown in Table 7.5. Note that in Table 7.5, the risk profile for each strategy is shown directly below that strategy. Also, note that even though there are only six possible strategies for this particular hypothetical situation, for decision trees associated with many actual situations, the number of potential strategies may well be too many to enumerate. For example, if for this situation we modeled four possible outcomes for the market (very good market, good market, bad market, and very bad market) rather than two, and the survey could reflect a prediction of any of these four, then there would be 10 possible strategies rather than only six. This proliferation of strategies is the reason why in many situations, the strategies analyzed in detail are typically specified in terms of just the initial decisions, followed by "the decisions that maximize EMV following the initial decisions."

TABLE 7.4

Risk Profile for the Strategy *Hire Firm–Follow Optimal Decisions after with Respect to Maximization of EMV* for the Hypothetical Example

Net Income (Payoff), $	Probability
440,000	0.7
−160,000	0.3

TABLE 7.5

All Strategies and Associated Risk Profiles for the Decision Tree of Figure 7.1

S1: Do not hire firm–develop product			
Net income (payoff)	500,000	−100,000	
Probability	0.7	0.3	
S2: Do not hire firm–do not develop product			
Net income (payoff)	0		
Probability	1		
S3: Hire firm–develop product (if predict good market) and develop product (if predict bad market)			
Net income (payoff)	440,000	−160,000	
Probability	0.7	0.3	
S4: Hire firm–develop product (if predict good market) and do not develop product (if predict bad market)			
Net income (payoff)	440,000	−60,000	−160,000
Probability	0.63	0.28	0.09
S5: Hire firm–do not develop product (if predict good market) and develop product (if predict bad market)			
Net income (payoff)	440,000	−60,000	−160,000
Probability	0.07	0.72	0.21
S6: Hire firm–do not develop product (if predict good market) and do not develop product (if predict bad market)			
Net income (payoff)	−60,000		
Probability	1		

Obviously, just on the face of the situation, some of the strategies would never be chosen; for example, you would not hire the firm if you were not going to develop the product no matter what the survey predicted. But with respect to choosing a particular strategy, there are many possible criteria that a decision maker might employ, depending on the amount of risk he or she is willing to take and the value he or she places on various payoffs. (This is where utility functions can be very useful.) However, if one did not want to use a utility function and wanted to use the criterion of "maximize the probability of receiving a payoff of at least $440,000," then there would be two "best strategies":

1. *Do not hire firm–develop product*
2. *Hire firm–develop product (if predict good market) and develop product (if predict bad market)*

(This would obviously not be a good criterion to employ unless any income over $440,000 is worthless to the decision maker.)

7.2.6 Cumulative Risk Profiles and Dominance

The *cumulative risk profile* for a strategy is just the cumulative probability distribution over the payoff associated with that strategy, that is, $F(x) = PR(X \leq x)$, where X is the payoff for the relevant strategy. Often, it is helpful to view the cumulative risk profile for a strategy as the graph associated with the cumulative probability distribution of the strategy.

Consider the cumulative risk profile for strategy S4 of Table 7.5: *hire firm–develop product (if predict good market) and do not develop product (if predict bad market).*

The cumulative risk profile associated with this strategy is shown in Table 7.6.

The graph associated with the cumulative risk profile of Table 7.6 is shown in Figure 7.8.

The cumulative risk profiles for each of the strategies shown in Table 7.5 are shown in Table 7.7.

TABLE 7.6

Cumulative Risk Profile for Strategy S4: *Hire Firm–Develop Product (If Predict Good Market) and Do Not Develop Product (If Predict Bad Market)*

Net Income (Payoff), $	Cumulative Probability
–160,000	0.09
–60,000	0.37
440,000	1

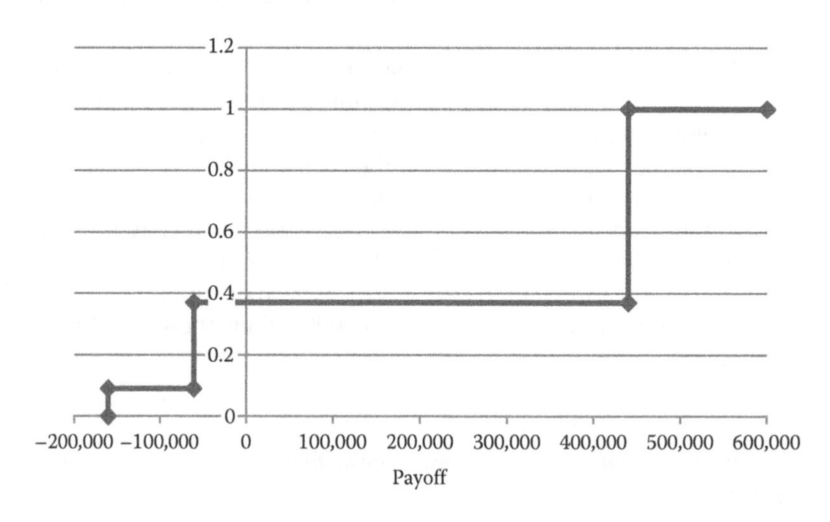

FIGURE 7.8
Graph associated with cumulative risk profile of strategy S4: *Hire firm–develop product (if predict good market) and do not develop product (if predict bad market).*

TABLE 7.7

Cumulative Risk Profiles for Each of the Strategies of Table 7.5

S1				
x	$x < -100{,}000$	$-100{,}000 \le x < 500{,}000$	$x \ge 500{,}000$	
F(x)	0	0.3	1	
S2				
x	$x < 0$	$x \ge 0$		
F(x)	0	1		
S3				
x	$x < -160{,}000$	$-160{,}000 \le x < 440{,}000$	$x \ge 440{,}000$	
F(x)	0	0.3	1	
S4				
x	$x < -160{,}000$	$-160{,}000 \le x < -60{,}000$	$-60{,}000 \le x < 440{,}000$	$x \ge 440{,}000$
F(x)	0	0.09	0.37	1
S5				
x	$x < -160{,}000$	$-160{,}000 \le x < -60{,}000$	$-60{,}000 \le x < 440{,}000$	$x \ge 440{,}000$
F(x)	0	0.21	0.93	1
S6				
x	$x < -60{,}000$	$x \ge -60{,}000$		
F(x)	0	1		

Strategy A *stochastically dominates* strategy B if the cumulative risk profile graph for B lies on or above the cumulative risk profile graph for A, that is, if $F_B(x) \ge F_A(x)$ for any value of x. Moreover, if as one is moving from left to right on the x-axis, the cumulative risk profile graph for B reaches a value of 1 while the cumulative risk profile graph for A is still at a value of 0, then strategy A *deterministically dominates* strategy B; that is, there is some value of x, say x′ such that $F_B(x') = 1$ and $F_A(x') = 0$. No rational decision maker will choose a strategy B over a strategy A if A stochastically dominates B. Note that if a strategy deterministically dominates another strategy, then, by necessity, the first strategy will stochastically dominate the second strategy; however, the reverse does not hold—that is, just because one strategy stochastically dominates another, this does not mean that the first strategy deterministically dominates the second.

Note also that just because A stochastically dominates B, this does not necessarily mean that the probability of A receiving a better payoff than B is 1, but this is the case if A deterministically dominates B. A summary of the effects associated with both stochastic and deterministic dominance is shown in Table 7.8.

Consider strategies S2 and S6 from Table 7.7. The cumulative risk profile graphs for these two strategies are shown in Figure 7.9. It is clear from

TABLE 7.8
Stochastic Dominance versus Deterministic Dominance

Type of Dominance	Definition	Effects	Comments
Stochastic dominance (A stochastically dominates B)	$F_B(x) \geq F_A(x)$	B may achieve a better payoff than A, but for any particular payoff, the probability of achieving less than or equal to that payoff is larger for B than for A.	The probability of a better payoff from A than from B is not necessarily 1.
Deterministic dominance (A deterministically dominates B)	There exists at least one value of x such that $F_B(x) = 1$ and $F_A(x) = 0$	The worst possible payoff for A is better than the best possible payoff for B.	The probability of a better payoff from A than from B is 1.

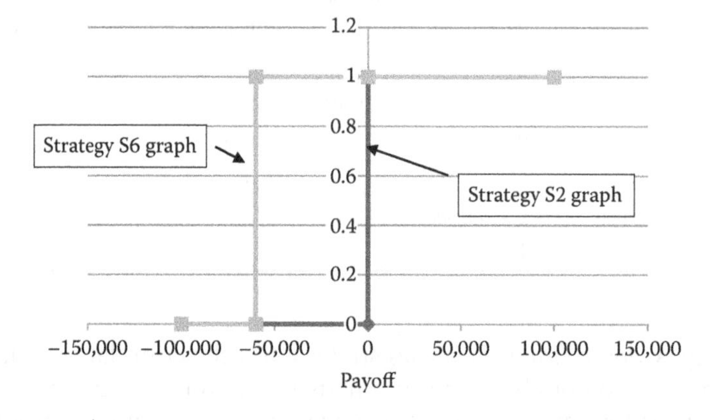

FIGURE 7.9
Cumulative risk profile graphs for strategies S2 and S6.

viewing these graphs that strategy S2 deterministically dominates strategy S6.

Conclusions that can be made from viewing Table 7.7 include the following:

S1 stochastically dominates S3
S4 stochastically dominates S5
S2 deterministically dominates S6

The key point is that any strategy that is dominated by another (either stochastically or deterministically) can be automatically eliminated from any further analysis. One must keep in mind, however, that if one strategy dominates a second strategy stochastically (but not deterministically), it is still possible to achieve a better payoff from the second strategy; it is not possible however to achieve a better payoff from a strategy that is deterministically dominated.

7.3 Influence Diagrams

7.3.1 Basic Concepts

An influence diagram is a model of a dynamic decision situation and can be thought of as being analogous to a decision tree; therefore, an influence diagram can be thought of as an evaluation or criterion model and a communication tool. In fact, because of its "higher level" perspective, an influence diagram typically works better as a communication tool (between an analyst and a group of decision makers) than a decision tree.

An influence diagram is a directed network, consisting of nodes and arcs. (Note that we use the term *branch* to indicate an arrow that is used in a decision tree and the term *arc* to indicate an arrow that is used in an influence diagram.) In its most basic form, an influence diagram has three types of nodes:

1. Decision nodes (represented as rectangles)
2. Chance event nodes (represented as ovals)
3. Outcome/consequence nodes (represented as rounded rectangles)

A decision node represents a set of mutually exclusive, collectively exhaustive alternatives just as with a decision tree. A chance event node typically represents a random variable—some quantity about which there is uncertainty such as a survey result, a medical test, or a sampling about something; in this sense, a chance event node is analogous to an outcome node in a decision tree. An outcome/consequence node is deterministic in nature and can represent a constant value or a function of values associated with other nodes in the diagram.

There are two types of arcs associated with influence diagrams:

1. Relevance arcs (arcs that enter a Chance Event Node or an Outcome/Consequence Node)
2. Sequence arcs (arcs that enter a decision node)

A sequence arc entering a decision node from another node indicates that the information associated with the previous node is known prior to the decision that is to be made for the decision node. In the case of the preceding node to a decision node also being a decision node, the prior decision is made first; in the case of the preceding node being a chance event node, the outcome of the chance event node is known prior to the decision having to be made.

Where most people seem to have trouble with influence diagrams is with the concept of relevance arcs joining two chance event nodes. A relevance arc entering a chance event node from another chance event node *does not mean* that the first chance event's outcome occurs prior to the second chance

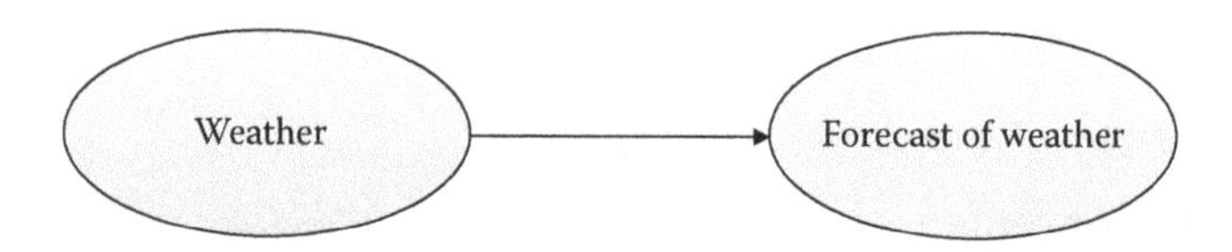

FIGURE 7.10
Portion of an influence diagram for two chance event nodes: weather and forecast of weather.

event's outcome, but only that the outcome of the first chance event has *relevance* for the outcome of the second chance event. For example, consider two chance events to be modeled as part of an influence diagram: *weather* (simply modeled with the possible outcomes of *sunshine* and *rain*) and *forecast of weather* (simply modeled with the possible outcomes of *forecast sunshine* and *forecast rain*). The portion of the influence diagram corresponding to these two chance event nodes would appear as in Figure 7.10.

The input to the portion of the influence diagram corresponding to Figure 7.10 would be the conditional probabilities of

1. P(weather forecast of rain|weather of rain)
2. P(weather forecast of sun|weather of sun)

The complementary probabilities of P (weather forecast of sun|weather of rain) and P (weather forecast of rain|weather of sun) would just be computed as 1 –P (weather forecast of rain|weather of rain) and 1 –P (weather forecast of sun|weather of sun), respectively. Bayes theorem would then be used to compute the conditional probabilities of P (weather of rain|weather forecast of rain) and P (weather of sun|weather forecast of sun), as well as their complementary probabilities and the unconditional probabilities of P (weather forecast of sun) and P (weather forecast of rain).

Consider the hypothetical example problem illustrated with the decision tree of Figure 7.1. The corresponding influence diagram associated with this problem is shown in Figure 7.11.

In the influence diagram of Figure 7.11, there are two *decision nodes*: "hire firm?" and "develop product?" There are two *chance event nodes*: "market" and "survey prediction." Finally, there is one *outcome/consequence node*: "payoff."

The arc leading from "survey prediction" to "develop product?" in Figure 7.11 is a sequence arc since the survey prediction will be known (given that the firm is hired) prior to the decision of whether or not to develop the product. The arc from "market" to "survey prediction" is a relevance arc. The information needed for input to the influence diagram related to the relationship between "market" and "survey prediction" would be the conditional probabilities of P(PGM|GM), P(PBM|GM), P(PGM|BM), and P(PBM|BM), where, as discussed earlier, PGM, PBM, GM, and BM represent the events of "predict good market," "predict bad market," "good market," and "bad market," respectively. In addition, the user (of the software

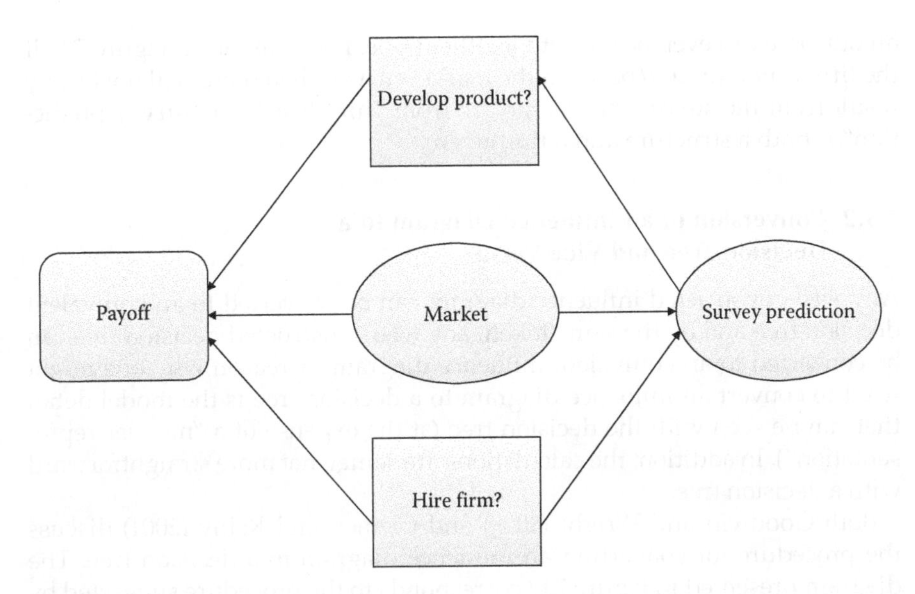

FIGURE 7.11
Influence diagram for the example problem.

package employed to "solve" the influence diagram) would need to input the marginal probabilities associated with the events of "good market" and "bad market": P(GM) and P(BM). The software program would then compute the "reversed probabilities" of P(GM|PGM), P(GM|PBM), P(BM|PGM), and P(BM|PBM), along with the marginal probabilities of P(PGM) and P(PBM) using Bayes' theorem, in order to proceed with the computations required for the influence diagram.

The reader should note that different software packages for constructing and solving decision trees and influence diagrams employ different sets of terminology. For example, Precision Tree, marketed by Palisades Corporation (www.palisade.com), calls the nodes in their influence diagrams chance nodes, decision nodes, calculation nodes, and payoff nodes. Each influence diagram in Precision Tree can have only *one* payoff node. In our terminology, we have used an outcome/consequence node to represent either a calculation node or a payoff node.

Also in Precision Tree, the arcs in the influence diagrams can be one or more of three types: timing, value, and structure. A timing type of arc is what we earlier called a precedence arc. Value arcs and structure arcs are in some sense related to relevance arcs. More specifically, a value arc pointing from one node to another just means that the value associated with the first node affects the value of the second. An example would be the arc pointing from the "market" node to the "payoff" node in Figure 7.11; the value of "market" affects the value of "payoff." A structure arc is used to indicate that the decision associated with a decision node can affect whether or not an outcome can even occur with another node. For example, in Figure 7.11 if the firm is not hired to perform the market survey, then there will not be any result from the survey; hence, the arc from "hire firm?" to "survey prediction" is both a structure and a timing arc.

7.3.2 Conversion of an Influence Diagram to a Decision Tree and Vice Versa

Any well-constructed influence diagram can be converted to an equivalent decision tree and by the same token, any well-constructed decision tree can be converted to an equivalent influence diagram. A reason why one might want to convert an influence diagram to a decision tree is the model detail that can be seen with the decision tree (at the expense of a "messier representation"). In addition, the calculations are somewhat more straightforward with a decision tree.

Both Goodwin and Wright (2009) and Clemen and Reilly (2001) discuss the procedure for converting an influence diagram to a decision tree. The diagram presented in Figure 7.12 corresponds to the procedure suggested by Goodwin and Wright (2009, p. 172) for converting an influence diagram to an equivalent decision tree.

Applying the procedure of Figure 7.12 to the influence diagram of Figure 7.11, we note that there are two nodes with no arrows pointing to it: "market" and "hire firm?" From these two, we would choose "hire firm?" since it is a decision node and place it at the beginning (i.e., on the left-hand side) of the decision tree, with the two appropriate decisions of "yes" and "no"; we would also remove the "hire firm?" node from the influence diagram.

The actual process is a little more complicated than what is depicted in Figure 7.12 since one may have to place multiple copies of a node in the decision tree; such would be the case with the decision node "develop product?" in Figure 7.11, which appears three times in the decision tree of Figure 7.1. For a more detailed description of the conversion process, the reader is directed to Clemen and Reilly (2001).

7.4 Sensitivity Analysis for Decision Trees and Influence Diagrams

Sensitivity analysis involves studying the sensitivity of a model's output (e.g., optimal performance measure value or optimal solution/policy) to changes to a model's input. For decision trees and influence diagrams, the sensitivity analysis process is relatively straightforward, as compared to sensitivity analysis for linear programs. Specifically, for decision trees and influence diagrams, one just varies the value(s) for the input parameter(s)

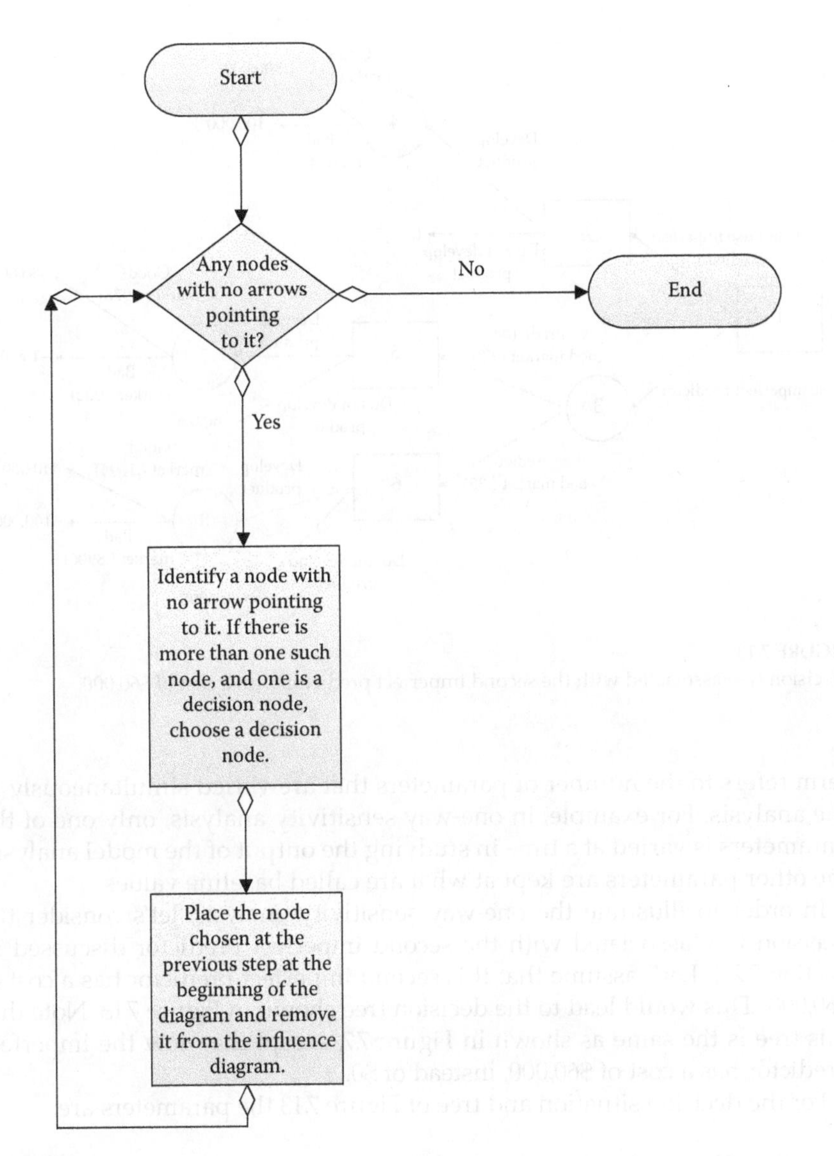

FIGURE 7.12
Flowchart of a procedure for converting an influence diagram to an equivalent decision tree. (Derived from Goodwin, P. and Wright, G., *Decision Analysis for Management Judgment*, 4th ed., John Wiley & Sons, Chichester, UK, 2009, p. 172.)

and determines any changes in the optimal performance measure values and optimal policy.

There are several categories of sensitivity analysis for decision trees and influence diagrams. These categories include one-way sensitivity analysis, two-way sensitivity analysis, and so on. The number in the category

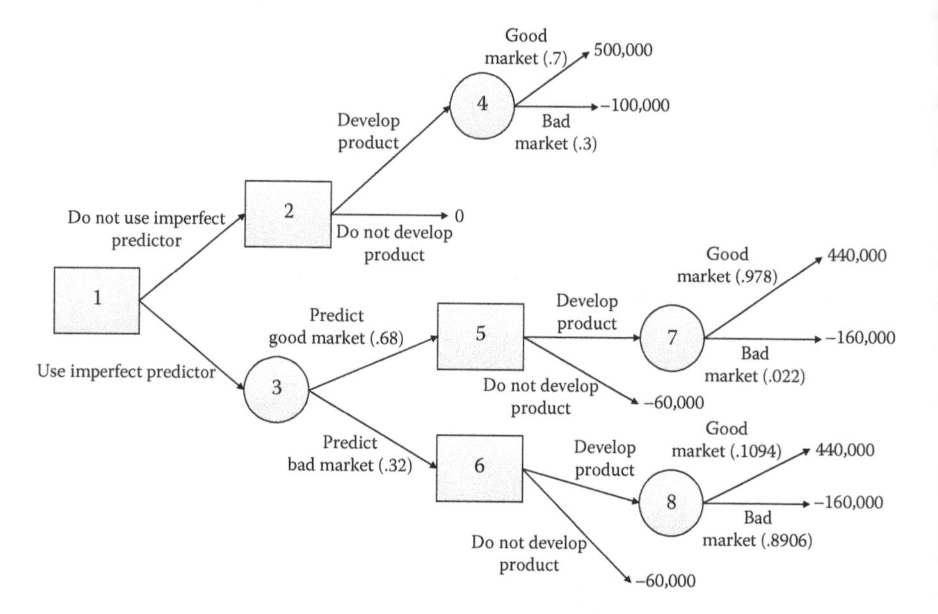

FIGURE 7.13
Decision tree associated with the second imperfect predictor with a cost of $60,000.

term refers to the number of parameters that are varied simultaneously in the analysis. For example, in one-way sensitivity analysis, only one of the parameters is varied at a time in studying the output of the model analysis; the other parameters are kept at what are called baseline values.

In order to illustrate the one-way sensitivity analysis, let's consider the decision tree associated with the second imperfect predictor discussed in Section 7.2.4. Let's assume that this second imperfect predictor has a cost of $60,000. This would lead to the decision tree shown in Figure 7.13. Note that this tree is the same as shown in Figure 7.7, except that now the imperfect predictor has a cost of $60,000, instead of $0.

For the decision situation and tree of Figure 7.13 the parameters are:

1. Payoff associated with the "good market" (baseline value of $500,000).
2. Payoff associated with the "bad market" (baseline value of −$100,000).
3. Cost of the "imperfect predictor" (baseline value of $60,000).
4. Probability associated with a "good market" (P(GM)—baseline value of .7).
5. Probability associated with a "bad market" (P(BM)—baseline value of .3).

6. Conditional probabilities associated with the "predictor," with the associated baseline values of P(PGM|GM) = .95, P(PBM|BM) = .95. Then, P(PBM|GM) = .05, and P(PGM|BM) = .05.

Note that, as discussed earlier, in order to obtain the values for many of the probabilities shown in the decision tree of Figure 7.13, we needed to apply Bayes' theorem to the input conditional probabilities.

Performing a one-way sensitivity analysis for any of the parameters involving one or more of the probabilities listed earlier would be more difficult than such an analysis not involving one of these probabilities. The reason for this is that the values for these probabilities are dependent on each other. For example, changing the probability of a good market, P(GM), also requires simultaneous changes to P(BM), P(PGM), and P(PBM).

Let's perform a one-way sensitivity analysis for each of the first three input parameters: payoff associated with a good market, payoff associated with a bad market, and cost of the imperfect predictor. As input for the sensitivity analysis, we need to provide baseline values, minimum values, and maximum values for each of these input parameters. These values are shown in Table 7.9.

Note that in Table 7.9 we are allowing payoffs associated with good and bad markets, respectively, to vary by 25% from their baseline values; however, we are allowing the cost of the imperfect predictor to vary by 100% from its baseline value of $60,000.

Now, probably the most well-known vehicle for portraying information about a one-way sensitivity analysis is the *tornado diagram*, sometimes called a *tornado graph*. This diagram has two axes: x and y. The x-axis portrays the expected value (or EMV) associated with an optimal sequence of decisions for the tree. The y-axis varies according to the input parameter analyzed; there is one bar for each input parameter, ranging from the worst expected value to the best expected value as the parameter is varied over its range. The tornado diagram associated with the parameters and associated values of Table 7.9 is shown in Figure 7.14.

TABLE 7.9

Baseline, Minimum, and Maximum Values for the Input Parameters for Sensitivity Analysis

Input Parameter	Minimum Value, $	Baseline Value, $	Maximum Value, $
Payoff associated with "good market"	375,000	500,000	625,000
Payoff associated with "bad market"	−125,000	−100,000	−75,000
Cost of imperfect predictor	0	60,000	120,000

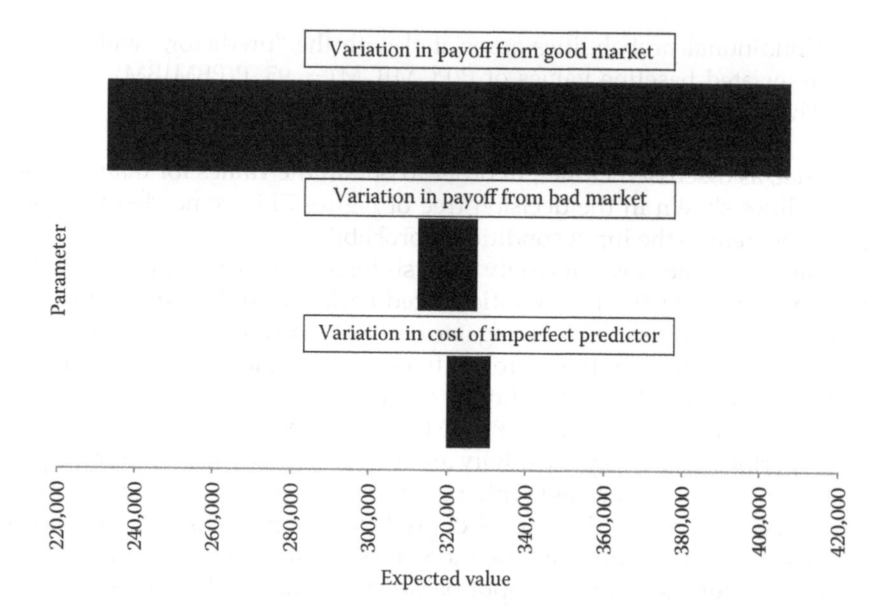

FIGURE 7.14
Tornado diagram for parameters shown in Table 7.9. (Output from Precision Tree.)

TABLE 7.10
Data Associated with the Tornado Diagram of Figure 7.14

Parameter	Minimum Value, $ (Associated Expected Value, $)	Maximum Value, $ (Associated Expected Value, $)
Payoff associated with "good market"	375,500 (232,500)	625,000 (407,500)
Payoff associated with "bad market"	−125,000 (312,000)	−75,000 (327,500)
Cost of imperfect predictor	0 (331,024)	120,000 (320,000)

The tornado diagram shown in Figure 7.14 corresponds to the numbers shown in Table 7.10.

Note that the parameters in the tornado diagram are arranged from top to bottom in order of decreasing effect on the expected value (or EMV)—hence the term *tornado diagram*. In this way, the DM can quickly see which parameter variation has the largest effect on the payoff. But in considering this variation, the DM should remember the upper and lower bounds used for each parameter. For example, in Table 7.10, the payoffs associated with both the good and bad markets were allowed to vary by 25% in each direction from their respective baselines; however, the cost of the imperfect predictor was allowed to vary 100% in each direction from its baseline value of $60,000.

Also, note that the expected (payoff) value varies as one would expect for each parameter. For example, as the payoff associated with a good market or with a bad market increases, the expected value increases; but as the cost of the imperfect predictor increases, the expected value decreases.

Since the tornado diagram contains information about several parameters, it is sometimes easy to forget that the diagram portrays only a *one-way* sensitivity analysis. That is, in looking at any bar on the graph, the DM needs to remember that the values for the parameters not associated with the relevant bar are kept constant at their respective baseline values throughout the variation of the relevant parameter.

The EMV associated with an optimal policy is not necessarily a linear function of a parameter value within the limits established for the tornado diagram. For example, consider the cost for the imperfect predictor; since the baseline optimal policy does not involve the use of the imperfect predictor, changing its value only slightly will not result in a change in the EMV. The relationship between the change in a parameter value and the optimal EMV value is seen in another important tool for one-way sensitivity analysis: the *sensitivity graph*. There is one sensitivity graph for each bar in the tornado diagram. For example, the sensitivity graph associated with the cost of the imperfect predictor is shown in Figure 7.15.

The reader will note that the expected value (the y-axis) in Figure 7.15 decreases in value from $331,024 to $320,000 as the cost for the imperfect predictor increases from 0$ to $11,024; once the cost increases past $11,024, the expected value remains at $320,000. Note from Table 7.2 that the expected value for this imperfect predictor was computed as $11,024.

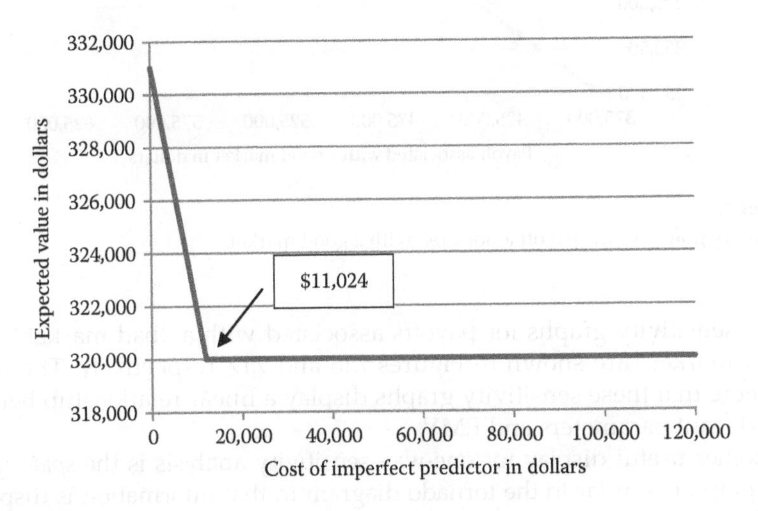

FIGURE 7.15
Sensitivity graph for the cost of an imperfect predictor.

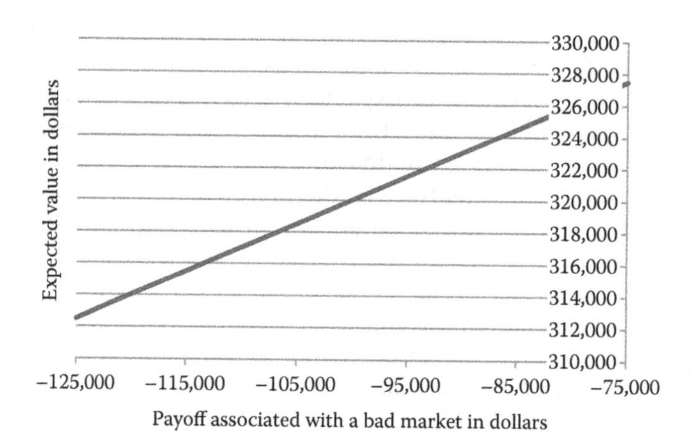

FIGURE 7.16
Sensitivity graph for the payoff associated with a bad market.

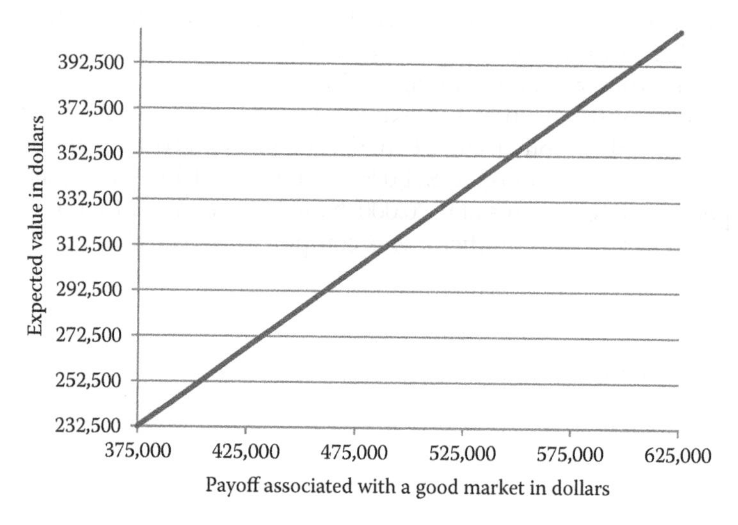

FIGURE 7.17
Sensitivity graph for the payoff associated with a good market.

The sensitivity graphs for payoffs associated with a "bad market" and a "good market" are shown in Figures 7.16 and 7.17, respectively. The reader will note that these sensitivity graphs display a linear relationship between the relevant parameters and EMV.

Another useful display for one-way sensitivity analysis is the *spider graph.* This graph is similar to the tornado diagram in that information is displayed for all of the relevant parameters on one graph. The x-axis of the spider graph displays the percent change from the baseline value for the parameter, while

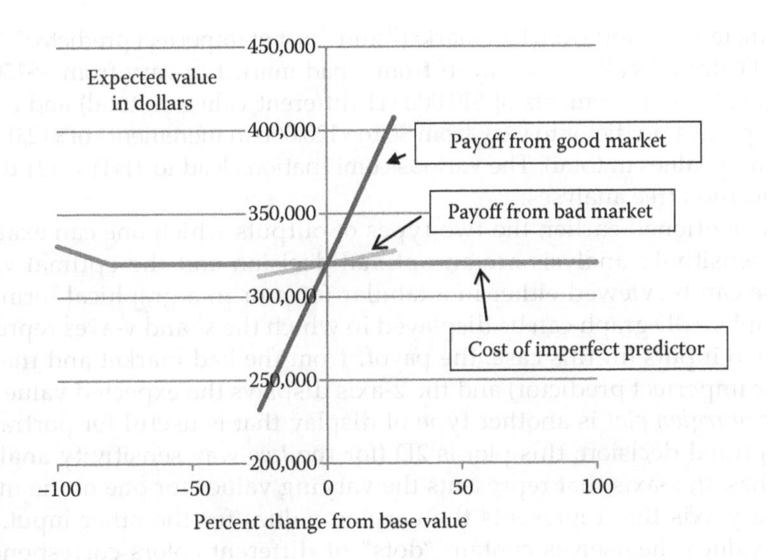

FIGURE 7.18
Spider graph corresponding to the tornado graph of Figure 7.14. (Output from Precision Tree.)

the y-axis gives the expected value as a function of this change. The spider graph corresponding to the tornado graph of Figure 7.14 is shown in Figure 7.18.

The reader will note that the slope of the plot in the spider graph for any parameter corresponds to the width of the corresponding bar in the tornado diagram—the wider the bar, the steeper the slope.

Each of the aforementioned constructs associated with a one-way sensitivity analysis of a decision tree/influence diagram was developed with the Precision Tree software, referred to earlier. The computational approach involved setting the values for all of the relevant parameters to their baseline values and then varying the studied parameter value in a stepwise fashion (e.g., 11 different values equally spaced from its minimum to its maximum value).

In using an Excel-based software package like Precision Tree, the user often has the choice of (1) placing the value for a cost or payoff directly at the node associated with the value, or (2) in a separate cell and then using a cell reference within a formula to refer to the parameter value. In general, it is better to use the second approach since a parameter value is typically employed in several places in a decision tree; for example, the payoff associated with a "good market" is employed at Nodes 4, 7, and 8 in the decision tree of Figure 7.1. This approach of using a cell reference within a formula for different places in the decision tree greatly simplifies the sensitivity analysis process.

A *two-way sensitivity analysis* involves changing two of the parameters simultaneously and determining how an optimal value and optimal solution changes. Let's apply a two-way sensitivity analysis to the input parameters: "payoff from bad market" and "cost of imperfect predictor." More specifically, let's allow the payoff from a bad market to vary from −$150,000 to −$50,000 in increments of $10,000 (11 different values in total) and cost of an imperfect predictor to vary from $0 to $120,000 in increments of $12,000 (11 different values in total). The various combinations lead to 11*11 = 121 different decision tree analyses.

As mentioned earlier, the two types of outputs which one can examine in a sensitivity analysis are an optimal decision and the optimal value. These can be viewed either in a tabular form or in a graphical form. For example, a 3D graph can be displayed in which the x- and y-axes represent the two inputs (in this case, the payoff from the bad market and the cost of the imperfect predictor) and the z-axis displays the expected value. The *strategy region plot* is another type of display that is useful for portraying an optimal decision; this plot is 2D (for the *two*-way sensitivity analysis) and has an x-axis that represents the varying values for one of the inputs and a y-axis that represents the varying values for the other input. The plot values themselves contain "dots" of different colors corresponding to an optimal decision to make at a decision node in the decision tree or influence diagram. A strategy region plot for our example, developed with the Precision Tree software, is shown in Figure 7.19.

The points in the strategy region plot of Figure 7.19 represent the optimal decision to make for hiring of the firm (and therefore use the imperfect predictor) for the various values of payoff from a bad market and cost of the imperfect predictor. The plot indicates that (1) if the payoff from the bad market is −$150,000 (i.e., $150,000 is lost) and the cost of the imperfect predictor is $24,000 or less, then the firm should be hired, and (2) if the payoff from the bad market is −$110,000 or less, and the cost of the imperfect predictor is $12,000 or less, then the firm should be hired.

Sensitivity analysis of decision trees and influence diagrams is usually conducted using a "brute force" procedure. That is, the standard "rollback procedure" for a decision tree is conducted for each set of input values used in the sensitivity analysis. Certainly, if one were doing the analysis manually, this would be impossible for most decision trees. However, with a software package, given the relatively minor computational requirements for a decision tree analysis, such a brute force approach is not difficult.

A more sophisticated approach for sensitivity analysis could involve treating the input parameters as an algebraic variables and finding the variable values that satisfy an equation or inequality.

For example, let's consider the decision tree of Figure 7.13, but with the cost of the imperfect predictor being denoted as CIP, instead of being set at $60,000. Hence, for example, the payoffs at Node 7 for good market and bad market, respectively, would be 500,000 − CIP and −100,000 − CIP instead of 440,000 and −160,000. Then the EMV for Node 7 is given by

$$.978(500{,}000 - \text{CIP}) + .022(-100{,}000 - \text{CIP}) = 486{,}800 - \text{CIP}.$$

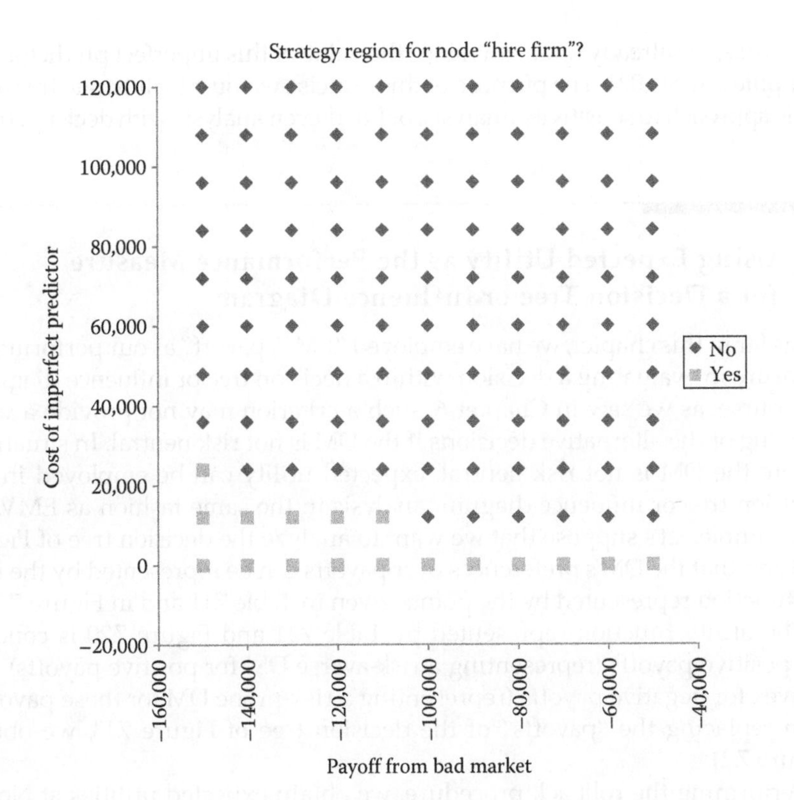

FIGURE 7.19
Strategy region plot for the decision of "hire firm?" (Output from Precision Tree.)

The EMV for Node 8 is given by

$$.1094(500{,}000 - \text{CIP}) + .8906(-100{,}000 - \text{CIP}) = -34{,}360 - \text{CIP}.$$

At Node 5, we would choose the decision of "develop product" since 486,800 – CIP > –CIP and this decision would have an EMV of 486,800 – CIP. At Node 6, we would choose the decision of "do not develop product" since –CIP > –34,360 – CIP, with an EMV of –CIP.

Now, the EMV for Node 3 is given by

$$.68(486{,}800 - \text{CIP}) + .32(-\text{CIP}) = 331{,}024 - \text{CIP}.$$

At Node 2, we had already computed an EMV of 320,000. Therefore, in viewing the decision from Node 1, we would use the imperfect predictor (i.e., hire the firm) if

$$331{,}024 - \text{CIP} > 320{,}000, \quad \text{or} \quad \text{if CIP} < \$11{,}024.$$

Of course, we already knew this since the value of this imperfect predictor was computed as $11,024. The purpose of this exercise was just to illustrate this algebraic approach to sensitivity analysis, or breakeven analysis with decision trees.

7.5 Using Expected Utility as the Performance Measure for a Decision Tree or Influence Diagram

Thus far in this chapter, we have employed "EMV/payoff" as our performance measure in evaluating a decision within a decision tree or influence diagram. Of course, as we saw in Chapter 6, such a criterion may not provide a valid ranking of the alternative decisions if the DM is not risk neutral. In situations where the DM is not risk neutral, expected utility can be employed in the decision tree or influence diagram analysis in the same fashion as EMV. As an example, let's suppose that we want to analyze the decision tree of Figure 7.13 and that the DM's preferences over payoffs can be represented by the utility function represented by the points given in Table 7.11 and in Figure 7.20.

The utility function represented by Table 7.11 and Figure 7.20 is concave for positive payoffs (representing a risk-averse DM for positive payoffs) and convex for negative payoffs (representing a risk-prone DM for these payoffs).

In replacing the "payoffs" of the decision tree of Figure 7.13, we obtain Figure 7.21.

Performing the rollback procedure, we obtain expected utilities at Nodes 4, 7, and 8 of

Expected utility at Node 4 = .7(1) + .3(.06) = .718

Expected utility at Node 7 = .978(.96) + .022(0) = .9388

Expected utility at Node 8 = .1094(.96) + .8906(0) = .105

TABLE 7.11

Points on Decision Maker's Utility Function for the Evaluation of the Decision Tree of Figure 7.13

x in Dollars	u(x)
−160,000	0
−100,000	0.06
−60,000	0.13
0	0.3
80,000	0.47
200,000	0.66
300,000	0.8
440,000	0.96
500,000	1

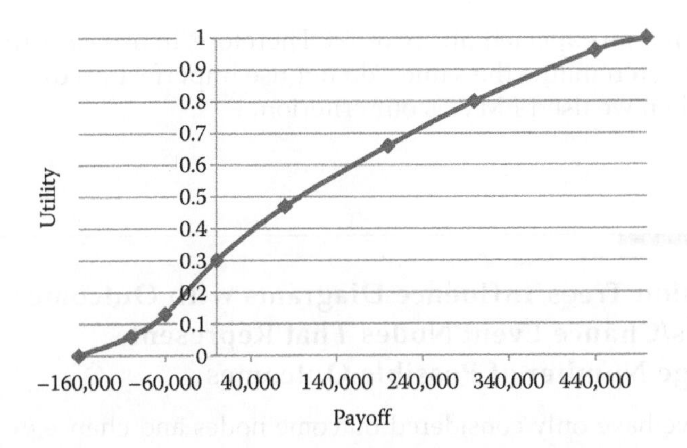

FIGURE 7.20
Utility function corresponding to the points of Table 7.11.

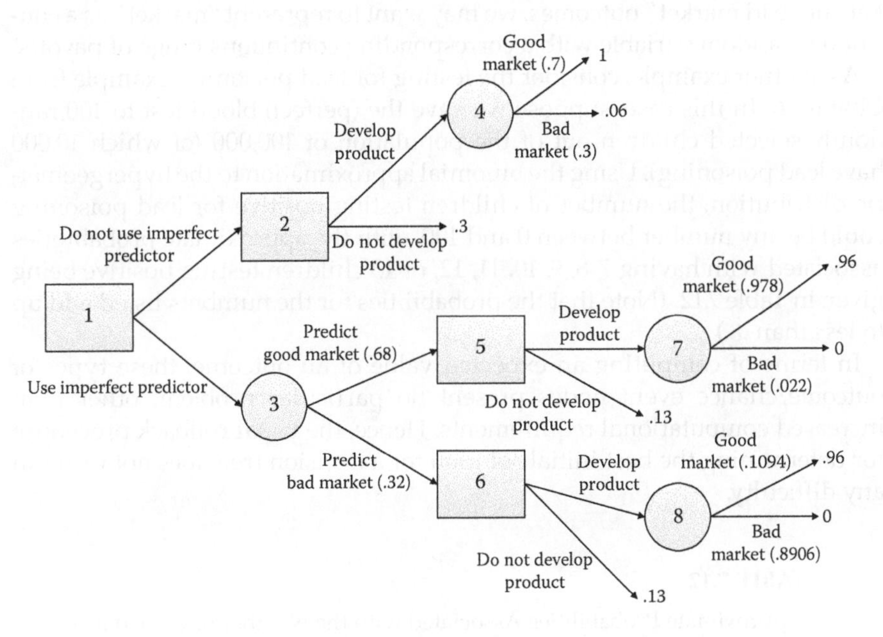

FIGURE 7.21
The decision tree of Figure 7.13 with the utilities in place of the payoffs.

Then the optimal decisions at Nodes 2, 5, and 6 are "develop product," "develop product," and "do not develop product," respectively. These optimal decisions have expected utilities of .718, .9388, and .13, respectively.

The expected utility associated with Node 3 is given by .68(.9388) + .32(.13) ≈ .68. Hence, in viewing the two decisions from Node 1, "do not use imperfect predictor" has an expected utility of .718, while "use imperfect predictor" has an expected utility of .68. Therefore, in this case, the optimal initial decision remains the same: "do not use imperfect predictor," as was the case when we used EMV as our criterion.

7.6 Decision Trees/Influence Diagrams with Outcome Nodes/Chance Event Nodes That Represent a Large Number of Possible Outcomes

Thus far, we have only considered outcome nodes and chance event nodes representing discrete random variables with a few (usually 2) outcomes. Often, we may want to consider situations involving many (e.g., 10 or more) or even an infinite number of outcomes. For example, instead of "good market" or "bad market" outcomes, we may want to represent "market" as a continuous random variable with a corresponding continuous range of payoffs.

As another example, consider the testing for lead poisoning example from Chapter 6. In this case, suppose we gave the (perfect) blood test to 100 randomly selected children out of the population of 100,000 (of which 10,000 have lead poisoning). Using the binomial approximation to the hypergeometric distribution, the number of children testing positive for lead poisoning could be any number between 0 and 100, with the approximate probabilities associated with having 7, 8, 9, 10, 11, 12, or 13 children testing positive being given in Table 7.12. (Note that the probabilities for the numbers listed add up to less than .8.)

In terms of computing an expected value of an outcome, these types of outcome/chance event nodes present no particular problem, other than increased computational requirements. Hence, the usual rollback procedure for determining the best initial solution for a decision tree does not result in any difficulty.

TABLE 7.12

Approximate Probabilities Associated with the Number of Children (Out of a Sample of 100) Testing Positive for Lead Poisoning

Number of Children Testing Positive	Approximate Probability
7	0.089
8	0.115
9	0.130
10	0.132
11	0.120
12	0.099
13	0.074

However, computation of risk profiles for the various strategies becomes problematic. In these situations, one may want to employ a Monte Carlo simulation approach, which will be discussed in Chapter 9.

With respect to representation of the distribution function, one has the option of approximating the function with a simpler distribution. In particular, Keefer and Bodily (1983) have suggested the use of the extended Pearson–Tukey approximation for a wide range of continuous distribution functions, especially for those continuous distributions with one peak that are "close" to symmetric in nature.

Example 7.1: Bidding for Construction of a University Dormitory

The A1 Construction Company is considering placing a bid for the construction of a new dormitory at State University. The cost of preparing the bid would be $5000. A1 estimates the lowest competing bid as a random variable with a triangular distribution with a minimum value of $10 million, a most likely value of $14 million, and a maximum value of $20 million. A1 estimates their costs for constructing the building as an uncertain quantity, represented as a discrete random variable with the following distribution:

Cost for Construction	Probability
Low ($12 million)	.2
Medium ($12.8 million)	.6
High ($13.6 million)	.2

If they decide to place a bid, A1 is considering one of three bid amounts: $13 million, $15 million, or $17 million. So, for example, if A1 places a bid for $15 million and wins the contract, they will receive a net profit of their bid, minus the cost of preparing the bid, minus their costs for the project; their project costs will be either $12 million, $12.8 million, or $13.6 million, according to the table mentioned earlier.

In order to present their problem in the format of a decision tree with a relatively small number of outcomes for each outcome node, A1 needs to compute the probability of their bid being the lowest given that their bid is either $13 million, $15 million, or $17 million. If X represents a random variable corresponding to the lowest competing bid, given that there is a competing bid, then we need to compute

$$P(\$13 \text{ million} < X), \; P(\$15 \text{ million} < X), \text{ and } P(\$17 \text{ million} < X),$$

given that X is a triangularly distributed random variable with parameters of $10, $14, and $20 million. These probabilities are given by the following integrals: $\int_{13}^{\infty} f(x)dx$, $\int_{15}^{\infty} f(x)dx$, and $\int_{17}^{\infty} f(x)dx$, where f(x) represents the distribution function for the triangular distribution with the specified parameter values in millions of dollars. Computing the values for the integrals, we determine that

$$P(\$13 \text{ million} < X) = .775,\ P(\$15 \text{ million} < X) = .208,$$
$$\text{and } P(\$17 \text{ million} < X) = .15.$$

This gives us the decision tree shown in Figure 7.22.

In rolling back the decision tree of Figure 7.22, the EMVs of Bid Amount 1 (bid \$13 million), Bid Amount 2 (bid \$15 million), and Bid Amount 3 (bid \$17 million) are \$104,750, \$407,600, and \$580,000, respectively. Hence, the optimal decision (if a bid is made) is to bid \$17 million, with an EMV of \$580,000. Comparing this to the decision of not to bid at all, the optimal decision is to bid, with a bid of \$17 million.

A one-way sensitivity analysis was performed with four parameters: the probability of winning the contract at Bid 1 (baseline of .775), the probability of winning the contract at Bid 2 (baseline of .208), the

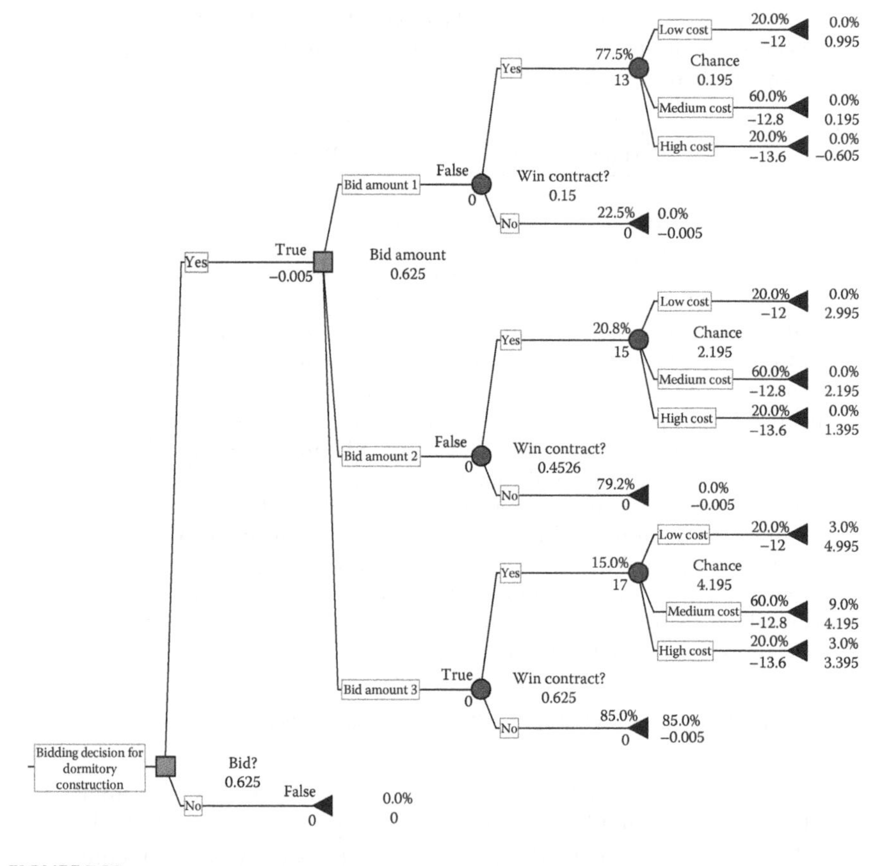

FIGURE 7.22
Decision tree for bidding example. (Reproduced from Precision Tree.) (Ouput from Precision Tree.)

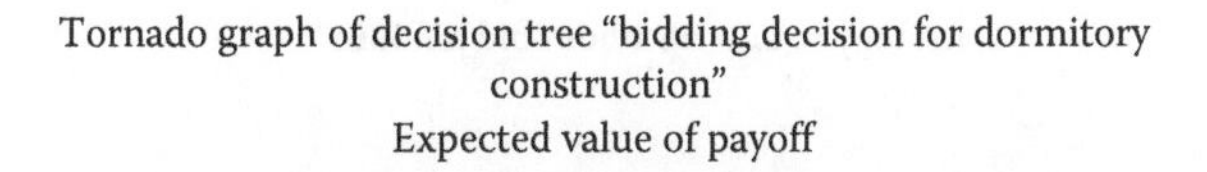

Tornado graph of decision tree "bidding decision for dormitory construction"
Expected value of payoff

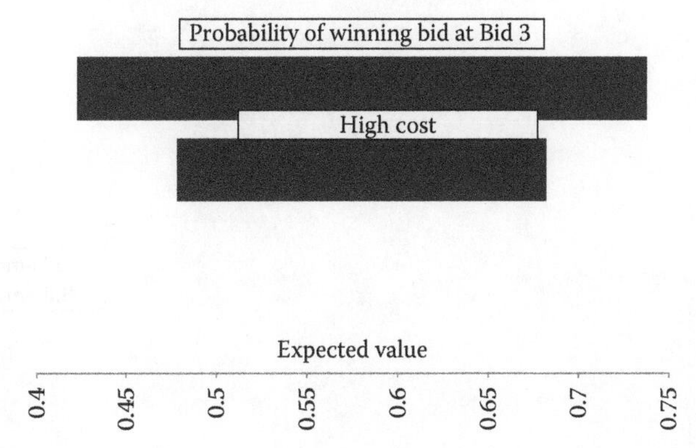

FIGURE 7.23
Tornado graph for one-way sensitivity analysis for construction bidding problem. (Output from Precision Tree.)

probability of winning the contract at Bid 3 (baseline of .15), and high cost for construction (baseline of \$13.6 million). The parameters were each allowed to vary from their baseline values by ± 25%. (Note that when "high cost for construction" became less than \$12.8 million, it actually became "low" or "medium" cost.)

The tornado diagram associated with this one-way sensitivity analysis is shown in Figure 7.23. The reader will note that two of the parameters, the probability of winning the contract at Bid 1 and the probability of winning the contract at Bid 2, do not even appear on the tornado diagram. This indicates that even when set at values of 25% over their baseline values, these probabilities do not affect the EMV; more specifically, even at their larger probability values, placing a bid at the Bid 1 or Bid 2 levels is suboptimal.

A two-way sensitivity analysis was performed using two parameters: the probability of winning the contract at Bid 2 and the probability of winning the contract at Bid 3. For this analysis, the probabilities were allowed to vary ±50% from their baseline values. In particular, A1 was interested in studying whether to place a bid at the Bid 2 or the Bid 3 levels according to their respective probabilities of winning the contract at these levels. The strategy region graph for this analysis is shown in Figure 7.24. This graph shows, for example, that if the probability of winning the contract at Bid 2 is .2912 or higher and the probability of winning the contract at Bid 3 is .15 or lower, then the optimal decision is to place a bid at Bid 2.

The risk profiles associated with the decisions of not bidding, bidding \$13 million, bidding \$15 million, and bidding \$17 million are shown in Tables 7.13 through 7.16, respectively.

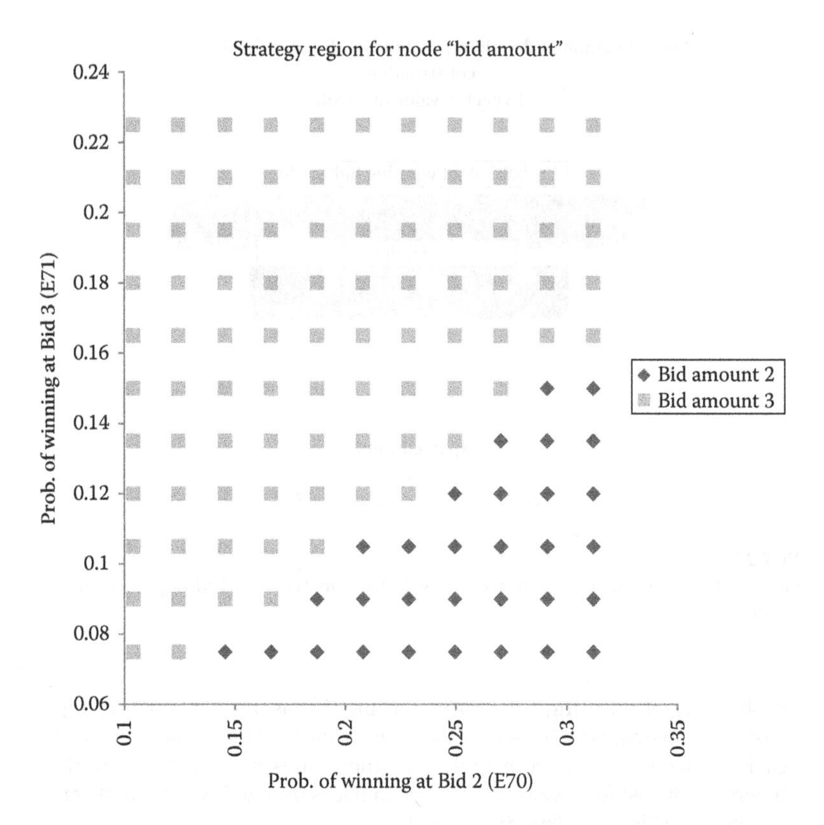

FIGURE 7.24
Strategy region for probabilities of winning contract at Bid 2 or Bid 3. (Output from Precision Tree.)

TABLE 7.13
Risk Profile for Decision of Not Bidding

Profit	Probability
$0	1.

TABLE 7.14
Risk Profile for Decision of Bidding $13 Million

Profit, $	Probability
–65,000	0.155
–50,000	0.225
150,000	0.465
950,000	0.155

TABLE 7.15

Risk Profile for Decision of Bidding $15 Million

Profit, $	Probability
−50,000	0.792
1,350,000	0.0416
2,150,000	0.1248
2,950,000	0.0416

TABLE 7.16

Risk Profile for Decision of Bidding $17 Million

Profit, $	Probability
−50,000	0.85
3,350,000	0.03
4,150,000	0.09
4,950,000	0.03

A1 notes that there is a high probability of losing money if they bid on the project; in particular, if they implement the alternative that maximizes EMV (i.e., bid $17 million), there is a .85 probability of losing $50,000. Of course, this is offset by a .15 probability of making $3.35 million or more.

One important fact to remember with respect to this decision by A1 is that, as was pointed out in Chapter 1 for most decisions, it does not exist in a vacuum. Specifically, A1 bids on a number of contracts. For example, suppose that A1 bids the high amount ($17 million) on five identical contract situations; the number of successful bids (assuming each bid situation is independent) would be a binomial random variable with parameters n = 5 (the number of bids) and P = .15 (the probability of success with the bid). The probabilities associated with the number of successful bids are shown in Table 7.17.

TABLE 7.17

Number of Successful Bids out of Five Independent Identical Bid Situations Given That the Probability of a Successful Individual Bid Is .15

Number of Successful Bids	Approximate Probability
0	0.44
1	0.39
2	0.14
3	0.03
4	0.002
5	0

Hence, the probability of having at least one successful bid would be greater than .5 since 1 – .44 = .56.

In Chapter 9, we will consider a generalization of this bidding decision situation in which the amount to be bid will be defined as a continuous decision variable and some of the outcomes will be defined as continuous random variables. Such a decision situation will be modeled as a Monte Carlo simulation.

Material Review Questions

7.1 Influence diagrams represent decision situations in a (more, less) compact form than a decision tree (choose one).

7.2 Which is the more natural way to represent a decision situation to a decision maker: an influence diagram or a decision tree?

7.3 In general terms, what does a decision tree represent?

7.4 A decision tree can contain "cycles" (*true* or *false*).

7.5 What are the two types of nodes in a decision tree?

7.6 What would be the error associated with having two separate outcomes emanating from an outcome node as (a) demand for product will be greater than 1000 items and (b) demand for product will be greater than 1200 items?

7.7 The typical basic objective associated with an initial analysis of a decision tree is the determination of the best (initial decision *or* all decisions) in the decision tree (choose one).

7.8 In a decision tree, one can have multiple branches entering a node (*true* or *false*).

7.9 The probabilities associated with the branches in a decision tree are either marginal or conditional probabilities (*true* or *false*).

7.10 The certainty equivalent (CE) for an outcome node in a decision tree is often represented by an expected monetary value (EMV) for that node (*true* or *false*).

7.11 In the "evaluation" of a decision tree, one will usually perform the calculations working from (choose one):

a. Right to left
b. Left to right

7.12 What important theorem is often used to determine the conditional probabilities employed in a decision tree?

7.13 The quality of a predictor is typically indicated by what type of a probability?

7.14 What are the two types of conflicting measures that need to be considered in deciding whether or not to obtain and use the information from a predictor?

7.15 Give several examples of things that can be employed as predictors.

7.16 What is another name for "expected value of a perfect predictor?"

7.17 What are the values for the relevant conditional probabilities for a "perfect predictor?"

7.18 Both the "expected value of a perfect predictor" and the "expected value of an imperfect predictor" can be computed with decision trees (*true* or *false*).

7.19 The "expected value of a perfect predictor" will always be greater than or equal to the "expected value of an imperfect predictor" for a given decision situation (*true* or *false*).

7.20 Give the functional relationship between the following set of variables: EVOPP, EVWPP, and EVWAI.

7.21 Give the functional relationship between the following set of variables: EVOIP, EVWIP, and EVWAI.

7.22 What is a strategy associated with a decision tree?

7.23 Give an example of a strategy that is not "well defined."

7.24 What is a risk profile associated with a strategy?

7.25 Why is it that strategies analyzed in detail are typically specified in terms of just the initial decisions, followed by "the decisions that maximize EMV following the initial decisions?"

Exercises

7.1 Consider the following risk profiles for two decision strategies: A and B. The payoffs are given in tens of thousands of dollars.

A

Payoff	4	11	14	18
Probability	.2	.3	.2	.3

B

Payoff	4	11	14	18
Probability	.2	.2	.3	.3

Circle each of the following true statements:

a. *A* stochastically dominates *B*.
b. *B* stochastically dominates *A*.
c. *A* deterministically dominates *B*.
d. *B* deterministically dominates *A*.
e. Neither *A* nor *B* stochastically dominates the other.
f. Neither *A* nor *B* deterministically dominates the other.

7.2 The payoffs associated with the decision tree as follows are given in thousands of dollars. Give the risk profile (in the form of a table) associated with strategy A–A1.

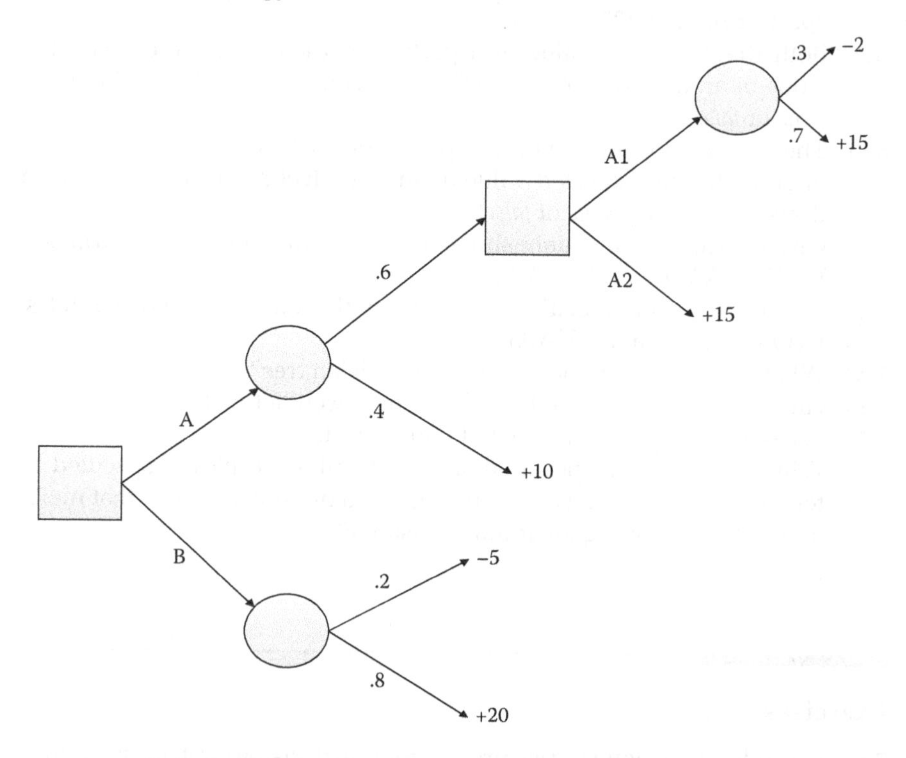

7.3 A particular test for a certain type of disease gives a false-positive reading with a probability of .03 and a false-negative reading with a probability of .02. (A false-positive reading means that the test says a person does have the disease when he or she does not. A false-negative reading means that the test says a person does not have the disease when he or she does, i.e., the sensitivity of the test is 1 − .02 = .98 and the specificity of the test is 1 − .03 = .97.) In Louisville, it is estimated that 1 person in 10,000 has the disease.
Sally, a person randomly selected from Louisville, has tested positive according to the test. What is the probability that she has the disease?

7.4 What are the two types of arcs that are used in an influence diagram and what are they used for? (Be as brief as possible in your answers)

7.5 What are the three types of nodes used in an influence diagram?

7.6 If strategy A deterministically dominates strategy B, then A also stochastically dominates B (*true* or *false*).

7.7 The executives of a particular company are considering the national launch of a new product. The launch will be either a "success" (resulting in a payoff of $8 million) or a "failure" (resulting in a loss of $2 million).

The initial estimates of the probabilities of success and failure are .8 and .2, respectively. The company has the opportunity to contract with a marketing firm in order to conduct a study to determine whether or not the product will be a success. What would be the maximum amount that the company should pay this marketing firm for the study, assuming that the study would be a perfect indicator of the success of the product?

7.8 For the company in Exercise 7.7, data have been gathered as to the accuracy of studies made by the marketing firm. These data indicate the following conditional probabilities: P(study forecasts success|product is a success) = .95, P(study forecasts failure|product is a failure) = .9. What is the maximum amount that should be paid for this (imperfect) marketing study?

7.9 Consider the decision situation as described in Exercise 7.8. Enumerate all of the explicit strategies associated with this situation and give the risk profile associated with each strategy. Assume that the cost associated with using the firm for its imperfect predictor is $20,000.

7.10 Consider the decision tree of Figure 7.22 for the bidding decision example. Develop an equivalent simplified decision tree with fewer nodes by "collapsing" the tree where appropriate.

7.11 Use the algebraic approach discussed in the chapter to perform a two-way sensitivity analysis on the parameters—(a) probability of winning at Bid 2 and (b) probability of winning at Bid 3 for the bidding decision example—thus conveying the same information as in Figure 7.24. More specifically, determine an inequality with two variables (the respective probabilities of winning at Bid 2 and winning at Bid 3), for which one can supply values for these variables and thus determine whether to place a bid at the Bid 2 value or the Bid 3 value.

8

Determining Probabilistic Inputs for Decision Models

8.1 Introduction

As the reader probably realizes from reading Chapter 7, random inputs are important for decision models. These inputs, usually represented by probability distribution functions, are used to represent system randomness or uncertainty. These functions, which can be obtained through analysis of data, through expert opinion, or through a combination of both, can be either discrete or continuous in nature. When the decision model is executed (or "run"), the values generated from these input distributions are called "random variates."

The set of distribution functions used for a decision model, in terms of both the form of the function (e.g., normal, triangular, binomial, and empirical) and the parameters (e.g., mean and standard deviation) of that function, can affect the ranking of the set of alternatives being considered. For example, the first-ranked alternative for staffing schedule for nurses in an emergency department at a hospital will obviously vary depending on the probability distribution used to represent patient arrivals. This is one of the main reasons for performing sensitivity analysis with a model—by determining which input distributions cause the greatest variations in the rankings of alternatives, one can spend additional time and effort in analysis to specify these distributions.

In decision trees and influence diagrams, the input distributions correspond to the outcome nodes (decision trees) and chance event nodes (influence diagrams). In particular, these could be distributions representing sales, percentage of population having a disease, and so on.

In more general simulation models used to aid in decision making, the input distributions can correspond to such things as the arrival pattern of customers/patients to a service system, the distribution associated with categorization of customers/clients/patients to a service system (e.g., percentages of callers to a 911 call center who require service from police, EMS, or fire department or some combination of these services), the time required to perform some activity (e.g., in a project simulation, the time required to perform each activity might be represented as a beta-distributed random variable), and so on.

The remainder of this chapter is separated into two main sections. The first section addresses methodologies to be used when data are available or can be obtained. The second section addresses the situation when data are not available and subjective estimates must be made.

8.2 Situation of Data Availability

In most cases, data will be available, or can be obtained, to aid in the estimation of random inputs for a decision model. In many situations, organizations will automatically collect the relevant data. For example, manufacturing organizations will collect data on machine failures (time between failures and time to repair) and call centers collect data on patterns of incoming calls (rates by day and time of day and type of caller).

If data are available, as noted by Law (2007, p. 279), they can be used to form probabilistic inputs for models in any of three different ways:

1. The data can be used directly as input to the model.
2. The data can be used to form an empirical probability distribution to input to the model.
3. A "theoretical" distribution, such as a triangular, normal, and binomial, can be fit to the data.

As an example of the first approach, if a simulation model is being built to determine staffing levels at a fast-food restaurant, data could be collected on the arrival times, size of groups, and menu items ordered by groups of customers arriving at the restaurant from 6 to 9 a.m. The exact data associated with these groups would then be input for the simulation model. This approach of using the exact data as input might be appropriate for model validation/accreditation purposes, but rarely for experimentation in determining policies and decisions since the data may not represent a typical morning in the operation of the restaurant.

An empirical distribution, to be discussed in Section 8.2.5.1, is constructed in such a way that it will most closely represent the data collected; as such, an empirical distribution will also represent irregularities in the data more accurately than a theoretical distribution. However, a theoretical distribution (1) allows the generation of values outside the bounds of the data collected, (2) is typically more compact to represent within the simulation model than an empirical distribution, and (3) (because of the second point) will typically be easier to change than an empirical distribution. In addition, there may well be a reason associated with the process itself to use a theoretical distribution. For example, as noted in Chapter 5, the number of defective items in a sample taken from a group of items will correspond to a hypergeometric distribution, and the sample mean of a random variable, which is the sum of a number of independent, identically distributed random variables, will approximate a normally distributed random variable; the larger the number in the sample, the more closely the sample mean will be distributed approximately as a normal distribution.

At least a few researchers, however, have suggested that more emphasis should be placed on the use of empirical distributions than there is currently for the modeling of real systems. The reason for this is that irregularities do exist in real systems and they should be modeled as such and not "smoothed out."

The bottom line is that the approach used out of the three mentioned earlier should correspond to the purpose of the study. For purposes of validation/accreditation of the model, the actual data could be used as input. But for analysis and experimentation, one should use either an empirical distribution or a theoretical distribution that is "fit" to the data.

8.2.1 Steps in Fitting a Distribution to the Data

If there are data available for analysis, the steps to be followed for determining the distribution function to represent those data for a decision model should be as follows:

1. Determine which data to collect and how to collect them.
2. Collect the data.
3. Examine the data for any anomalies and/or outliers and take appropriate action with respect to these anomalies and/or outliers (e.g., removing data points collected in error, taking action with respect to correcting a bad situation, or treating data as if from different processes).
4. Determine either an empirical or a theoretical distribution function to represent the data in the decision model.

8.2.2 Step 1: Determine Which Data to Collect and How to Collect Them

There are two types of input to a decision model: input that is qualitative in nature and input that is quantitative in nature. Typically, the qualitative input is determined first; examples of this type of input would be things like the alternative decisions under consideration, the performance measures, the potential outcomes associated with decisions, the number and types of scenarios under consideration, the steps associated with any processes being modeled, and the resources required to perform the activities of a process. The quantitative input would include things like the probability distributions over various economic/demand scenarios, process times, time between failures, time to repair, categories of entities, and arrival rates by the type of customer/patient/client. Through interaction with experts/stakeholders involved with the system, the analyst gradually builds a conceptual decision model; as part of this development, the analyst should be able to determine the data to gather in order to form the probabilistic inputs required for the decision model. In many cases, this determination is straightforward, while in other cases, this determination is not straightforward.

Assume that the data collected for a particular process associated with the system are represented as $X_1, X_2, \ldots, X_n$. These numbers could be such quantities as (1) times between arrivals of customers to a restaurant or of arrivals of patients to an emergency department of a hospital, (2) processing times associated with an activity, (3) "indicators" of whether a forecast of some discrete variable is correct or not (where a 0 would represent an incorrect forecast and a 1 would represent a correct forecast), (4) the number of customers in a group entering a fast-food restaurant, (5) times required to repair a machine, or (6) the number of defective items in lots of part received from a vendor.

8.2.2.1 *Independent Samples*

When deciding which data to collect, the analyst should remember that the sample data collected should represent "independent samples" of the underlying distribution. One important reason for this requirement is that many of the statistical techniques (such as chi-square goodness of fit tests and maximum likelihood estimation) used to analyze the data assume this independence. Even when the samples are not independent though, the use of histograms is still valid.

Examples of situations involving *dependent* samples would include collection of sequential data on (1) times to perform an activity from a person who at the end of the day is getting more tired, (2) times between arrivals for patrons of a restaurant when some of those patrons arrive by bus and others arrive by car (see Example 8.2), and (3) processing times on a machine that is slowly deteriorating.

One way to check that the samples are independent is to examine the sample correlations between the pairs of numbers that are one, two, three, and so on, apart in the sequence of numbers. This can be accomplished through the use of a "scatter diagram" (to examine independence for samples that are one apart in the sequence) or a "correlation plot" (to examine independence for samples that are 1, 2, 3, etc., apart in the sequence).

A scatter diagram is just a plot of the data points: (X_i, X_{i+1}) for $i = 1, \ldots, n-1$, where X_i is the x-coordinate value and X_{i+1} is the y-coordinate. If the data values are all positive and are uncorrelated, then the points should appear to be "well dispersed" through the first quadrant. For example, Figure 8.1

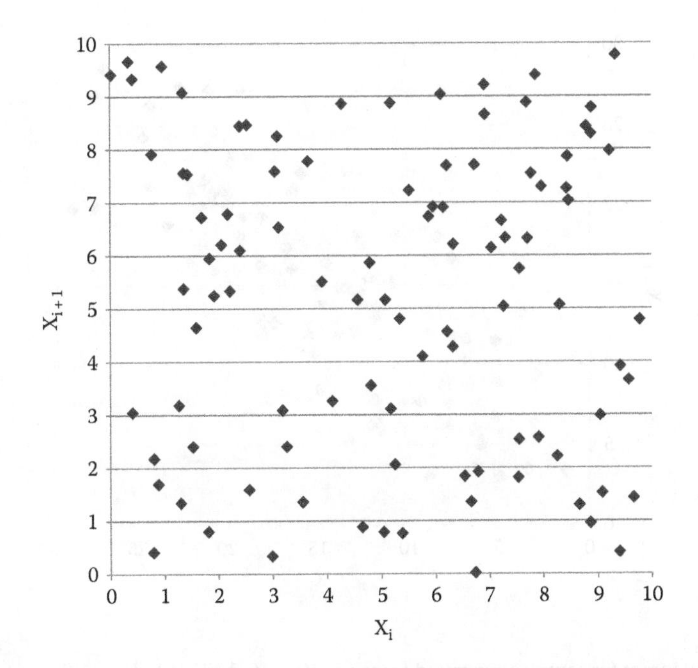

FIGURE 8.1
A scatter diagram for independent samples generated from a uniform (0, 10) distribution.

shows a scatter diagram for 99 points derived from a series of 100 values: $X_1, X_2, \ldots, X_{100}$ representing independent samples for a uniform (0, 10) distribution. The points on the diagram are given by $(X_1,X_2),(X_2,X_3),\ldots,(X_{99},X_{100})$.

The points do appear to be dispersed appropriately through the first quadrant. Note that in order for the scatter plot to be correctly interpreted, its two axes should have the same scales.

If a positive correlation exists between sequential data points, then the points should appear along a line with a positive slope. For example, Figure 8.2 shows a scatter diagram for samples generated from 100 points for a uniform (a, b) distribution with a trend, as defined by an increment given to a and b for each sequential point sampled. In particular, a = 0 + .1i and b = 10 + .1i to generate the X_i data point for i = 1,…,100.

A correlation plot shows the values for the sample correlations, denoted as $\hat{\rho}_j$ for j = 1,2,…,l (l is a positive integer) where $\hat{\rho}_j$ is an estimate of the true correlation ρ_j between two observations that are j observations apart in time, j is the x-axis coordinate, and $\hat{\rho}_j$ is the y-axis coordinate. Note that the sample correlation values for samples $X_1, X_2, \ldots, X_n$ are given by

$$\hat{\rho}_j = \frac{\hat{C}_j}{S^2(n)}.$$

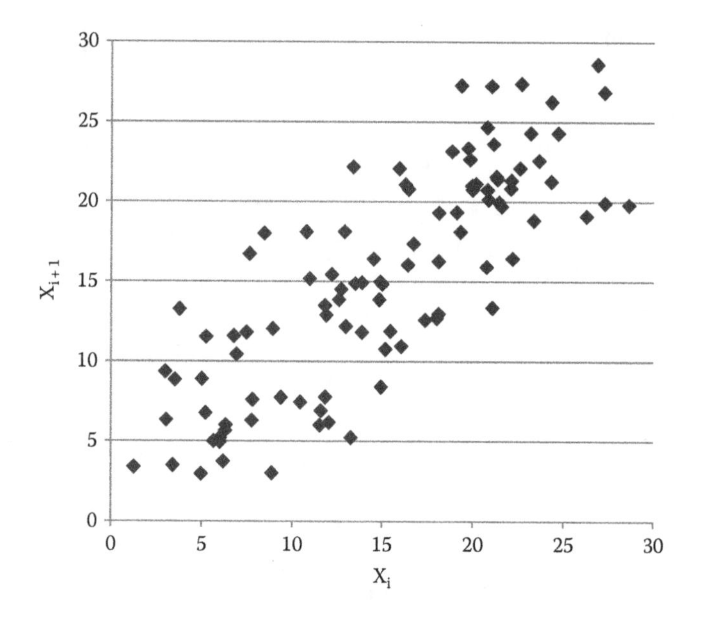

FIGURE 8.2
A scatter diagram for samples generated from a uniform (a, b) distribution with a trend, where a = 0 + .1i and b = 10 + .1i for X_i.

where

$S^2(n)$ is the sample variance for the data

$\hat{C}_j$ is the sample covariance between two observations that are j apart

$\hat{C}_j$ is given by

$$\hat{C}_j = \frac{\sum_{i=1}^{n-j}\left[\left(X_i - \bar{X}(n)\right)\right]\left[\left(X_{i+j} - \bar{X}(n)\right)\right]}{n-j}.$$

Note that the $\hat{\rho}_j$'s will attain a value of between −1 and +1. The ideal values, to indicate independence, would be 0. However, since these are just estimates of the actual correlations between two observations that are j observations apart, one should only expect their values to be "close to" 0 to indicate independence.

Example 8.1: Sample Covariances for Data from a Changeover Operation

Suppose that the following samples for changeover times (in minutes) have been collected for a coating operation.

30.6, 24.5, 31.8, 44.3, 34.1, 25.4, 28.1, 35.5, 23.7, 28.6, 39.1, 33.4, 26.8, 27.2, 31.1, 35.3, 25.6, 29.8, 33.9, 37.6, 26.5, 28.2, 22.1, 29.8, 39.3.

TABLE 8.1
Sample Covariances and Correlations for Data of Example 8.1

j	$\hat{C}_j$	$\hat{\rho}_j$
1	1.333	0.043812
2	−17.347	−0.56995
3	−4.902	−0.16109
4	6.361	0.209007
5	−2.487	−0.08171
6	−4.924	−0.16177
7	6.636	0.218011

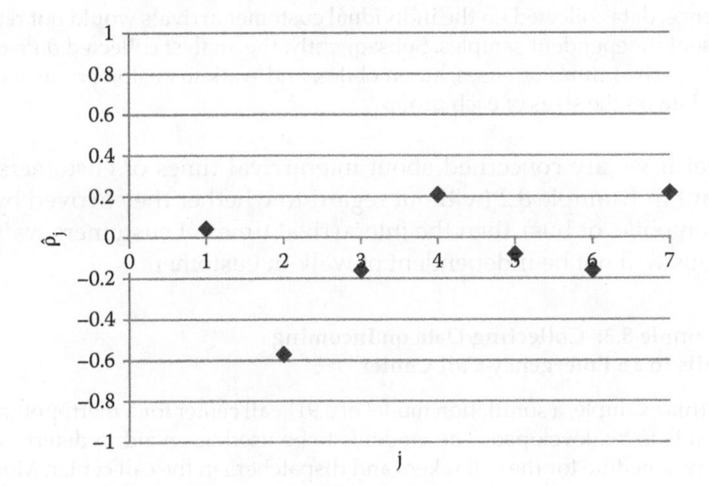

FIGURE 8.3
Correlation plot for data of Example 8.1.

Hence, we have 25 samples (n = 25) and

$$\bar{X}(25) = 30.892, \quad S^2(25) = 30.4366.$$

The values for the sample covariances are given in Table 8.1.

A correlation plot, corresponding to the $\hat{\rho}_j$ values shown in Table 8.1, is shown in Figure 8.3.

Except possibly for the value associated with $\hat{\rho}_j$ for j = 2, these correlations would be acceptable for the independence assumption.

In addition to being independent, the observations collected should represent the process correctly. These considerations (independence and correct representation) can often be related to such modeling aspects as categorization of entities and identification of time periods. Examples 8.2 and 8.3 illustrate these considerations.

Example 8.2: Collecting Data on Interarrival Times for Customers of a Fast-Food Restaurant

An analyst has been asked to construct a simulation model of a fast-food restaurant in order to aid in determining workforce staffing policies for the restaurant. In the construction of the model, the analyst needs to collect data on the interarrival times of the restaurant's customers. Initially, the analyst collected data over specific time periods with respect to the interarrival times of individual customers. However, upon further reflection, the analyst determined that the customers arrived in groups, as determined by buses, automobiles, or groups of walk-in customers. Hence, data collected on the individual customer arrivals would not represent independent samples. Subsequently, the analyst collected data on interarrival times of buses, automobiles, and walk-in customers, as well as data on the sizes of each group.

Note that if we are concerned about interarrival times of customers to the restaurant in Example 8.2 (without regard to whether they arrived by walking, automobile, or bus), then the interarrival times of customers walking in from a bus will not be independent of walk-in customers.

Example 8.3: Collecting Data on Incoming Calls to an Emergency Call Center

In this example, a simulation model of a 911 call center for a metropolitan area is to be developed. The model is to be used as an aid in determining a schedule for the call takers and dispatchers in the call center. More specifically, the center's management would like to know the number of call takers and dispatchers to assign to each of several predetermined shifts in order to minimize the number of staff hours while satisfying a constraint on the fraction of callers who wait longer than 12 seconds to have their call answered.

The calls received by the call center are initially answered by the call takers. The call taker then determines whether or not a response is required by emergency services (either police, fire, ambulance, or some combination of these). If a response is required, the call taker then gathers some basic information from the caller, such as the location and nature of the emergency. This information is then relayed to a dispatcher (or dispatchers if more than one type of emergency service is required) who communicates with the emergency responders.

A major input to the simulation model to be developed is the pattern of calls coming into the system. Typically, this pattern is represented as a nonstationary Poisson arrival process, but the analyst must still determine how to collect the data in terms of (1) categories of calls and (2) the time periods to be used in terms of unique arrival rates.

For example, should the analyst collect arrival rate data for each separate category of call (police, fire, ambulance, and every combination of these three) by time period? This approach would require much effort, especially when one considers that there would be seven different sets of arrival rates for which to gather data, since there are seven categories of calls corresponding to the combinations of responder types.

An alternative approach would be to assume that the fraction of calls associated with each combination of service required remains constant over time, and then the data could be gathered over the arrival rate "of all calls" by time period. This approach could represent an approximation to the actual system since, for example, the fraction of callers of any particular type might vary depending on when "rush hour" occurs; that is, the fraction of calls corresponding to traffic accidents could be expected to increase during rush hour.

As mentioned earlier, another consideration would be how the planning horizon is divided into time periods, for example, should the rate of incoming calls be allowed to vary by 30-minute periods, by 1-hour periods, or even by larger time periods? If, for example, the analyst builds a model that assumes that incoming calls arrive at a constant rate between 7 and 8 a.m. and therefore collects data during this time period over several days to estimate this rate, then the model may not accurately represent the system (for its purpose) if the rate varies greatly from the (7–7:30 a.m.) time frame to the (7:30–8 a.m.) time frame.

Certainly, information obtained from the system, decision makers, and stakeholders would be helpful to the analyst in making these decisions related to independence and correct representation. In Example 8.3, the system management of the emergency call center might specify that the fraction of calls that require only police as compared to those that require police and fire department varies greatly between 7 and 8 a.m. on a Monday morning as compared to 10–11 p.m. on a Friday night, *and* the typical amounts of time required of the call taker to handle these two different call types are very different.

Obviously, the larger the number of states and entity categories used in a model, the more accurate that model will be. The question that the analyst needs to consider is whether that additional accuracy is worth the time and effort in terms of data collection and model building. This is where the experience of the analyst comes into play, not to mention sensitivity analysis performed after the initial model is constructed.

8.2.3 Step 2: Collect the Data

The data themselves could be collected from any of several different formats/procedures. The examples would be from reports, observation, surveys, or interviews. In the case of observation, the analyst needs to be aware of the observational effect on the data themselves; for example, people may work faster if they know that they are being observed. In the case of surveys or interviews, the analyst should be aware of any inherent biases on the part of those participating in the survey/interview. And, of course, the comments regarding independence and representative nature made previously should also be considered.

Reports may not be initially understandable to the analyst and hence may require interpretation from the management of the system.

8.2.4 Step 3: Examine the Data for Any Anomalies or Outliers and Take the Appropriate Action

Prior to the use of any "sophisticated" procedures for fitting a distribution to the data, the analyst should spend some time "looking at" the data in its more or less raw form. This would imply such things as considering "outliers" in the data and viewing histograms of the data to determine any other anomalies in addition to the outliers found.

An outlier is loosely defined as a data point that is distant from other observations (Grubbs, 1969). The analyst should determine first if the outlier is a true data point or if it resulted in an error in the data collection process. For example, in Example 8.3 involving the emergency call center, the data for the processing time by the call taker were automatically collected by the system. For a data point, which corresponded to a call that began just prior to midnight and lasted past midnight, the program designed for computing the process time initially computed a time that was much longer than the actual time for this call (several hours for a call that came in a few minutes prior to midnight and finished a few minutes after midnight), an obvious error.

A second consideration with outliers is whether or not the outlying data point in question is representative of what actually happens in the system, even if the data point is a true data point. Consider a situation in which an analyst is collecting data on the arrival rates for the diners of a restaurant between 12 noon and 1 p.m. On a particular afternoon during which data are being collected, two busloads of diners arrive and the buses are operated by a particular tour operator. Instead of using these data without additional thought, the analyst should probably consult with the manager of the restaurant to determine how often this situation would occur. If the situation of two busloads of customers arriving almost simultaneously almost never occurs, the analyst (and restaurant management) might decide to delete the relevant data from analysis. On the other hand, a separate model might be developed just for this situation where two buses arrive simultaneously and contingency plans developed based on experimentation with this separate model.

In addition to discovering outliers, a rudimentary analysis of the data might reveal some opportunities for system improvement that would not require the development of a sophisticated decision model. Consider Example 8.4, involving data collected for the lead time on the refurbishment for a major component of the space shuttle.

Example 8.4: Collecting Data on Time to Refurbish a Major Component of the Space Shuttle

The Space Shuttle (with the official name of the Space Transportation System) was one of the most complex pieces of machinery ever developed. The various individual ships flew 135 missions from 1981 until its retirement in 2011. The maintenance and refurbishment of an individual shuttle between flights required thousands of individual tasks, thus resulting in a complex problem in project management.

An industrial engineer was developing a simulation model to aid in the design of an inventory control policy for major components of the space shuttle. The data collected for the lead time (in days) for one of these components are shown in the following: 222, 171, 924, 210, 188, 902, 218, 183, 882, 865, 215, 167, 205, 245, 915, 885, 226, 198, 216, and 891.

Ordering these data from the smallest to largest value gave the following data points: 167, 171, 183, 188, 198, 205, 210, 215, 216, 218, 222, 226, 245, 865, 882, 885, 891, 902, 915, and 924.

Arranging the data into intervals of 20-day width, as shown in Table 8.2 and Figure 8.4, the engineer noted that that there were two groupings of the data: those data points within the 161–260-day range and those within the 861–940-day range. Rather than just trying to fit a theoretical distribution function to the raw data, she decided to spend additional time by interviewing the relevant personnel involved to determine if the instances involving refurbishment within the 861–940-day range could be eliminated. When it was discovered that these instances could not be eliminated, she decided to fit two separate distribution functions to the data, one function for the days in the 161–260-day range and one function for the 861–940-day range, with 65% of the points generated within the first range and 35% of the points generated for the second range, as suggested by the data.

Example 8.4 illustrates the value of a "stand-alone" simple analysis of the raw data, prior to any sophisticated analysis.

As seen from Example 8.4, part of a rudimentary analysis of the data would involve forming histograms corresponding to that data. A histogram allows

TABLE 8.2

Lead Time to Refurbish a Major Component of the Space Shuttle

Lead Time Interval (Days) (Inclusive Endpoints)	Number of Occurrences
161–180	2
181–200	3
201–220	5
221–240	2
241–260	1
861–880	1
881–900	3
901–920	2
921–940	1

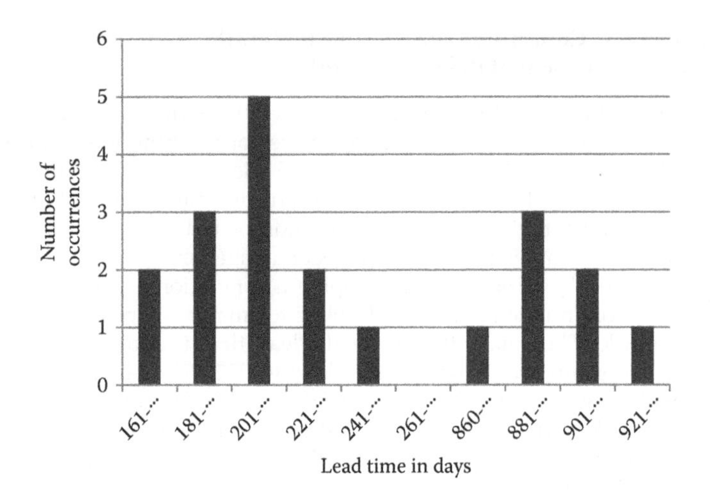

FIGURE 8.4
Lead times (in days) for time to refurbish a major component of the space shuttle.

the decision maker to view the data in a more coherent way than just viewing the raw data as numbers. Once a histogram is formed, anomalies can be more easily identified; in addition, the histogram can be easily compared to graphs of theoretical density functions.

A histogram can be described as a plot of the number of data points that fall into each of a series of equal-width cells. Therefore, in forming a histogram, the decision maker must make decisions with respect to the number and width of the histogram's cells. Sturges's rule (see Law, 2007, p. 319) for specifying the number of cells, k, to use in a histogram for continuous data collected is as follows:

$$k = [1 + \log_2 n], \text{ rounded to the nearest integer,} \tag{8.1}$$

where n is the number of data points collected. The specific values for k as a function of n for Sturges's rule are shown in Table 8.3. Note that the values for k in Table 8.3 are rounded to the nearest integer.

In using Sturges's formula given in Equation 8.1, the assumption is that any data points associated with anomalies and outliers have been removed. For example, in applying Sturges's formula to the data given in Example 8.4, the formula would be applied first to the set of data corresponding to the lead times within the interval of 167–245 days and then the lead times within the interval of 865–924 days. Since there are 13 data points with values within the [167, 245], the number of cells in the histogram for these data points according to Sturges's rule should be

$$k = [1 + \log_2(13)] = 4.7 \approx 5.$$

TABLE 8.3
Suggested Values for the Number of Cells (k) for a Histogram as a Function of the Number of Data Points (n) from Sturges's Rule

n	k
10	4
50	7
100	8
1000	11

Dividing the width of the entire interval by the number of cells gives the width of the cells in the histogram:

$$\frac{(245-167)}{5} = 15.6.$$

Therefore, the cells for the histogram for the first set of data points in Example 8.4 would be [167, 182.6), [182.6, 198.2), [198.2, 213.8), [213.8, 229.4), and [229.4, 245].

Note that the cells are set up so that they cover the entire interval in which the data points are contained and that the intersection of any two cells is the null set, thereby not allowing for the possibility of any data point falling in more than one cell.

With the cells given earlier, the number of data points in each of these cells are given as 2, 3, 2, 5, and 1, respectively; the resulting histogram appears in Figure 8.5.

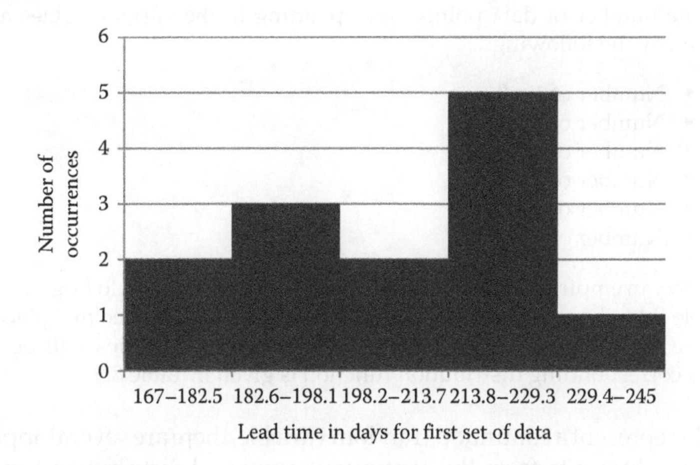

FIGURE 8.5
Histogram formed using Sturges's rule for the first set of data in Example 8.4.

8.2.5 Step 4: Determine Either an Empirical Distribution or a Theoretical Distribution Function to Represent the Data in the Decision Model

Suppose that steps 1 through 3 have been completed and therefore any outliers have been removed; in addition, the data collected are "independent" and represent a single process, as opposed to multiple processes as we had in Example 8.4. The data collected from a process, denoted as $X_1, X_2, \ldots, X_n$, are now to be "fit" with a probability distribution that represents the process that generated the data in the first place. As suggested earlier, there are two possibilities to consider at this point: an empirical distribution or a theoretical distribution.

8.2.5.1 Fitting an Empirical Distribution to the Data

Let's consider the use of an empirical distribution first. If the process represents a discrete random variable, the approach is very intuitive—just use the proportion of data points with a particular value as the relevant probability. Example 8.5 illustrates this approach.

Example 8.5: Fitting an Empirical Distribution Function for the Number of Customers in a Group Entering a Fast-Food Restaurant

Suppose that an analyst wants to estimate the distribution function for the number of customers in a group that enters a fast-food restaurant from 7 to 8 a.m. on a typical weekday morning. The analyst has collected the following observations from several representative weekdays over the 7 to 8 a.m. time period for the number of customers in a group entering the restaurant: 1, 3, 2, 2, 1, 1, 1, 4, 3, 2, 2, 3, 6, 4, 2, 1, 3, 2, 2, 3, 5, 3, 2, 1, 1, 1, 2, 2, 2, 4, 2, 2, 3, 2, and 4.

The number of data points corresponding to the various values are given by the following:

- Number of 1's: 8
- Number of 2's: 14
- Number of 3's: 7
- Number of 4's: 4
- Number of 5's: 1
- Number of 6's: 1

Hence, an empirical density function used for this case would be given as Table 8.4 (where, since the probabilities are given to two decimal places, some are rounded off so that the sum of the probabilities will be 1). The corresponding distribution function is given in Table 8.5.

If the data represent a continuous random variable, there are several approaches that one might use to form the continuous empirical distribution function to represent these data. Intuitively, the distribution function, F(x), should be one

TABLE 8.4

Empirical Discrete Density Function for the Number of Customers in a Group Entering a Fast-Food Restaurant between 7 and 8 a.m. on a Typical Weekday Morning

x	Pr(X = x)
1	8/35 = 0.23
2	14/35 = 0.4
3	7/35 = 0.2
4	4/35 = 0.11
5	1/35 = 0.03
6	1/35 = 0.03

TABLE 8.5

Empirical Discrete Distribution Function for the Number of Customers in a Group Entering a Fast-Food Restaurant between 7 and 8 a.m. on a Typical Weekday Morning

x	F(x) = Pr(X ≤ x)
1	0.23
2	0.63
3	0.83
4	0.94
5	0.97
6	1

that increases most rapidly in the regions where there are many data points. Also, in order to be a valid distribution function, F should be monotonically increasing, with a minimum value of 0 and a maximum value of 1.

Suppose that the data $X_1, X_2, \ldots, X_n$ are reordered from smallest to largest value so that $X_{(i)}$ represents the ith smallest value for i = 1,…,n. Law (2007) suggests three different functions that can be used as an empirical distribution function, as shown below. Note that (8.2) through (8.4) is the first approach, while (8.5) and (8.6) represent the second and third approaches, respectively.

$$F(x) = 0, \quad \text{for } x < X_{(1)}, \tag{8.2}$$

$$= \frac{(i-1)}{(n-1)} + \frac{(x - X_{(i)})}{((n-1)(X_{(i+1)} - X_{(i)}))}$$

$$\text{for } X_{(i)} \le x < X_{(i+1)}, \tag{8.3}$$

$$= 1, \quad x > X_{(n)}, \tag{8.4}$$

$$F(x) = \frac{(\text{number of } X_i\text{'s} \le x)}{n}, \tag{8.5}$$

$$F(X_{(i)}) = \frac{(i - .5)}{n} \quad \text{for } i = 1, \ldots, n. \tag{8.6}$$

Note that each of these three functions behaves in a way that one would intuitively expect in order to represent the data—that is, the functions increase most rapidly in value in regions where there are many data points.

The functions given in (8.2) through (8.5) provide values for F for any value of x, while the function given in (8.6) only provides values for F at the data points. In addition, the functions shown in (8.2) through (8.4) and (8.6) assume that each of the data points is distinct. This is not a major difficulty if the data represent a continuous random variable.

The advantage of the third function, given by (8.6), is that it allows the possibility of generating random variate values larger than the largest data point; the other two functions do not allow for this.

Finally, the function F given in (8.2) through (8.4) assumes a linear interpolation between the data points, while the function F given in (8.5) provides a step function for the relevant distribution.

A numerical example for these functions is shown in Example 8.6.

Example 8.6: Empirical Distribution Functions for Representing Changeover Times

Consider the data used for Example 8.1, involving the changeover times for a coating operation. In order to have each of the data points unique, let's suppose that the second value of 29.8 minutes was actually 29.9 minutes. The modified data set is shown as 30.6, 24.5, 31.8, 44.3, 34.1, 25.4, 28.1, 35.5, 23.7, 28.6, 39.1, 33.4, 26.8, 27.2, 31.1, 35.3, 25.6, 29.8, 33.9, 37.6, 26.5, 28.2, 22.1, 29.9, and 39.3.

Since the values are distinct, we can employ any of the three functional representations shown earlier. Reordering the data points from smallest to largest value gives us 22.1, 23.7, 24.5, 25.4, 25.6, 26.5, 26.8, 27.2, 28.1, 28.2, 28.6, 29.8, 29.9, 30.6, 31.1, 31.8, 33.4, 33.9, 34.1, 35.3, 35.5, 37.6, 39.1, 39.3, and 44.3.

The values for the three different formulations of F for a few of the data points are shown in Table 8.6.

A graph for the first representation of F, as given by Equations 8.2 through 8.4, is shown in Figure 8.6.

Note that the graph rises more rapidly in regions with more data points (and conversely less rapidly in regions with fewer data points). For example, the slope of the graph from $X_{(24)}$, 39.3, to $X_{(25)}$, 44.3, is very small.

Note also that these representations do not necessarily fully specify the distribution functions. For example, for the second representation given in Table 8.6, the value for x at which F(x) = 0 must be specified, and

TABLE 8.6

Values for Various Representations of the Empirical Distribution Function

Data Point	F Value from Equations 8.2 through 8.4	F Value from Equation 8.5	F Value from Equation 8.6
$X_{(1)}$, 22.1	0	0.04	0.02
$X_{(5)}$, 25.6	0.166666667	0.2	0.18
$X_{(9)}$, 28.1	0.333333333	0.36	0.34
$X_{(13)}$, 29.9	0.5	0.52	0.5
$X_{(17)}$, 33.4	0.666666667	0.68	0.66
$X_{(21)}$, 35.5	0.833333333	0.84	0.82
$X_{(25)}$, 44.3	1	1	0.98

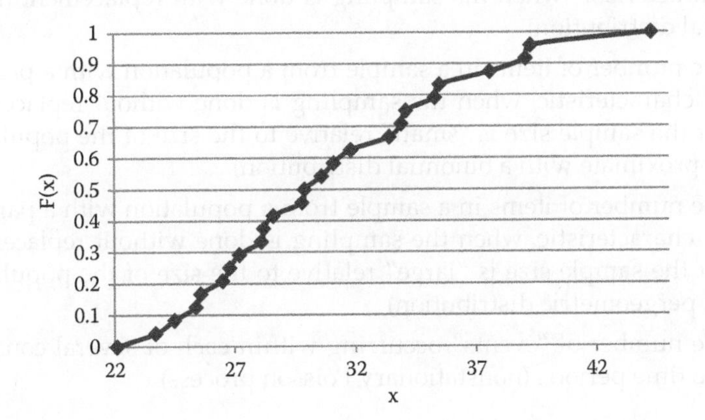

FIGURE 8.6
Empirical distribution function graph resulting from Equations 8.2 through 8.4.

for the third representation given in Table 8.6, the values for x at which F(x) = 0 and F(x) = 1 must both be specified.

The second representation for F allows for the generation of random variate values that are less than the smallest data point; in particular, $Pr(x \leq 22.1) = .04$, so as long as $Pr(x = 22.1) \neq .04$, such random variate values will be possible. By the same token, the third representation allows for the generation of random variate values that are smaller than the smallest data value or larger than the largest data value.

As the number of data points (n) becomes larger, the three representations for F will more closely approximate each other.

The representation of the continuous empirical distribution function varies according to the simulation software package employed. For example, the Arena simulation software package employs the representation given by Equations 8.2 through 8.4, involving the linear interpolation between data points.

8.2.5.2 Fitting a Theoretical Distribution to the Data

As noted earlier in this chapter, it is usually preferred to fit a theoretical distribution function as opposed to an empirical distribution function to the data. There are two aspects of fitting a particular theoretical distribution to the data: (1) choosing a correct family of distributions (e.g., normal, exponential, uniform) and (2) choosing the best parameter values associated with the distribution. We will discuss the choice of a particular family of distributions first.

In some situations, there may be an underlying reason to employ (or to not employ) a particular theoretical distribution function. Examples of situations that would call for a particular discrete distribution would include

1. The number of items in a sample from a population with a particular characteristic, when the sampling is done with replacement (binomial distribution)
2. The number of items in a sample from a population with a particular characteristic, when the sampling is done without replacement but the sample size is "small" relative to the size of the population (approximate with a binomial distribution)
3. The number of items in a sample from a population with a particular characteristic, when the sampling is done without replacement but the sample size is "large" relative to the size of the population (hypergeometric distribution)
4. The number of "events" occurring within each of several consecutive time periods (nonstationary Poisson process)

With respect to the second situation, Burr (1953) notes that if the size of the population is at least eight times the size of the sample, the binomial is an appropriate approximation to the hypergeometric. One of the main reasons for employing a binomial distribution instead of a hypergeometric would be its simplicity.

An example of the "events" referred to in the fourth situation would be customer arrivals to a service system such as a hospital emergency department, a branch bank office, and a restaurant.

For situations involving a continuous random variable, Law (2007, Chapter 6), in his book notes several examples in which a particular distribution might be used. For example, the uniform distribution can be appropriate for situations in which little is known about the underlying process, the gamma or Weibull distributions often represent the time to complete a task, and the normal distribution can be used to represent the sum of a large number of independent, identically distributed quantities (by the central limit theorem).

By the same token, there are certain distributions that should not be used in particular situations. For example, one should not employ a distribution that can result in negative random variate values for nonnegative quantities.

For this reason, instead of using an unmodified normal distribution with say a mean of 4 and a standard deviation of 10, one might employ a truncated normal distribution for representing task duration, which obviously cannot be negative.

In addition to the use of prior knowledge, one might employ summary statistics of the data to help identify an appropriate distribution function. For example, for a symmetric distribution, the mean must equal the median; hence, if the sample mean associated with the data is not at least "close" to the sample median, a symmetric distribution is probably not appropriate.

Once a theoretical distribution has been identified as being appropriate, then one only has to identify the appropriate parameter values. The parameters associated with a particular family of distributions depend upon the family. For example, for a normal distribution function, the parameters that will completely define its density and distribution functions are its mean (μ) and standard deviation (σ), while for the uniform distribution, these parameters are its minimum (a) and maximum (b) possible values.

As an example, for the distribution associated with the number of defective items in a sampling (with replacement) of items, the parameters would be (1) the probability that a randomly selected item is defective and (2) the sample size.

Any of several different methods can be used to select the parameter values for a specific distribution and data set. One frequently used approach is maximum likelihood estimation. (See Myung, 2003, for a tutorial on maximum likelihood estimation.) This approach relies on the choice of a parameter value that maximizes a likelihood function of that parameter given the set of data collected. More specifically, the likelihood function represents the "likelihood" that a particular parameter value is correct, given a particular set of data. For a discrete distribution, this would be just the product of the probabilities associated with the individual data values; that is, the likelihood function is given by

$$L(\theta \mid X_1, X_2, \ldots, X_n) = P_\theta(X_1)P_\theta(X_2)\ldots P_\theta(X_n),$$

where

θ represents the set of parameters

$X_1, X_2, \ldots, X_n$ is the set of data points

$P_\theta(X_i)$ is the relevant density function value associated with the data point X_i

Note that we can express the likelihood as a product of the individual probabilities since the data are independent samples.

If the distribution function under consideration represents a continuous random variable, then the likelihood function is just the product of the density function values for the data collected.

Example 8.7 illustrates the calculation of the likelihood function values for a small set of data for a Poisson distribution.

Example 8.7: Computation of Likelihood Function Values for a Poisson Distribution

Suppose that three data points have been collected to represent the number of customers in a group entering a fast-food restaurant, as given by $X_1 = 2$, $X_2 = 3$, and $X_3 = 2$. The Poisson distribution function (with parameter λ, representing the mean of the random variable) is given by

$$\begin{aligned} p(x) &= \frac{e^{-\lambda}\lambda^x}{x!} \quad \text{for } x = 0,1,2,\ldots, \\ &= 0 \quad \text{otherwise.} \end{aligned}$$

Therefore, the likelihood function, for this set of three data points, is given by

$$\begin{aligned} L(\lambda \mid X_1 = 2, X_2 = 3, X_3 = 2) &= \frac{e^{-\lambda}\lambda^2}{2!} * \frac{e^{-\lambda}\lambda^3}{3!} * \frac{e^{-\lambda}\lambda^2}{2!} \\ &= \frac{e^{-3\lambda}\lambda^7}{2!3!2!} \\ &= \frac{e^{-3\lambda}\lambda^7}{24}. \end{aligned}$$

And various likelihood function values for respective values of λ are shown in Table 8.7.

Note that the value of λ in Table 8.7 that maximizes the likelihood function is the one that corresponds to the sample mean for the data. This will always be the case—that is, the maximum likelihood estimator for the Poisson distribution will always be the sample mean value for the data.

As with the Poisson distribution, the maximum likelihood estimates for the parameter values will often correspond to the respective values for sample statistics.

TABLE 8.7

Likelihood Function Values for Various Values of λ

λ	$L(\lambda)$
1	0.002075
2	0.0132
2.3333	0.014307
3	0.011246

For at least a few density functions, the maximum likelihood estimates based on data collected may not be the best parameter values to use. For example, the maximum likelihood estimates for the parameters of the uniform distribution are the minimum and maximum data values. Yet if one chooses these values for the parameters for a fitted uniform distribution to be used in a decision model, then that model would not be able to generate any values for the relevant variable that are less/greater than the minimum/maximum data value. Since such extreme values can greatly affect the model's behavior, the use of a minimum that is less than the minimum data value and a maximum that is greater than the maximum data value makes sense. Hence, sometimes, the software package used for fitting a distribution to the data will employ a heuristic approach to generate the parameters for specific distributions.

In order to test how well a particular distribution fits the data, any of several procedures can be applied. For example, the decision maker might just visually compare a histogram for the data to a graph of the distribution function. If this procedure is used, the decision maker should employ several different histograms with varying widths for the cells of the histogram.

A more sophisticated approach would involve the use of the square error criterion. The square error is given by

$$\sum_{i=1}^{k}(f_i - f(x_i))^2,$$

where

f_i is the relative frequency of the data within the ith interval of a histogram for the data

$f(x_i)$ is the probability that the random variable associated with the hypothesized distribution would fall in the ith interval of the histogram

k represents the number of cells of the histogram

Finally, one could employ a "goodness of fit" procedure such as the chi-square goodness of fit test, the Kolmogorov–Smirnov (K–S) goodness of fit test, or the Anderson–Darling test. These tests rely on an approach called "hypothesis testing" from the field of statistics. Basically, one is testing the null hypothesis (H_0) that the X_i data points are independent, identically distributed random variables corresponding to the hypothesized distribution function. A problem with hypothesis tests of this type is that failure to reject the null hypothesis should not be interpreted as accepting it as true, but only that there is not enough evidence to reject it. In addition, as noted by Law (2007, p. 340), when the amount of data are not very large, these tests are not very good at recognizing differences between the data and the hypothesized distribution, which means that H_0 will not be rejected; on the other hand, if there are a lot of data, H_0 will almost always be rejected since it will never be exactly true. For these reasons, it is usually a good idea to just view the hypothesized distribution function superimposed over a histogram of the data and to also look at the value of the square error in evaluating various hypothesized distributions.

The chi-square test involves computing the number of data points in each of several adjacent intervals and comparing these numbers to what they would be if hypothesized distribution is correct. More specifically, the steps of the procedure are as follows:

1. Divide the entire range of the fitted distribution into k adjacent intervals, $[b_0,b_1),[b_1,b_2),\ldots,[b_{k-1},b_k)$, where b_0 could be $-\infty$ and/or b_k could be $+\infty$.
2. Compute N_j = number of X_i's in the jth interval, $[b_{j-1},b_j)$, for $j = 1,2,\ldots,k$. (Note that $n = \sum_{j=1}^{k} N_j$ is the number of data points.)
3. Compute p_j as the expected probability of having a data point occur in the jth interval if the fitted distribution is correct. Note that p_j will be computed with an integral for continuous data and with a summation sign for discrete data.
4. Compute the chi-square test statistic, $\chi^2 = \sum_{j=1}^{k}((N_j - np_j)^2/np_j)$.

The smaller the value of χ^2, the better the fit of the distribution to the data. More specifically, the null hypothesis should be rejected at the $1 - \alpha$ level if the test statistic, χ^2, is greater than $\chi^2_{k-1,1-\alpha}$, the critical value from the chi-square distribution with $(k - 1)$ degrees of freedom. (Note that k is the number of intervals used in the process.)

A major difficulty with conducting the chi-square test is how to choose the intervals. A typical approach involves choosing the intervals so that $p_1 \approx p_2 \approx p_3 \approx \cdots \approx p_k$ (called the equiprobable approach). Example 8.8 illustrates this approach.

Example 8.8: Choosing Intervals for the Chi-Square Test Using the Equiprobable Approach

Suppose that the fitted distribution is the normal distribution with a mean of 10 and a standard deviation of 2. Suppose also that we have n = 100 data points and that we want to have k = 10 intervals. Now, using the equiprobable approach, we would have $p_j = 1/k = 1/10 = .1$ for $j = 1,2,\ldots,10$. Set $b_0 = -\infty$; then b_1 must satisfy that the integral of the hypothesized normal distribution evaluated from $-\infty$ to b_1 must have a value of .1 or $\int_{-\infty}^{b1} \hat{f}(x)dx = 0.1$, or PR($\eta$ (mean = 10, std. dev. = 2) $\leq b_1$) = .1, or PR($Z <= (b_1-10)/2$) = .1, or $(b_1-10)/2 = -1.28$, or $b_1 = 7.44$.

Similarly, $b_2 = 8.3$. So the first two intervals are $[-\infty,7.44),[7.44,8.3)$. And then one would follow the same approach for determining the remaining intervals.

Another general guideline as stated in Law (2007, p. 344) is that in the equiprobable case, if k (the number of intervals) is greater than or equal to 3 and np_j (over all intervals) ≥5, then the test will be approximately valid for the all parameters known case, as shown through extensive testing by Yarnold (1970).

The K–S tests basically compare an empirical distribution function (derived directly from the data) to the hypothesized distribution function. Its main advantage, as compared to the chi-square tests, is that these tests do not require a grouping of the data (into intervals), which therefore eliminates the troublesome problem of interval estimation.

The K–S tests tend to be more powerful than chi-square tests for many alternative distributions (Stephens, 1974); however, the range of applicability for the K–S tests for various distributions is limited, and for discrete data, they must use a complicated set of formulas as compared to the chi-square tests.

There are several software packages available for fitting a distribution to a set of data, including the EasyFit™ (EasyFit, 2013), Arena Input Analyzer™ (Kelton et al., 2015), StatTools™ (Palisade StatTools, 2013), and Stata™ (Stata, 2013). These software packages typically will allow a user to either specify a particular distribution function and then find the set of parameter values for that distribution function that will allow a "best fit" to the data; or the software will perform a "complete enumeration" of the various possible families of distributions, along with a procedure (heuristic or exact) to choose the parameter values that best fit the data for that particular family. Various tests will be made over each of the distributions fit in order to determine the distribution that best fits the data.

Example 8.9 illustrates the some of the calculations involved in evaluating how well a hypothesized distribution fits a set of data.

Example 8.9: Evaluating the Fit of a Hypothesized Distribution to Process Time Data

Suppose that the data on a process time (in minutes) have been collected. Twenty-five data points have been collected: 5.43, 4.23, 4.46, 6.18, 7.36, 7.28, 4.45, 6.14, 4.75, 7.89, 7.55, 7.36, 6.97, 5.84, 4.91, 5.64, 6.63, 7.41, 4.52, 7.18, 5.79, 4.33, 7.81, 5.21, and 7.60.

The hypothesized distribution is a uniform distribution with a minimum of 4 minutes and a maximum of 8 minutes. (Note that the parameter value of 4 is smaller than any of the data points and that the parameter value of 8 is larger than any of the data points, as should be the case in choosing the best parameter values for a uniform distribution.)

Let's choose our intervals for the histogram as [4, 5), [5, 6), [6, 7), and [7, 8). This choice satisfies the equiprobable condition, with $p_j = .25$ for $j = 1,\ldots,4$. This choice of intervals also satisfies the conditions given by Yarnold (1970) since $np_j = 25(.25) = 6.25$ for all j, which is greater than the value of 5.

Now, the number of data points in each of the four respective intervals of the histogram is given by 7, 5, 4, and 9. Hence, the square error is given by $(7/25-.25)^2 + (5/25-.25)^2 + (4/25-.25)^2 + (9/25-.25)^2 = .0236$.

The chi-square test statistic is given by

$$\chi^2 = \frac{\sum_{j=1}^{k}(N_j - np_j)^2}{np_j}$$
$$= \frac{(7-6.25)^2}{6.25} + \frac{(5-6.25)^2}{6.25} + \frac{(4-6.25)^2}{6.25} + \frac{(9-6.25)^2}{6.25}$$
$$= 2.36.$$

Now, we have k = 4 since we have four intervals for our histogram. If we let $\alpha = .05$ (i.e., we are conducting the test at the $1 - \alpha = .95$ or 95% level), then we have $\chi^2_{k-1,1-\alpha} = \chi^2_{3,.95} = 7.815$ (from a table of values). Since the test statistic is not greater than the critical value from the chi-square distribution, we fail to reject the null hypothesis. Therefore, we would accept the uniform distribution with a minimum value of 4 and a maximum value of 8 as a good fit for the data.

Example 8.10 illustrates the use of a software package for fitting a distribution to a set of data.

Example 8.10: Fitting a Distribution to the Number of Calls Made to an Emergency Call Center within an Interval of Time

A simulation model was constructed for the 911 call center of a major metropolitan area. The purpose of the model was to determine workforce staffing levels over a typical weekly period. Out of several types of input data for the model, one of the most important was the pattern of incoming calls. Through interaction with the manager of the system, the analyst charged with building the model determined that the probability distributions representing the incoming calls could be treated as varying by 2-hour periods according to the day of week. In particular, there were 48 different probability distributions that needed to be determined in order to represent the incoming calls: 12 midnight–2 a.m., 2–4 a.m.,…, 10 p.m.–12 midnight for any Monday through Thursday and 12 midnight–2 a.m., 2–4 a.m.,…, 10 p.m.–12 midnight for each of Friday, Saturday, and Sunday.

The analyst wanted to check to see if the probability distribution for the number of calls in a 2-hour period was approximately normally distributed. Of course, this would not be exactly true since the number of calls must be integer, but since the number was large, the normal distribution might be good enough through the use of a rounding process. In particular, a normal distribution is a good approximation to the Poisson distribution if the mean for the Poisson is greater than 10 and the variance of the normal is approximately equal to the mean. For the data analyzed here, the sample mean was approximately 191, and the sample variance was approximately 185. Hence, the normal distribution was taken as a good approximation to the Poisson.

Data were collected on the number of calls received from 10 a.m. to 12 noon on 44 different Fridays. The 44 data points were as follows:

TABLE 8.8

Fitness Measures for Various Hypothesized Distributions

Distribution (with Parameter Values)	Square Error	Chi-Square p Value	K–S Test p Value
Normal (191, 13.4)	0.018345	0.0849	>0.15
155 + 65*Beta(2.59,2.04)	0.026387	0.0311	>0.15
Triangular (155,199, 220)	0.020414	0.157	>0.15
155 + Weibull (39.1, 2.15)	0.049815	<0.005	0.0937

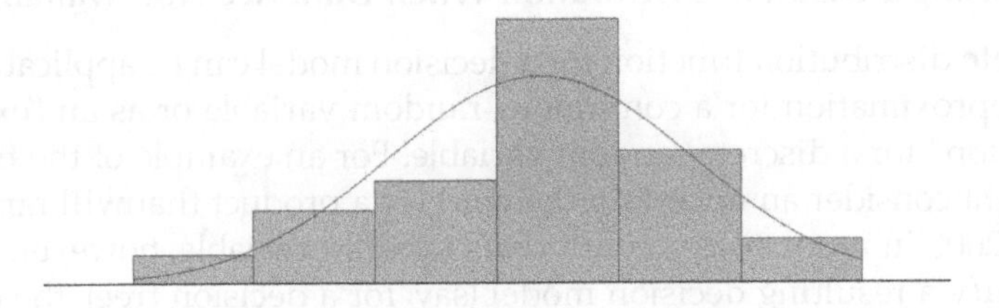

FIGURE 8.7
A histogram of the number of calls as compared to a graph of the fitted normal distribution. (Output from Arena's Input Analyzer.)

> 174, 198, 182, 205, 213, 202, 168, 172, 176, 162, 190, 197, 194, 181, 196, 201, 198, 220, 195, 195, 184, 191, 217, 181, 205, 191, 208, 196, 195, 175, 194, 197, 181, 191, 188, 201, 184, 200, 155, 188, 200, 204, 190, and 185.
>
> A number of different distributions were hypothesized, and estimates (using maximum likelihood estimation) were made of the relevant parameters. Results for this analysis, which were generated from Arena's Input Analyzer are shown in Table 8. A graph of the hypothesized normal distribution, superimposed over a histogram of the data, is shown in Figure 8.7.

In Table 8.8, the p values for the chi-square and K–S tests are the largest values of type I error probability that allows the respective distribution to fit the data; specifically, the higher the p value, the better in terms of the distribution's fit.

The reader will note that the normal distribution provides the best fit in terms of square error. However, the rankings of the various distribution functions vary according to the evaluation test used. For example, whereas the normal distribution performs best under the square error criterion, the triangular distribution ranks first according to the chi-square test.

8.3 What to Do If Data Are Not Available

There will be situations where data are not available to form the probability distributions needed for a decision model. Examples of these situations include cases where either there is not enough time available to collect and analyze the data or the decision situation is so novel that data are not available. In these situations, expert opinion may be employed to form the required distributions.

There are two cases to consider with respect to obtaining estimates from subject matter experts in order to estimate input probability distributions for a decision model—the first to be addressed will be the situation involving discrete random variables and the second situation involves continuous random variables.

8.3.1 Fitting a Discrete Distribution When Data Are Not Available

A discrete distribution function for a decision model can be applicable either as an approximation for a continuous random variable or as an "exact representation" for a discrete random variable. For an example of the first case, one might consider an uncertain demand for a product that will range from 1000 to 2000 in number as a continuous random variable; however, in order to simplify a resulting decision model (say, for a decision tree), the demand for the product might be represented as the following: high (1950), moderately high (1700), medium (1500), moderately low (1300), or low (1050).

With respect to the second case in which the uncertain quantity to be represented would be naturally a discrete random variable, examples would include (1) the number of customers in a group entering a fast-food restaurant (1, 2, 3, 4, 5, 6, or larger), (2) the outcome associated with a medical test (positive or negative), (3) the number of defects found in a part (0, 1, 2,...), (4) the outcome associated with inspection of a part (good or defective), and (5) the number of defective items in a sample of 5 randomly selected items from a lot of 10 items (0, 1, 2, 3, 4, 5).

Another way to classify the situation involving the fitting of a discrete distribution to the data involves whether the discrete distribution to be determined is empirical or theoretical in nature. Examples of the former would be cases 1, 2, 3, or 4 mentioned in the previous paragraph, while an example of the latter would be case 5 in the previous paragraph, involving the number of defective items in a sample taken from a larger lot. More specifically, as mentioned in Chapter 5, the number of defective items randomly sampled from a homogeneous lot of items will be a random variable with a hypergeometric distribution.

If the discrete distribution to be determined is empirical in nature, then specific probabilities associated with the various outcomes need to be estimated. This occurs, for example, in the case of a discrete chance node for a decision tree or influence diagram. Let's consider the hypothetical example of the decision tree in Chapter 7, specifically Figure 7.1 and outcome node 4. The outcomes associated with this node are "good market" and "bad market." The probabilities that need to be determined are probability of a good market and probability of a bad market.

One very commonly used approach that might be employed would be to just ask an expert or a group of experts the probability of the occurrence of a good market. Then, the probability of a bad market would, by default, just be (1 - the probability of a good market). If there are more than two branches emanating from the relevant outcome node, then the subject matter experts would assess the probabilities associated with all of the branches but one, with the probability associated with the last branch being implicitly determined.

A second way to assess the probabilities associated with an empirical discrete distribution involves an implicit approach in which hypothetical "bets" are placed by the subject matter experts. In this way, the experts provide amounts to bet (deterministic quantities) rather than probabilities.

The basic idea would be to first ask the subject matter experts to rank the various outcomes in terms of decreasing probabilities of occurrence. Then taking the most likely outcome first, the subject matter expert is asked to determine the specific amount of money to win or lose such that he or she is indifferent as to which side of a hypothetical bet to take. Then, an equation can be formulated such that the expected values for each side of the bet are equal; the equation has one unknown, the probability being sought. This approach assumes that the subject matter expert(s) is (are) risk neutral.

As an example, consider a hypothetical example in which a major defense contractor has a contract to refurbish a new weapons system for the U.S. Navy. In order to estimate manpower requirements, costs, and so on, a simulation model of the refurbishment process has been developed. One part of the model involves the possible repair of a major component of the system. Since the system and component are new, there is no directly applicable data that can be used to estimate the condition of the major component. However, the subject matter experts do have familiarity with similar systems and components; this familiarity allows them to make informed estimates with respect to the condition of the system component.

The model requires the probabilities associated with the following outcomes:

1. No repair required
2. Minor repair required
3. Major repair required

Suppose that the subject matter expert thinks that of these three outcomes, minor repair is most likely, with no repair next, and finally major repair least likely. The subject matter expert would be asked to set values for X and Y such that he or she would be indifferent between the following two options:

Option 1—Bet \$X to win \$Y if a randomly selected unit of the component requires minor repair; lose the \$X if the component does not require minor repair.

Option 2—Do not bet.

Now, the subject matter expert would be expected to "hone in" on the values for X and Y such that he or she is indifferent to these two options;

in addition, the values for X and Y should be set in such a way that, even though the situation is hypothetical in nature, the subject matter expert would "care" about the two options. Finally, the value for Y may be "set" at a particular value, while the value for X would be adjusted until indifference is achieved.

Let's suppose that if Y is set at $1400, the value for X must be $700 in order for the subject matter expert to be indifferent between the two options. Then, assuming that the subject matter expert is risk neutral over the range of outcomes, the expected value associated with option 1 must be equal to the expected value associated with option 2; or, if $P_{\text{minor repair}}$ denotes the probability that a randomly selected unit of the component requires minor repair, then

$$EV(\text{Option 1}) = EV(\text{Option 2}), \text{ or}$$

$$P_{\text{minor repair}}(1400 - 700) - (1 - P_{\text{minor repair}})(700) = 0, \text{or}$$

$$P_{\text{minor repair}}(1400) - P_{\text{minor repair}}(700) - 700 + P_{\text{minor repair}}(700) = 0, \text{or}$$

$$1400P_{\text{minor repair}} - 700 = 0, \text{or}$$

$$P_{\text{minor repair}} = .5.$$

A similar scenario would be set up to determine $P_{\text{no repair}}$, which must be less than .5 but greater than 0.

If the discrete distribution to be determined is theoretical in nature, only the parameters for that distribution need to be estimated. For example, for the case of a hypergeometric distribution, the parameters required to determine the distribution function would be the number of items in the lot, the number of items in the sample, and the number of defective items in the lot. Only the last quantity would have to actually be estimated by the expert, as the first two would be determined from the situation.

8.3.2 Fitting a Continuous Distribution When Data Are Not Available

As mentioned earlier, in some cases, a discrete quantity might be approximated by a continuous random variable—for example, this might be done in cases where a discrete quantity can attain very large values, such as item demand in thousands of units or cost in thousands of dollars.

As in the case with determining a discrete distribution in the absence of data, two situations are considered for continuous distributions: determination of an empirical continuous distribution and determination of a theoretical continuous distribution. We will consider the former situation first.

Determination of an empirical continuous distribution in the absence of data involves just the specification of points (x and y values) on the cumulative distribution function graph for the uncertain quantity and then interpolating between these points. The interpolation might be linear in nature or nonlinear if a more exact representation is desired.

Consider the example of specifying the probability associated with your favorite college basketball team winning x games in the upcoming season. (Here, we are approximating a discrete quantity, number of games, with a continuous variable.) Basically, you want to assess F(x), the distribution function for this quantity, for different values of x. Suppose that your team will win 40 games, if it wins all of the games it plays, including the NCAA tournament. Then you could set two initial points for the following function:

$$F(0) = 0 \quad \text{and} \quad F(40) = 1.$$

In other words, the probability of your team winning less than or equal 0 games is 0, (assume your team will win at least one game) and the probability of your team winning less than or equal to 40 games is 1. Then, F would be assessed for various values of x through the answers to the following questions:

What is the probability that your team will win less than or equal to 5 games?

Answer: 0

What is the probability that your team will win less than or equal to 10 games?

Answer: .05

What is the probability that your team will win less than or equal to 15 games?

Answer: .3

What is the probability that your team will win less than or equal to 20 games?

Answer: .5

What is the probability that your team will win less than or equal to 25 games?

Answer: .6

What is the probability that your team will win less than or equal to 30 games?

Answer: .8

What is the probability that your team will win less than or equal to 35 games?

Answer: .95

This would lead to the distribution function graph shown in Figure 8.8.

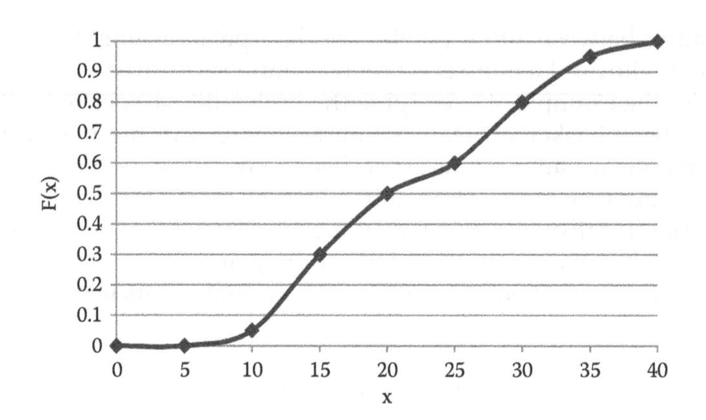

FIGURE 8.8
Distribution function for the number of games won by your favorite college basketball team.

Of course, the questions could be easily rephrased to something like: "What is the probability that your favorite team will win more than 25 games?" Then F(25) would be set equal to 1 minus this quantity.

In addition to the "direct approach" for assessing individual probabilities on F, one could also set up betting scenarios to determine points on the graph for F, as described in Section 8.3.1 for discrete distributions.

8.4 Biases and Heuristics in a Subjective Probability Assessment

Of course, one should always employ the best experts available in assessing probabilities in the absence of data. However, even the best experts can give biased estimates, resulting from the use of less than desirable heuristic approaches when providing these estimates. Being aware of these biases and heuristics can result in more accurate estimates.

Tversky and Kahneman (1974) are the major researchers in this field. One of the heuristics they identified is the "availability heuristic." This heuristic involves judging the probability of an event according to how easily these or similar events can be recalled. As an example, one of the important inputs for a simulation model of an emergency (911) call center was a probability distribution over the number of calls made for the same event. A recent event involved an outdoor shooting in a well-to-do neighborhood. Thirty-nine calls were made to the emergency call center as a result of the gunshots heard from the shooting. This salient event would have resulted in an overestimation of the number of calls associated with a single event if the data (relating to this and other events) had not been available for these situations.

The *representativeness heuristic* involves assessing the probability of an event or a sample "by the degree by which it is (1) similar in essential properties to its parent population, and (2) reflects the salient features of the process by which it is generated" (Kahneman and Tversky, 1972). A situation illustrating this heuristic involved an experiment in which a character named Tom W. was described to a group of subjects as a graduate student with the following characteristics: (1) a need for order and clarity, (2) a strong drive for competence, (3) mechanical in nature, and (4) self-centeredness, but with a deep moral sense. The experimental subjects were then asked to rank the likelihood of Tom W. being a grad student in each of the following fields: business administration, computer science, engineering, humanities and education, law, library science, medicine, physical and life sciences, and social science and social work.

Most people would only consider a stereotypical "representation" of the various people in these fields in developing their rankings of likelihood. In other words, Tom W. might be considered as "representative" of someone in engineering or computer science. However, a better approach would be to consider the unconditioned probability of someone being in each of these fields and then thinking about the conditional probabilities of someone being in each of these fields given that they have Tom W.'s characteristics. The bottom line is that most people would overestimate the likelihood of Tom W. being a graduate student in engineering or in computer science since they do not consider the marginal or unconditioned probabilities.

The *anchoring and adjusting heuristic* is typically applied when people have an initial estimate of some quantity (an anchor) and then make an adjustment to the estimate for a new, but related, situation. Often, the adjustment is not large enough because the estimator is anchored to the initial estimate. A good example might be in estimating the lead time for a new vendor of a product. The estimate of the new lead time might be anchored to the lead time of the previous vendor.

Finally, and probably, one of the most obvious things to consider in subjective estimation is "motivational bias." This occurs when experts provide estimates that they know are incorrect but they have a motivation do so. A good example of this occurs in the estimation of resource requirements and task durations in project management. Upper-level management will tend to underestimate these requirements and durations in order to provide a winning bid for the project. Lower-level management will tend to overestimate these quantities in order to make sure that they will achieve their goals.

Material Review Questions

8.1 What are the two basic approaches that can be used to obtain input probability distributions for decision models?

8.2 Why is it important to have accurate input probability distributions for decision models?

8.3 What types of nodes in a decision tree and an influence diagram, respectively, require input probability distributions?
8.4 Give three examples of elements of a simulation model that would require input probability distributions.
8.5 When data are available, what are the three basic ways that they can be used to represent probabilistic inputs for decision models?
8.6 In what situation would one want to use the actual data as input for a simulation model for decision making?
8.7 What is an advantage associated with the use of an empirical distribution function as opposed to a theoretical distribution function?
8.8 What are the four basic steps associated with fitting a probability distribution to a set of data?
8.9 What are the two types of input for a decision model?
8.10 Give two examples of qualitative input used for a decision model.
8.11 Why is it important that independent samples be collected in the data collection process?
8.12 Give two examples of situations that would result in dependent data samples being collected.
8.13 What are the two vehicles that one could use in determining whether the sample data collected are independent?
8.14 How would a scatter diagram appear for independent samples?
8.15 Give an example of bias that can result when an analyst collects data via observation.
8.16 Define an "outlier" with respect to a set of data that have been collected.
8.17 Why would a simple analysis of data collected (without the additional development of a sophisticated decision model) be useful in improving a system?
8.18 What is Sturges's rule?
8.19 What is the advantage of the representation given by (8.6) for an empirical distribution function over that given by (8.2) through (8.4)?
8.20 As the number of data points (n) becomes larger, the three representations for an empirical distribution function, given by (8.2) through (8.6), will more closely approximate each other (*true* or *false*).
8.21 What are the two aspects associated with fitting a theoretical distribution to a set of data?
8.22 Give two examples of situations where an analyst might, a priori, assume a particular distribution function for a discrete random variable to represent a process?
8.23 Under what condition might one safely assume that the binomial distribution would be an appropriate approximation to the hypergeometric distribution in a sampling without replacement process?
8.24 What would be an advantage of approximating a hypergeometric distribution with a binomial distribution?

8.25 Give two examples of situations where an analyst might, a priori, assume a particular distribution function for a continuous random variable to represent a process.

8.26 What is a typical method used to estimate the parameters for a distribution function based on data collected?

8.27 What is the maximum likelihood estimator value for the mean for a Poisson distribution for a set of data that have been collected?

8.28 What are three basic approaches that one can use to determine how well a hypothesized distribution function fits a set of data?

8.29 When the number of data points is very large, goodness of fit tests such as the chi-square or the K–S tests will almost always reject the null hypothesis of the data points being independent, identically distributed data points from the hypothesized distribution (*true* or *false*).

8.30 What is the main advantage of the K–S goodness of fit test over the chi-square test?

8.31 The three criteria, that is, square error, p value for the chi-square test, and p value for the K–S test, will always give the same ranking with respect to the quality of fit of a collection of density functions to a set of data (*true* or *false*).

8.32 Name four different software packages that one can employ to fit a distribution function to a set of data.

8.33 Subjective probability estimates can often be made through the use of a betting scenario (*true* or *false*).

8.34 Give at least one example (not in the book) of the application of each of the four heuristic approaches for subjective assessments of a probability or an unknown value: availability, representativeness, anchoring and adjustment, and motivational.

Exercises

8.1 Suppose that you have collected the following data for the interarrival times for customers arriving to a service system (in minutes): 3.2, 4.1, and 3.8. Suppose that you are assuming exponential interarrival times.

a. What is the likelihood function for this situation?

b. Give the likelihood function values for $\beta=2$, $\beta=3$, $\beta=3.7$, $\beta=4$, and $\beta=5$, respectively, where the density function for the exponential is given by $(1/\beta)e^{-x/\beta}$.

c. Which value for β gives the largest likelihood function value? Why?

8.2 In groups of one to three students, collect data (independent samples) on a process that might be used as input to a simulation model.

Analyze the data through the use of correlation plots(s) and/or scatter diagrams, histograms, empirical distributions, and a software package for fitting distributions. In your analysis with the software package, try histograms with different numbers of cells and try fitting different distributions. Consider how you might handle any anomalous data. Write a brief report describing your results.

8.3 Suppose that the following sequential interarrival times (given in minutes) have been collected for arrivals of walk-in customers to a fast-food restaurant:

1.5, 2.1, 1.3, 1.1, 2.3, 2.1, 1.7, 4.2, 3.3, 3.5, 1.9, .6, 5.4, 3.4, 2.5, 6.3, 1.4, 3.4, 3.9, 1.8, 4.7, 2.9, .5, 4.6, 3.5, 4.4, .8, 6.6, 4.9, 2.8

Construct a scatter diagram to help determine whether or not these observations are independent in nature.

8.4 Suppose that the following data have been collected on the amounts of time (in minutes) spent by the staff of a call center with individual callers. Arrange the data in an increasing order of values, and using Sturges's rule, construct histogram(s) for these data. Determine whether the data represent more than one type of caller. If so, estimate the number of different types of callers, the percentage of calls coming from each type of caller, and the sample average time taken by each type of caller. Determine any "outliers" in these data:

6.95, 2.83, 9.41, 4.26, 3.68, 6.75, 9.48, 9.36, 6.98, 3.17, 8.89, 4.34, 4.72, 7.79, 3.38, 1.33, 7.22, 7.77, 9.77, 8.06, 8.55, 8.49, 3.33, 8.12, 6.96, 2.14, 6.27, 8.36, 4.04, 9.96, 6.87, 7.09, 7.14, 8.43, 4.86

8.5 Calculate an empirical distribution function of type (8.2) through (8.4) for the data given in Exercise 8.4. According to this function, what would be the probability associated with generating a random variate with a value less than 6.87 minutes?

8.6 Consider the data in Problem 8.4, after the removal of any anomalous data. Using an appropriate software package, fit several different probability distributions to the data generated from each of the respective underlying processes. Discuss the meaning of the various fitness criterion values associated with each hypothesized distribution function.

8.7 Suppose that you wish to employ the chi-square goodness of fit test in order to test the goodness of fit for a particular distribution function for a set of data. You have six intervals for the categorization of the data. The *number* of data points in each of the six intervals is in the order (ranging from the interval with the smallest left-side endpoint to the largest) 75, 87, 68, 85, 57, and 95.

You have used the equiprobable approach to determine the intervals. Determine the value of the chi-square test statistic and, based on this, make the decision of whether or not to reject H_0 (at the 95% level): the data points are IID random variables corresponding to the distribution function being tested.

8.8 Consider the Tom W. case discussed in this chapter. Suppose that there are about 2.4 million graduate students in the United States and about 200,000 of these students are enrolled in engineering. Suppose also that you estimate that approximately 40% of these engineering graduate students have the characteristics described for Tom W. and that only 10% of all graduate students have the characteristics described for Tom W. What would be your estimate of the probability that Tom W. is an engineering graduate student?

8.9 Discuss a personal situation in which an acquaintance of yours made an estimate of (1) a probability (either a number or a verbal estimate such as "highly likely") or (2) a quantity in which you think bias affected their estimate. What do you think was the source of the bias? What type of heuristic did they use? What could you have suggested to improve their estimate?

9

Use of Simulation for Decision Models

9.1 Introduction

This chapter presents the use of simulation for decision modeling. The basic idea is to provide a model that will allow a mapping from the decision space to the outcome space. More specifically, we want to allow for a representation or a computer program that will "map" a decision and then compute a probability distribution over the outcome space.

The word *simulation* means different things to different people. In its broadest sense, a simulation is just an imitation or representation of something real, usually a system or a process. In this sense, a simulation can involve people and hardware as well as software. In this chapter though, we are concerned with simulations that are software programs.

In the context presented here, we are interested in representing a decision situation and what happens in that situation when a decision is made. Most industrial engineers are familiar with the concept of *discrete event simulation*, in which the state of the system (as measured by important system variables such as number of entities in respective queues and status of resources) changes instantaneously at particular points in time called *events*; in particular, these state variables are discrete in nature. This is as opposed to *continuous simulation* in which the state variables (which could be such things as flows or velocities) are continuous in nature. Both of these types of simulation represent a system over time and are thus called time-based or *dynamic simulation models*.

Although certainly many decision situations involve the use of dynamic simulation models, the emphasis in the field of decision analysis has been in the area of *static simulation models*. In particular, influence diagrams and decision trees are usually evaluated through the use of static simulation models, which use *Monte Carlo simulation*. The focus in Monte Carlo simulation is the generation of random variates, that is, values associated with random variables.

Most introductory books on simulation provide a categorization of simulation models; for example, see Law (2007, Chapter 1), which provides a categorization according to static versus dynamic simulation models, deterministic versus stochastic simulation models, and continuous versus discrete simulation models. The focus in this chapter is in studying the probabilistic output associated with stochastic, static simulation models; however, we also present a few examples involving dynamic simulation.

The various results can be categorized according to (1) the number of attributes (or objectives) being considered and (2) the number of alternatives being considered. In particular, we can consider

1. One attribute
2. Multiple attributes

With respect to the number of alternatives, we can consider

1. One alternative
2. Two alternatives
3. A few alternatives (say 3–100)
4. A "large number" of alternatives

The focus in most books on simulation is on the case involving one attribute. In a sense, if we employ a multiattribute utility function as our main output measure, we are considering the "single attribute" case, where the attribute is "utility"; the real attributes are considered only implicitly. Therefore, all of the results on output analysis, except for one category of approaches, can be directly applied if expected utility (as computed via a multiattribute utility function) is the performance measure. The one category that requires a modification is the category of techniques called *ranking and selection*.

Another approach that can be used for the multiple attribute case would involve optimization of one attribute while constraining the other attributes; by varying the right-hand sides of the constraints, we can give the decision maker (DM) a variety of outcomes to consider.

With respect to the number of alternatives to consider, understanding the output associated with the single alternative case gives us a starting point to expand upon. The last category, involving a "large number" of alternatives, requires the use of a sophisticated optimization procedure since not all of the alternatives can be explicitly evaluated.

As a prelude to this chapter, we presented methodologies associated with the determination of probabilistic inputs for simulation models in Chapter 8. In this chapter, we briefly mention methodologies for the generation of random variates in Section 9.2. In Section 9.3, we discuss the characteristics of the output from a simulation model.

In Section 9.4, we briefly discuss variance reduction techniques, while in Sections 9.5 and 9.6, we discuss the comparison of two alternatives and several alternatives (more than two), respectively, using simulation. In Section 9.7, we present the use of optimization methods as interfaced with simulations to allow for the analysis of decision situations with many, or even an infinite, number of alternatives. In each of these three sections we present the use of multiattribute utility functions to allow for analysis of situations involving multiple objectives and risk.

9.2 The Generation of Random Variates

The key to a Monte Carlo simulation is the generation of random variates. A *random variate* value is an independent sample from a prescribed probability distribution, such as a normal distribution or a binomial distribution. These values are required in order to execute a simulation model with random aspects, such as a Monte Carlo simulation. For example, in the influence diagram for the lead poison testing decision example presented in Chapter 6, random variates are needed to represent the number of children with lead poisoning in the sample tested; these random variates should correspond to the hypergeometric distribution, as noted in the example.

Methods to generate random variates are typically programmed into simulation software packages. Even so, the user of one of these packages should have a general understanding of these methods. In this section, we just provide a brief discussion of this area. For more details, see any of several different books on simulation, such as Law (2007) or Banks et al. (2005).

The most popular method for generating random variates is the inverse transform method. Other methods include the acceptance–rejection method, the composition method, and the convolution method. The specific method used in a particular case is dependent on the distribution function that the random variate is supposed to represent. More specifically, the choice of a method for a particular distribution is based upon (1) how accurately the method represents the distribution function, (2) its efficiency (computing time and storage requirements), and (3) its complexity.

With respect to accuracy, the methods can vary depending on the parameters for the distribution.

With respect to efficiency, the techniques discussed in Chapter 8 involving fitting an input distribution to a set of data are of interest, except in this case the process is reversed; that is, we are interested in how well a set of data represents a particular distribution function.

9.3 Characterizing and Analyzing Output for a Single Alternative

The basic idea in the use of a simulation model for ranking alternatives is to execute the model with differing inputs (corresponding to the respective alternatives under consideration) and then rank the alternatives based on analyses of the outputs from the simulation. The difficulty is that the outputs from a simulation are both multidimensional (corresponding to the multiple attributes of the decision situation) and random in nature. In particular, an output from a simulation model will be an *estimate* of a particular performance measure or attribute value; this estimate is itself a realization of a random variable.

Examples of the outputs from a run of a simulation model for a decision situation would be the following:

1. A "discrete outcome," including successful/unsuccessful project or product, a positive/negative medical test, a successful/unsuccessful surgery, a project that finishes on schedule or behind schedule, and so on. (Note that there can be more than two outcomes, e.g., successful, moderately successful, moderately unsuccessful, and unsuccessful, depending upon how the model is constructed.)
2. A singular "continuous outcome" such as a profit or a cost.
3. A mean value such as mean utilization of a resource, mean time to complete a process, or mean waiting time for customers. Note that there are two types of mean values that one might have as output—time persistent means such as mean utilization or mean number in a queue or means based on observations from individual entities in the model such as mean waiting time over all customers. Monte Carlo models would never have time-persistent variables as output; hence, these models would never output time-persistent means.
4. A proportion (or a probability), such as proportion of customers who must wait longer than X minutes.

Note that a proportion might result from a model that was formulated in such a way that many independent trials were run with one replication of the model—in this case, each trial would provide a "discrete outcome," the first type of output listed. The analyst therefore has an option to formulate his or her model to represent a single trial or multiple trials within each replication. Note that these multiple trial–type models would only be formulated for Monte Carlo simulations, in which each trial simulation would "start fresh," with no entities in the system, often referred to as "empty and idle."

9.3.1 Sample Means, Sample Variances, Sample Proportions, and Confidence Intervals

With respect to mean values, which are computed from the outputs of several independent replications of a simulation model, each individual output itself can represent a sample mean. These sample means could be time persistent or based on observation. For example, let x(t) be a time-persistent variable from any particular independent replication of the simulation model, run from time 0 to time T; that is, x(t) is the value of the variable at time t. Such a variable might represent the number of customers in a system, the number of busy servers, or the status of a particular server (with a value of 0 for idle and 1 for busy). Then, the sample mean value and the sample variance for this variable would be given by $\bar{X}_T$ and s^2_{xT}, respectively,

$$\bar{X}_T = \frac{\int_0^T x(t)dt}{T} \tag{9.1}$$

$$s^2_{xT} = \frac{\int_0^T x^2(t)dt}{T} - \bar{X}_T^2 \tag{9.2}$$

In many cases, computing an integral of a function is a relatively complex procedure; however, for a discrete event simulation, this is not so since the x(t) function graph over time can be represented as a sequence of horizontal lines, connected by vertical lines. Hence, computing an integral (i.e., the area under a curve) just corresponds to summing the areas of rectangles. As mentioned earlier, these time-persistent variables would not be relevant for a Monte Carlo simulation model.

For the sample mean within a replication for a variable based on observation, examples would be mean time in the system by customers, mean time spent waiting by customers, and so on. One thing to note about this computation is that the number of customers considered in the computation can vary between replications. This is not a problem as long as the analyst identifies the variable properly. For example, the analyst might be interested in the mean time spent waiting by customers in a restaurant from 11 a.m. to 1 p.m.; the number of customers might vary over five replications according to 45, 56, 42, 59, and 48. If x_i represents the value of the variable for the ith observation for any particular replication of the model and I is the number of observations, the sample mean would be computed by $\bar{x}_I$ and the sample variance would be computed by s^2_X, respectively:

$$\bar{x}_I = \frac{\sum_{i=1}^{I} x_i}{I}, \tag{9.3}$$

$$s^2_X = \frac{\left(\sum_{i=1}^{I} x_i^2 - I\bar{x}_I^2\right)}{(I-1)}. \tag{9.4}$$

A summary of these computations for a "within replication" run for a simulation model is shown in Table 9.1.

Each type of output represents a sample value. In order to obtain multiple, independent samples, multiple (and independent) replications of the model must be made. These replications are typically run so that an independent stream (or substream) of random numbers is used for each replication. This is typically handled by the software package used in building the model. For example, the Arena simulation software allows the model builder to specify

TABLE 9.1

Summary of Computations for a "Within Replication" Run of a Simulation

Notation	Description/Definition	Expression for Computation
$\overline{X}_T$	Sample mean of a time-persistent variable, x(t), from a single independent run of a simulation	$\int_0^T x(t)dt/T$
$s^2_{X_T}$	Sample variance of a time-persistent variable, x_t, from a single independent run of a simulation	$\int_0^T x^2(t)\,dt/T - \overline{X}^2_T$
$\overline{x}_I$	Sample mean of an observation-based variable, x_i, from a single independent run of a simulation	$\sum_{i=1}^{I} x_i/I$
s^2_X	Sample variance of an observation-based variable, x_i, from a single independent run of a simulation	$\left(\sum_{i=1}^{I} x_i^2 - I\overline{x}_I^2\right)\Big/(I-1)$

different random number streams to be used for different respective purposes in the model (e.g., different outcome nodes for a decision tree), and then sequential substreams within a stream are used for the model replications. For example, if you specified that stream number 3 was to be used to generate the random variates for an outcome node representing market demand (as a continuous variable) for a product and that 10 replications were to be made for the simulation model representing the decision tree, then the ith replication of the model would employ the random variates generated from the ith substream of the third stream of the random number generator.

Let's consider these outputs and their characteristics in more detail. First of all, an analyst/decision maker should be very specific about the type of output desired from a simulation model and therefore not be confused about similar, though different, output measures. For example, if a decision maker is concerned about the *fraction* or *proportion* of callers to a call center who must wait longer than 12 seconds for their call to be answered, then an estimate of the *mean* waiting time for callers is probably not appropriate for this purpose.

Let X_i denote an output from one independent replication of a simulation model, say, replication i. So, if n replications are made, the outputs are denoted as $X_1, X_2, \ldots, X_n$. As noted earlier, this outcome could be a discrete outcome, a mean value, a proportion, or just a number such as payoff or cost. In the case of a discrete outcome, the outcomes could be coded in order to provide a numerical output such as 0 for failure and 1 for success. These X_i are independent and identically distributed (iid) since they arise from independent replications of the same model. Denote the population mean and variance of X_i as μ and σ^2, respectively.

As an example, suppose that you had a simulation model of a project that resulted in two possible outcomes when the model is run: success or failure.

Suppose also that success is coded with a value of 1, failure is coded with a value of 0, and the underlying probability of project success is 0.7 (which of course means that the underlying probability of project failure is 0.3). Then the X_i are independent and identically distributed according to a Bernoulli distribution corresponding to a 0.7 probability of a value of 1 and a 0.3 probability of a value of 0; moreover, $\mu = 0.7$ and $\sigma^2 = .7(1 - .7) = .21$. (Refer to the variance associated with a Bernoulli random variable [Law, 2007, p. 302].) Of course, we would not know the value of μ but would have to estimate its value by experimenting with the model.

An unbiased estimator of μ is given by the sample mean of the X_i's, denoted as $\bar{X}(n)$ in the following equation:

$$\bar{X}(n) = \frac{\sum_{i=1}^{n} X_i}{n}. \tag{9.5}$$

An estimator of the variance of the X_i's is denoted as $S^2(n)$ and is computed as follows:

$$S^2(n) = \frac{\sum_{i=1}^{n} \left[X_i - \bar{X}(n)\right]^2}{n-1}. \tag{9.6}$$

Now, $\bar{X}(n)$ is itself a random variable. The variance of $\bar{X}(n)$, denoted as $\mathrm{Var}[\bar{X}(n)]$, is given by σ^2/n; since we typically do not know the value of σ^2, we estimate $\mathrm{Var}\,\bar{X}[(n)]$, denoted as $\widehat{\mathrm{Var}}[\bar{X}(n)]$:

$$\widehat{\mathrm{Var}}[\bar{X}(n)] = \frac{S^2(n)}{n}. \tag{9.7}$$

Substituting for $S^2(n)$ from (9.6) into (9.7), we obtain an expression for $\widehat{\mathrm{Var}}[\bar{X}(n)]$ in terms of the outputs from the replications of the model, as shown:

$$\widehat{\mathrm{Var}}[\bar{X}(n)] = \frac{\sum_{i=1}^{n} [X_i - \bar{X}(n)]^2}{n(n-1)}. \tag{9.8}$$

In considering the values for $S^2(n)$ and $\widehat{\mathrm{Var}}[\bar{X}(n)]$ as a function of n, the reader should note that the value for $S^2(n)$ should stay relatively the same as n increases since its expected value remains the same for any value of $n \geq 2$. However, the value for $\widehat{\mathrm{Var}}[\bar{X}(n)]$ will usually, though not always, decrease as n increases. For a small increase in n, $\widehat{\mathrm{Var}}[\bar{X}(n)]$ might actually increase in value, though usually not.

When we compute $\bar{X}(n)$, we have no way of knowing how close it is to the actual value of μ. On one set of n replications, $\bar{X}(n)$ may be close to μ, while on another set of replications (i.e., using a different sequence of random numbers), there may be quite a large difference. We do know however that $[\bar{X}(n)]$ is an *unbiased* estimator since its expected value is μ.

Also, via the central limit theorem (Walpole and Myers, 1993, p. 216), we can say that as n becomes larger, $(\bar{X}(n) - \mu)/(\sigma/\sqrt{n})$ is distributed according to the standard normal distribution as n approaches infinity. This allows us to form confidence intervals for μ based on the outputs from the independent replications of the simulation model if we know the value of σ. In particular, an approximate (1 – α) * 100% confidence interval for μ is given as n becomes "large":

$$\bar{X}(n) \pm z_{1-\alpha/2}(\sigma^2/n)^{.5} \tag{9.9}$$

where $z_{1-\alpha/2}$ is the upper (1 – α) critical point for the standard normal random variable (the normal random variable with mean equal to 0 and variance equal to 1). In other words, $z_{1-\alpha/2}$ is the value of Y such that PR $(Z \le 1-\alpha/2) = Y$, where Z is the standard normal random variable.

In (9.9), the confidence interval is basically given as an interval on the real number line with a left-hand side of $\bar{X}(n) - z_{1-\alpha/2}(\sigma^2/n)^{.5}$, a "center point" of $\bar{X}(n)$, and a right-hand side of $\bar{X}(n) + z_{1-\alpha/2}(\sigma^2/n)^{.5}$. The quantity $z_{1-\alpha/2}(\sigma^2/n)^{.5}$ is called the *half-width* of the confidence interval.

Typically, one uses a 95% confidence interval, so α is set to a value of .05; correspondingly, $z_{1-\alpha/2} = z_{1-.05/2} = z_{.975}$ (and $z_{.975} = .196$). These "z values" can be attained through the use of a "lookup table," or they are accessible through the use of statistical software packages.

A confidence interval will either contain or not contain the actual mean value once it has been computed. Hence, the way to think about confidence intervals is that if a large number of 95% confidence intervals are computed from a simulation model for the same situation (using respective different streams of random numbers), then approximately 95% of these confidence intervals will contain the true mean.

Typically, one will not know the value of σ^2, the actual variance of the X_i's. However, one can replace σ^2 with its sample value, $S^2(n)$, computed from the X_i's, as computed from (9.6). If the X_i are normally distributed (note that this is not the same thing as saying that $\bar{X}(n)$ is normally distributed), then one can use the expression for a confidence interval for μ, as given by (9.9). This expression (given by Equation 9.9) is an *exact* confidence interval for any value of $n \ge 2$, as opposed to (9.10), which represents an *approximate* 100 * (1 – α)% confidence interval for μ:

$$\bar{X}(n) \pm t_{n-1,1-\alpha/2}(S^2(n)/n)^{.5} \tag{9.10}$$

In (9.10), $t_{n-1,1-\alpha/2}$ represents the upper $1 - \alpha/2$ critical point for the t distribution with $n - 1$ degrees of freedom.

Even if the X_i values are not normally distributed, one can still employ this equation as a good approximation for a confidence interval for μ as long as the distribution of the X_i's is "bell shaped" (Walpole and Myers, 1993, p. 248).

Even if a bell-shaped distribution cannot be assumed for the X_i random variables, one can substitute $S^2(n)$ for σ^2 in (9.9) as long as n, the sample size, is at least 30, and obtain a reasonable approximation for a confidence interval for μ as shown in (9.11). The argument here is that if $n \geq 30$, $S^2(n)$ will be a good approximation for σ^2. The confidence interval shown in (9.11) is often referred to as a *large-sample confidence interval*:

$$\overline{X}(n) \pm z_{1-\alpha/2}(S^2(n)/n)^{.5} \tag{9.11}$$

One can also note that as n increases, $t_{n-1,1-\alpha/2}$ approaches the value of $z_{1-\alpha/2}$ so that for $n \geq 30$, there is usually almost no difference between these two quantities; hence, (9.10) gives approximately the same confidence interval as (9.11).

The confidence interval typically computed by simulation software packages (for the estimate of an expected value as derived from independent replications of a simulation model) is the one given by (9.10).

Another type of confidence interval of interest, especially for outputs from simulation models of decision trees or influence diagrams, is one involving a confidence interval for a probability of a particular outcome. (This probability might also be termed as the proportion of time that a specific outcome occurs.) Examples would be the probability of a treatment's success/failure, the probability that a customer must wait longer than 10 minutes, and the probability that a project's duration will be longer than 70 days.

Suppose that we have a simulation model that has as one of its outputs any of a number of discrete, mutually exclusive, but exhaustive, outcomes, numbered as 1, 2,..., O; for example, the outcomes could be success or failure (in which case, O = 2), or the outcomes could be average waiting time for customers that was (1) less than 2 minutes, (2) between 2 and 4 minutes inclusive, or (3) more than 4 minutes. In the second case, the decision maker might be interested in the respective probabilities that the average customer waiting time is less than 2 minutes, between 2 and 4 minutes, and more than 4 minutes.

Now, suppose that n independent replications of the simulation model are made and that outcome i occurs n_i times for i = 1, 2,..., O, so that $n = n_1 + n_2 + \cdots + n_o$. Then a point estimate of the probability of outcome i occurring (denoted as p_i) is given by $\hat{p}_i$ in the following equation:

$$\hat{p}_i = n_i/n \tag{9.12}$$

If n is "large," then an approximate 100(1 – α)% confidence interval for p_i is given by (9.13):

$$\hat{p}_i \pm z_{1-\alpha/2}(\hat{p}_i(1-\hat{p}_i)/n)^{.5} \quad (9.13)$$

The reader will recognize that the relevant random variable in this case has a Bernoulli distribution. Since an estimate for the variance in this case would be $\hat{p}_i(1-\hat{p}_i)$, the confidence interval of (9.13) is actually just a special case of (9.11).

A summary of the notation and the confidence intervals discussed in this section is shown in Table 9.2.

Example 9.1: The Thief of Baghdad Problem

(This example is derived from an exercise given by A. Alan B. Pritsker in his textbook on simulation [Pritsker, 1986, problem 2–7, p. 48].)

A thief has been placed in a cell with three doors. One door leads immediately to freedom. A second door leads to a short tunnel, which takes 2 hours to travel but returns the thief to the jail cell. A third door leads to a long tunnel, which takes 6 hours to travel; but this tunnel also returns to the cell. The thief has an equal chance of choosing any of the three doors; and once back in the cell (assuming the thief chooses one of the doors that leads to a tunnel), the thief again has an equal chance (i.e., one-third probability) of choosing any of the three doors. (As noted by Pritsker, this is a Markov, or forgetful, thief.) Given that the thief continually keeps choosing a door until he or she gets to freedom, what is the expected time until the thief reaches freedom?

Upon first viewing the problem, the reader might think that there is no reasonable answer—that is, the thief could just continually choose a door that leads to one of the tunnels and therefore never reach freedom. However, there is an analytical solution to the problem, which is determined by solving the following equation:

$$E(X) = 1/3\ (0) + 1/3\ (2 + E(X)) + 1/3\ (6 + E(X)),$$

where

X denotes the time to freedom in hours
E(X) is the expected time to freedom

The argument for this equation goes as follows. If the thief chooses the door to freedom on the first try (probability of 1/3), then the time to reach freedom is just 0. If the thief chooses the door to the short tunnel or the long tunnel, respectively, with a probability of 1/3 for each, then the time to reach freedom is 2 hours or 6 hours, respectively, plus the expected time to freedom (since the thief starts all

TABLE 9.2

Summary of Notation for Simulation Model Outputs Associated with Several Independent Replications

Notation	Description/ Definition	Expression for Computation If Applicable	Comments If Applicable
n	Number of independent replications of the simulation	—	—
X_i	Output from the ith replication	—	The X_i are independent and identically distributed.
μ	Mean of the X_i's	—	μ is a constant but is usually unknown.
σ^2	Variance of the X_i's	—	σ^2 is a constant but is usually unknown.
$\bar{X}(n)$	Sample mean of the X_i's	$\frac{\sum_{i=1}^{n} X_i}{n}$	$\bar{X}(n)$ is a random variable.
$S^2(n)$	Sample variance of the X_i's	$\frac{\sum_{i=1}^{n} [X_i - \bar{X}(n)]^2}{n-1}$	$S^2(n)$ is a random variable; its expected value remains constant as n increases.
$\widehat{Var}[\bar{X}(n)]$	Sample variance of $\bar{X}(n)$	$S^2(n)/n$	$\widehat{Var}[\bar{X}(n)]$ is a random variable, but its expected value will decrease as n increases.
$z_{1-\alpha/2}$	The upper $(1-\alpha)$ critical point for the standard normal random variable	$z_{1-\alpha/2}$ is the value of Y such that $PR(Z \le 1-\alpha/2) = Y$, where Z is the standard normal random variable	$z_{1-\alpha/2}$ is typically obtained from a table.
$t_{n-1,1-\alpha/2}$	The upper $1-\alpha/2$ critical point for the t distribution with $n-1$ degrees of freedom	—	$t_{n-1,1-\alpha/2}$ is typically obtained from a table.
$\bar{X}(n) \pm z_{1-\alpha/2}(\sigma^2/n)^{.5}$	Z confidence interval for μ, σ^2 known	—	Approximate confidence interval for μ with σ known.

(Continued)

TABLE 9.2 (*Continued*)

Summary of Notation for Simulation Model Outputs Associated with Several Independent Replications

Notation	Description/ Definition	Expression for Computation If Applicable	Comments If Applicable
			Results from the central limit theorem. Typically not used because σ^2 is usually not known. Only an approximate CI; becomes more accurate as n increases.
$\bar{X}(n) \pm t_{n-1,1-\alpha/2}(S^2(n)/n)^{.5}$	t confidence interval for μ, σ^2 unknown	—	Exact confidence interval for μ if the X_i are normally distributed. Can be used as a good approximation if the distribution of the X_i is "bell shaped."
$\bar{X}(n) \pm z_{1-\alpha/2}(S^2(n)/n)^{.5}$	Z confidence interval for μ, σ^2 unknown; also called "large-sample" confidence interval	—	Provides a good approximation for a confidence interval for μ even if the distribution of the X_i is not "bell shaped," as long as $n \geq 30$. Approximately the same as the t CI as long as $n \geq 30$.
p_i	Probability of outcome i	—	
$\hat{p}_i$	An unbiased point estimate for p_i	n_i/n	n_i is the number of times that outcome i occurs in the n replications
$\hat{p}_i \pm z_{1-\alpha/2}(\hat{p}_i(1-\hat{p}_i)/n)^{.5}$	Confidence interval for p_i	—	An approximate CI.

over being back in the cell). Now solving this equation, one obtains the following:

$$E(X) = 2/3 + 1/3\ E(X) + 2 + 1/3\ E(X)$$

or

$$1/3\ E(X) = 8/3$$

or

$$E(X) = 8 \text{ hours}$$

Besides the expected time to freedom, additional outputs in which one might be interested from an analysis of this situation would include such measures as the probability that the time to freedom would be longer than Y hours, where Y is some constant. Since this situation can be represented with an analytic (in particular, a Markov) model, one could derive the exact values for these measures analytically (i.e., without the analysis of a simulation model). (For a description of how to solve Markov chain models of the type to represent this situation, see Hillier and Lieberman, 2010, Chapter 16.)

In general, given that an analytic model is an accurate representation for the purposes of the study, one would want to use such a model. One reason for this is that an exact value for the performance measure of interest can be attained, as in this case. Another reason is that experimentation with such a model, for example, for sensitivity analysis, is much simpler than with a simulation model. However, there are advantages to a simulation model. For example, if complications arise in the actual system, these may very well be relatively easy to account for with a simulation model. Second, a simulation model may be easier to explain to a decision maker than an analytic model.

In order to illustrate the computations presented in this section, a simulation model of the Thief of Baghdad problem was developed, using the Arena simulation software package (Kelton et al., 2015). The model was run initially for five replications, using random number stream 3 to generate a sequence of random numbers to determine which door was chosen for each trip. If the random number was less than 1/3, then the door leading to the short tunnel was chosen; if the random number was between 1/3 and 2/3, then the long tunnel was chosen; finally, if the random number was between 2/3 and 1, then the door to freedom was chosen. The times to freedom for the five replications are shown in Table 9.3.

The *sample mean* and *sample variance of the mean* of the time to freedom for these five independent replications are given by

$$\bar{X}(5) = \sum_{i=1}^{5} X_i/5 = 12,$$

$$\widehat{Var}(\bar{X}(5)) = \sum_{i=1}^{5} (X_i - \bar{X}(5))^2/5(4) = 43.6.$$

TABLE 9.3

Outputs from the Thief of Baghdad Simulation for Five Replications

Replication Number, i	Estimate of Time to Freedom, X_i
1	0
2	10
3	36
4	0
5	14

So, since $Z_{.975} = 1.96$ (note that $Z_{.975} = Z_{1-\alpha/2}$ for $\alpha = .05$), a 95% confidence interval for the mean time to freedom computed from these five replications would be

$$12 \pm 1.96\,(43.6)^{.5}, \text{ or } 12 \pm 12.942, \text{ or } (-.942, 24.942).$$

Note that since $t_{4,.975} = 2.776$, the 95% t confidence interval is given by

$$12 \pm 2.776\,(43.6)^{.5}, \text{ or } 12 \pm 18.33, \text{ or } (-6.33, 30.33).$$

Obviously, these are not very good confidence intervals. Very rarely would one want to use only 5 replications for simulation model experimentation; in general, at least 30 replications are needed for accurate estimates, or, as noted by Walpole and Myers, for large-sample confidence intervals (Walpole and Myers, 1993, p. 249). Hence, the model was run for 5, 20, 50, 100, 500, and 10,000 replications, respectively, using first random number stream 3 and second random number stream 6. The results for these experiments are shown in Table 9.4.

The reader will note several things from Table 9.4:

1. Every confidence interval computed contains the true value for the mean time to freedom of 8 hours.
2. Except for the experiments involving only five replications, the half-widths for the confidence intervals are approximately the same for each of the two different random number streams for a given set of replications.
3. The sample standard deviations for the means decrease as n increases, while the sample standard deviations for the X_i's remain relatively constant, as is to be expected. This results in a decrease in the half-width of the confidence interval for the mean as the number of replications increases.
4. The sample means do not necessarily become closer to the true value for μ as n increases. However, for the cases of 10,000 replications, the sample values are very close to the true value of 8 hours.

With respect to the third comment earlier, in general one can *almost* always expect the half-width of the confidence interval for the mean to decrease with an increase in the number of replications of the model.

TABLE 9.4

Experimental Results for Time to Freedom from the Thief of Baghdad Simulation

Number of Replications, n	Random No. Stream	Sample Mean	Sample Standard Deviation of the Mean	Sample Standard Deviation of the X_i	Half-Width for 95% Z CI	Half-Width for 95% t CI
5	3	12.	6.6	14.765	12.94	18.33
20	3	8.8	2.4	10.735	4.91	5.24
50	3	6.72	1.27	8.97	2.49	2.55
100	3	7.6	1.02	10.18	2.00	2.02
500	3	8.1	0.47	10.44	0.915	0.92
10,000	3	8.05	0.01	10.2	0.2	0.2
5	6	8.4	3.6	8.05	7.05	9.99
20	6	10.	2.39	10.7	4.69	5.01
50	6	8.32	1.34	9.46	2.62	2.69
100	6	8.52	1.03	10.33	2.03	2.05
500	6	8.744	0.46	10.22	0.896	0.9
10,000	6	8.12	0.01	10.2	0.2	0.2

If the increase in n is only slight, however, the confidence interval half-width might actually increase since the increase in n might actually be offset by an increase in the estimate for the variance of the X_i, denoted as $S^2(n)$.

Consider a different confidence level (90% rather than 95%) for the confidence interval for the experiment with 500 replications using random number stream 3. The 95% t confidence interval for μ is given by

$$8.1 \pm .92, \text{ or } (7.18, 9.02).$$

The 90% t confidence interval would use the t value given by $t_{499,.95} = 1.648$. Hence, one would obtain the confidence interval given by

$$8.1 \pm t_{499,.95}(S^2(n)/n)^{.5}, \text{ or } 8.1 \pm 1.648 * (109/500)^{.5},$$

or

$$8.1 \pm .77, \text{ or } (7.34, 8.87),$$

which is smaller than the 95% confidence interval. At first thought, this seems counterintuitive. However, if one returns to the operational definition given earlier for a confidence interval, 90% of the 90% confidence intervals will contain the true mean, while 95% of the 95% confidence intervals generated will contain the true mean, such a smaller width confidence interval makes sense. Hence, it will always be the case that an X% confidence interval will have a larger half-width than a Y% confidence interval for X greater than Y (e.g., for 95% larger than 90%).

With a slight modification, the Thief of Baghdad simulation can be used to estimate the probabilities associated with various outcomes related to the amount of time for the thief to reach freedom. In particular, one could estimate the probabilities associated with reaching freedom in X hours or in ≥Y hours where X and Y are just specific values. As before, since this situation can be represented as a Markov process, these values could be computed analytically. The purpose here is to compare the exact (analytic) results to the results from the simulation and also to illustrate the computations required from the simulation output.

Suppose that we wanted to compute the probability associated with time to freedom being greater than 2 hours. The exact probability can be computed just by using a little common sense. In particular, if we let X be a random variable representing time to freedom, then

$$
\begin{aligned}
\text{PR}\,(X > 2) &= 1 - \text{PR}\,(X \le 2) \\
&= 1 - \text{PR (Door 1 (door to freedom) chosen on first try} \\
&\qquad \text{or Door 2 (to 2-hour tunnel) chosen on first try and} \\
&\qquad \text{Door 1 chosen on second try)} \\
&= 1 - (1/3 + ((1/3) * (1/3))) \\
&= 1 - (1/3 + 1/9) \\
&= 5/9 \\
&\approx .55555\ldots
\end{aligned}
$$

Now, the simulation was set up to record the number of outcomes in which the time to freedom was more than 2 hours, out of a specific number of "trials." This number was divided by the total number of trials in order to obtain a point estimate of the time to freedom being greater than 2 hours. Random number stream 3 was used to determine which door was chosen on each try by the thief. The simulation was run for the following number of trials: 20, 50, 100, 1,000, and 10,000. The point estimates and corresponding 95% confidence intervals using (9.13) are shown in Table 9.5.

TABLE 9.5

Experimental Results for the Probability of Reaching Freedom in More Than 2 Hours from the Thief of Baghdad Simulation

Number of Replications, n	Sample Value for Probability of Time to Freedom >2 Hours, $\hat{p}$	Half-Width for a 95% Confidence Interval for Probability
20	0.50	0.22
50	0.60	0.14
100	0.57	0.097
1,000	0.536	0.031
10,000	0.5593	0.01

As an example of the calculations in Table 9.5, consider the case of 100 replications. This run resulted in a point estimate of .57 for the probability, which in turn led to a half-width for a 95% confidence interval of

$$z_{.975}[\hat{p}(1-\hat{p})/100]^{.5} = 1.96\ [.57\ (1-.57)/100]^{.5} = .097,$$

which gives a 95% confidence interval of .57 ± .097 or (.473, .667).

The reader will note that every confidence interval in Table 9.5 contains the true value for the probability of .555.... However, "good accuracy" of a .031 (or less) half-width for a 95% confidence interval required at least 1000 independent trials/replications of the simulation.

9.3.2 Achieving a Specified Accuracy with a Confidence Interval

Sometimes, the execution of a simulation model of a decision situation requires much computational effort for each replication of the model. Hence, the trade-off between computational effort and accuracy (i.e., width of a confidence interval) in the estimation of a performance measure for an alternative may be an issue. In these cases, an analyst may want to have an idea of the relationship between the number of replications of a model and the half-width of a confidence interval for a performance measure.

In looking at (9.10) for the t confidence interval, one can see that the half-width for this confidence interval, denoted as HW, is given by (9.14).

$$HW = t_{n-1,\,1-\alpha/2}(S^2(n)/n)^{.5}, \tag{9.14}$$

which can be rewritten as

$$n = t^2_{n-1,\,1-\alpha/2}(S^2(n)/HW^2). \tag{9.15}$$

Now, we can estimate the value for $S^2(n)$ for any value of n by making just a few replications, say, 5. (Remember that the *expected* value of $S^2(n)$ remains constant for any value of n.) Let this approximation be denoted as just S^2. However, we still have n on both sides of (9.15) since it remains a parameter for the t distribution. But we can replace $t_{n-1,\,1-\alpha/2}$ with the critical value from the standard normal distribution, $z_{1-\alpha/2}$, as an approximation. (Remember that as n approaches infinity, the value of $t_{n-1,\,1-\alpha/2}$ approaches the value of $z_{1-\alpha/2}$, and for $n \geq 30$, these values are almost the same.) So we have

$$n \approx z^2_{1-\alpha/2}(S^2/HW^2) \tag{9.16}$$

Suppose that we make a few initial replications, say, n_0, in order to estimate S^2. Let HW_0 be the half-width for the confidence interval from these initial replications. Then, since we have $HW_0 = z_{1-\alpha/2}(S/n_0^{.5})$, and therefore $z^2_{1-\alpha/2}S^2 = n_0 HW_0^2$, we can rewrite (9.16) as a simpler approximation for the number of replications needed to achieve a specified half-width for a confidence interval, as shown in the following equation:

$$n \approx n_0(HW_0^2/HW^2) \quad (9.17)$$

Let's consider the results for 5 replications for random number stream 6 for the Thief of Baghdad problem, as shown in Table 9.4. For this run, we have $n_0 = 5$, $HW_0 = 9.99$ (for the 95% t confidence interval). Let's suppose that we wanted to improve the accuracy of our estimate for the mean time to freedom by achieving a half-width of 2 for the confidence interval. How many total replications would be needed to achieve this? According to our first approximating formula, given by (9.16), we would need

$$n \approx z_{1-\alpha/2}^2(S^2/HW^2) = 1.96^2(8.05^2/2^2) = 62.2 \approx 62$$

total replications or 57 additional replications in order to achieve the desired half-width of 2. According to our second, less accurate, formula given in (9.17), we would need

$$n \approx n_0(HW_0^2/HW^2) = 5\,(9.99^2/2^2) = 124.75 \approx 125$$

total replications, or 120 additional replications in order to achieve the desired half-width of 2. Note that from looking at Table 9.4, the actual value for the total number of replications required to achieve a half-width of 2 is approximately 100.

9.4 Variance Reduction Techniques

Reduction of the variance of estimates associated with simulation output is useful in improving the accuracy of the estimates. More specifically, the widths of the confidence intervals of the respective estimators can be reduced with these variance reduction techniques. This in turn can allow one to more easily distinguish between the various alternatives under consideration.

There are many variance reduction techniques available for use in simulation studies, including common random numbers, antithetic variates, control variates, indirect estimation, and conditioning. In this section, we will focus briefly on common random numbers, which is probably the easiest and most commonly used of the various techniques. The reader is referred to any of several books on simulation such as Law (2007, Chapter 11) for additional information on common random numbers and the other techniques.

One important fact to remember is that these variance reduction techniques do not always work as intended. In particular, sometimes, a trial study (in which a few replications of the simulation model is made) is sometimes employed to see if the method will work.

Common random numbers, which have also been called correlated sampling, differs from some of the other techniques in that it can be applied when two or more alternatives are being compared. That is, while some of the other techniques can be applied when obtaining a confidence interval associated with the output (such as expected utility) for a single alternative, common random numbers can be used to reduce the confidence interval associated with the difference between the expected utilities of respective alternatives—see Section 9.5.1 on paired-t confidence intervals.

The basic procedure for using common random numbers is very simple: specific random number streams are dedicated to specific respective purposes in the model. For example, stream 1 might be dedicated to the interarrival times for customers, stream 2 might be dedicated to the generation of the process times for a specific activity, stream 3 might be dedicated to the generation of discrete random variates associated with the type of customer entering the system, and so on. For the model associated with Example 9.2 (given below) involving project management, streams 1, 2, and 3 might be dedicated to generating the random variates associated with generating the task durations for existing data transition, sales master data acquisition, and global master data acquisition, respectively. In this way, the random numbers used will be "matched up" (also called synchronization) across the different alternatives.

Most simulation software packages are easily set up to handle this type of procedure. For example, in the Arena simulation package, the stream number is appended at the end of the list of arguments representing the parameters of the distribution. Each time a new replication of the model starts execution, the random numbers being used will be taken from the start of the next substream of the relevant stream (see Kelton et al., 2015, pp. 524–525).

9.5 Comparing Two Alternatives

Although one of the emphases in this book is the analysis of decision situations with multiple performance measures, there are situations where one wants to optimize over a single measure, albeit with uncertainty/risk involved. As mentioned previously, it could be a situation where the other performance measures/attributes are considered through the use of constraints or all measures are "consolidated" through the use of a multiattribute utility/value function or some combination of these two approaches, where some of the performance measures are constrained while the rest are considered through the use of a utility/value function.

As seen in the previous section, output from a simulation model of a decision situation will be represented as a random variable through the use of multiple replications of the model. If one is considering two alternatives, then two separate simulations are used, one for each alternative. Often, these simulations will differ only in the values associated with one or more of their input/control variables.

9.5.1 Paired-t Confidence Intervals

Consider the output from two alternative simulation models, representing two alternatives over n_1 independent replications for alternative 1 and n_2 replications for alternative 2. Let $X_{1,j}$ for $j = 1,\ldots, n_1$ be the outputs associated with the first alternative and $X_{2,j}$ for $j = 1,\ldots, n_2$ be the outputs associated with the second alternative. As noted previously, these outputs could represent expected profits, expected utilities, individual discrete outcomes, and so on.

Initially, we will consider an approach called the *paired-t confidence interval* in which the number of replications run for each alternative are the same (i.e., $n_1 = n_2$). Let $n = n_1 = n_2$, and let

$$Y_j = X_{1,j} - X_{2,j} \quad \text{for } j = 1,\ldots, n.$$

Since the individual replications are independent, the Y_j are independent and identically distributed (so each Y_j has the same expected value). Let $\varepsilon = E(Y_j)$ be the expected value of Y_j; then

$$\bar{Y}(n) = \sum_{j=1}^{n} Y_j/n$$

is an unbiased estimator of ε and

$$\widehat{Var}\,[\bar{Y}(n)] = \sum_{j=1}^{n} [Y_j - \bar{Y}\,(n)]^2/n(n-1)$$

is an estimate of the variance of $\bar{Y}(n)$. Then an approximate $(1 - \alpha) * 100\%$ confidence interval for ε, the true mean, is given by (9.18).

$$\bar{Y}(n) \pm t_{n-1,\,1-\alpha/2}(\widehat{Var}[\bar{Y}(n)])^{.5}. \quad (9.18)$$

If the Y_i are normally distributed, then (9.18) is an exact confidence interval. Even if the Y_i are not normally distributed, then via the central limit theorem, (9.18) approaches an exact confidence interval as n approaches infinity. In most cases if $n \geq 30$, sufficient accuracy will result.

If the paired-t confidence interval does not contain 0, we say that there is a *statistically significant difference* between the two alternatives.

A second approach for comparing two alternatives is called the two-sample t-test. This test, which will not be discussed in detail here, is used less frequently than the paired t-test. It does not have the requirement that the number of samples from each alternative be the same. However, it does have the disadvantage of not being capable of incorporating the variance reduction technique (to improve the accuracy of the estimate of the difference between the two means) of common random numbers. The two-sample t-test can be useful for the validation of a simulation model where one may be able to obtain only a few samples from the actual system and many more samples from the model.

Example 9.2: Using a Paired-t Confidence Interval to Compare Two Alternatives for Project Management

This example is derived from an unpublished report involving a decision problem in project management encountered by a German company. The company produces beer trailers. The company used an outdated enterprise resource planning (ERP) system that supported only about 25% of the required deliverables of the process. ERP systems support data collection, database management, and management decision making in the areas of production planning, inventory control, marketing, personnel management, shipping, and others. The company wanted to install a new ERP system that would allow for a new, more comprehensive approach for management decision making in these areas.

As with any large project involving many tasks performed by a variety of personnel, the techniques associated with project management were useful for this ERP installation project. A project management approach involves subdividing a project into various separate tasks through the use of a project network. Implicit in the project network are the precedence relationships among these tasks—that is, certain tasks cannot be started until other tasks are completed. Through proper analysis of the project network, one can determine the project schedule and also allocate resources for the project over time.

The tasks, projected task durations, direct costs, and precedence relationships for the ERP installation project for this German company are shown in Table 9.6.

An activity-on-node project network illustrating the precedence relationships shown in Table 9.6 is shown in Figure 9.1.

Each of the tasks shown in Table 9.6 and in Figure 9.1 can be divided into "subtasks." For example, employee training can be divided into training of sales staff, training of purchasing staff, training of manufacturing personnel, and so on; preinstallation tasks include current data backup, manual data gathering, and server preparation. In order to keep this example relatively simple, however, we will keep the number of tasks at 12.

Several of the tasks have uncertain durations, which are represented by triangular distributions, as noted in Table 9.6.

Because of the uncertainties associated with the task durations, a Monte Carlo simulation model was developed for the project. The simulation was developed using the Arena simulation software package,

TABLE 9.6

Tasks, Projected Durations, and Precedence Relationships for Enterprise Resource Planning Installation Project

Task Number	Task Name	Task Duration (Days)	Task Cost (Euros/Day)	I.P.'s (Task Numbers)
1	Project initialization	2	600	—
2	Quality documentation	8	1000	1
3	Existing data transition	TRIA (16, 22, 30)	2000	1
4	Preinstallation tasks	7	1500	1
5	System installation	3	3000	4
6	Sales master data acquisition	TRIA (9, 12, 16)	2000	5
7	Global master data acquisition	TRIA (10, 14, 20)	1500	5
8	Layout definition	4	1600	5
9	Employee training	4	4000	6, 7
10	Master data acquisition	TRIA (38, 45, 58)	1500	9
11	Introduction into productive operations	1	1000	2, 3, 8, 10
12	Postproject review	1	1500	11

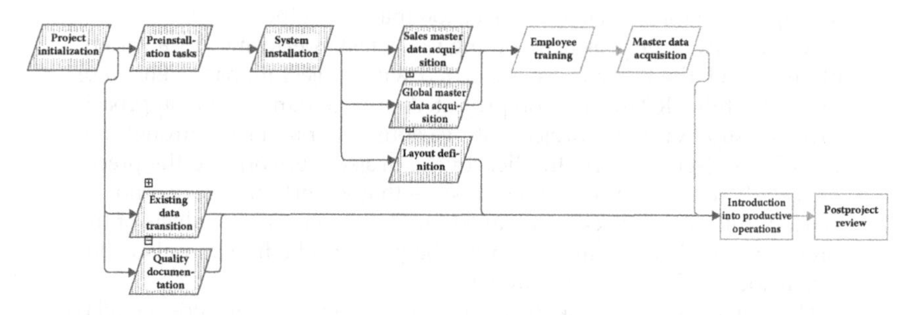

FIGURE 9.1
Activity-on-node project network for an enterprise resource planning installation project. (Output from MS Project™.)

but any of several different software packages could have been used. A simulation model of a project network is relatively easy to build in a process-oriented simulation software package such as Arena. In particular, for the Arena model that was constructed for this project, the following modules are given:

1. A *Create module* was used to create a single entity to represent the starting point of the project.
2. *Process modules* of the delay type were used to represent the individual tasks of the project.
3. For a task that is an immediate predecessor of more than one succeeding tasks, one or more *Separate modules* were used to

create multiple entities that will arrive at the *Process modules* representing those succeeding tasks. An example in this project is the "project initialization" task, which has three succeeding tasks: "quality documentation," "existing data transition," and "preinstallation tasks."

4. For a task that has more than one immediate predecessor, a *Batch module* was used to batch multiple entities, one entity from each of the immediate predecessor tasks, thereby assuring that each immediate predecessor task is finished before the successor task can begin. An example of this situation is the "employee training" task, which has "sales master data acquisition" and "global master data acquisition" as immediate predecessors.
5. A *Dispose module* was used to represent project termination when one entity reaches it.

Variable modules are used to input the cost per week for each of the tasks, and *record modules* are used toward the end of the network to record project duration and project cost for each replication of the model. Figure 9.2 illustrates the Arena simulation model for this project.

The reader should note that this particular model employs only a few of the features that simulation as a methodology can use to represent a project. In particular, this particular model does not address resource allocation or usage, probabilistic dependence of various task durations, or multiple project outcomes (e.g., project success and project failure). However, these aspects associated with real projects could be easily modeled with simulation as a tool.

Running the simulation for 30 replications resulted in the following outputs associated with project duration and cost:

Estimated expected project duration = 79.4 days
Maximum project duration (over 30 replications) = 88.6 days
Minimum project duration (over 30 replications) = 71.6 days
Estimated expected project cost = 214,559 euros
Maximum project cost (over 30 replications) = 234,129 euros
Minimum project cost (over 30 replications) = 199,915 euros

Upon viewing these results, the project manager realized that there would be a problem with the project as planned. Specifically, orders for the beer trailers occurred throughout the year but varied on a seasonal basis; in particular, the order rate was expected to increase sharply beginning in 74 days. Although allowing the duration of the ERP installation project to be greater than 74 days was feasible, such project duration would be undesirable.

Various alternatives were available for decreasing the project duration. Specifically, by contracting with the vendor of the ERP system, both the duration of the "existing data transition" task and/or the duration of the "master data acquisition" task could be reduced in duration, but at an increased cost per day for these tasks. These durations would still be modeled as uncertain in nature though through the use of triangular probability distributions; the parameters for these distributions would

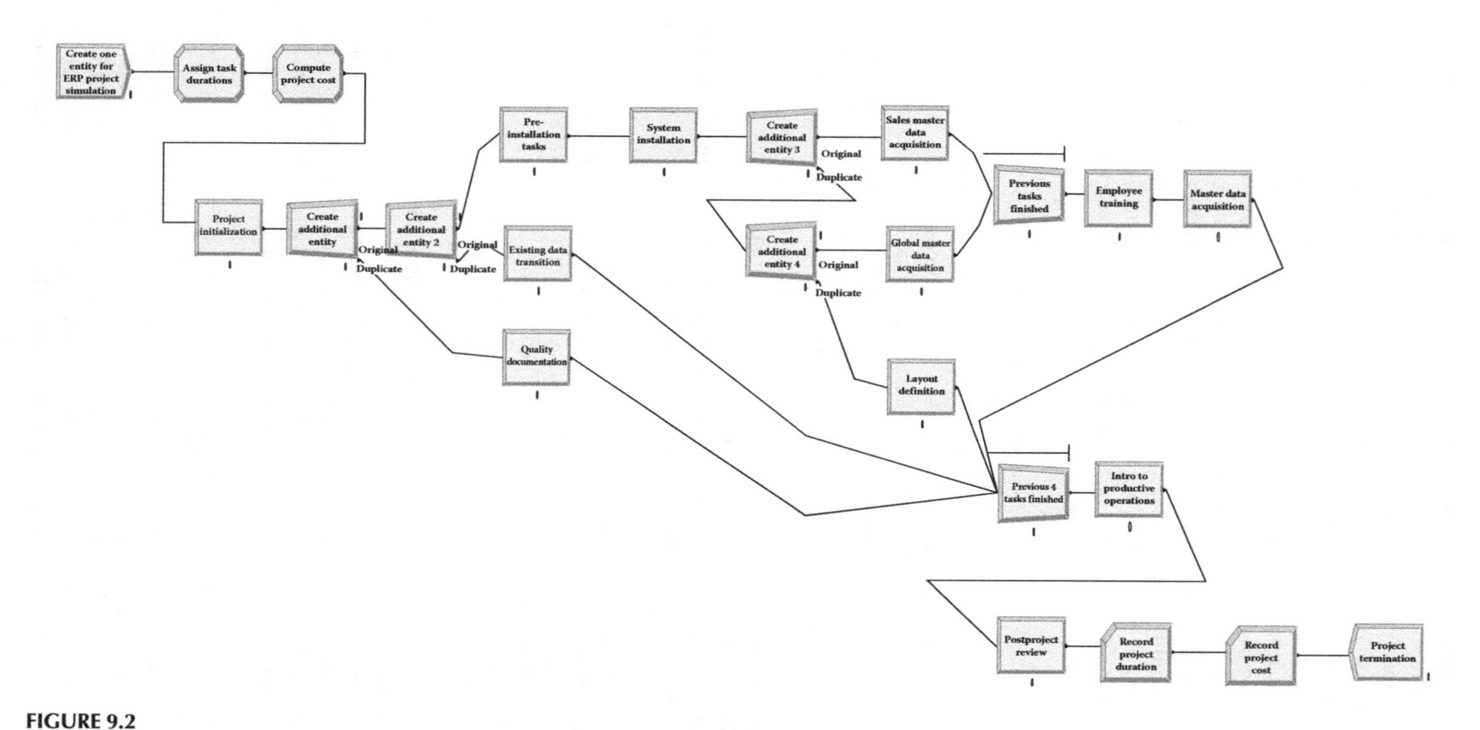

FIGURE 9.2
Arena simulation model of enterprise resource planning simulation project. (Output from the Arena simulation software package.)

TABLE 9.7

Investigated Alternatives for an Enterprise Resource Planning Installation Project

Alternative Number	Description	Results in Terms of Changes to Task Durations and Costs
1	Baseline—do not contract with software vendor.	—
2	Contract with vendor to reduce duration of "existing data transition."	Reduce duration of "existing data transition" to TRIA (14, 16, 22), with new cost per day of $4000.
3	Contract with vendor to reduce duration of "master data acquisition."	Reduce duration of "master data acquisition" to TRIA (30, 36, 46), with new cost per day of $3500.
4	Alternatives 2 and 3 are combined.	Changes to durations and cost associated with "existing data transition" and "master data acquisition" as noted earlier for alternatives 2 and 3.

TABLE 9.8

Results from the Simulation of the Alternatives Listed in Table 9.7

Alternative Number (as Shown in Table 9.7)	Estimated Expected Cost (Minimum, Maximum, over 30 Replications), in Thousands of Euros	Estimated Expected Project Duration (Minimum, Maximum, over 30 Replications), in Days
1	214.6 (200, 234.1)	79.4 (71.6, 88.6)
2	238.3 (222.7, 257)	79.4 (71.6, 88.6)
3	274.1 (251.9, 300.2)	69.8 (63.1, 77.9)
4	297.8 (274.7, 323.1)	69.8 (63.1, 77.9)

be provided through expert estimates. Consideration of these possibilities led to four mutually exclusive alternatives, as shown in Table 9.7.

Running the simulation model for 30 independent replications for each of the four alternatives listed in Table 9.7 led to the results shown in Table 9.8.

In viewing the results from Table 9.8, the project manager was initially surprised that alternative 2 performed no better than alternative 1 and that alternative 4 performed no better than alternative 3 on the project duration measure, even though they each performed worse on the cost measure. Upon further reflection, however, he realized that since the "existing data transition" task was not on the critical path for the network (even when the "master data acquisition" task duration was reduced), this was not surprising.

Hence, two alternatives, as listed in Table 9.7, were considered for further analysis: alternatives 1 and 3. Alternative 1, the baseline plan, had a lower cost than alternative 3, but alternative 1 resulted in longer project duration than alternative 3. However, there was uncertainty associated with both measures for each alternative. Hence, the project manager's utility function over two attributes, X_1 (cost in thousands of euros) and X_2 (project duration in days), was constructed.

Using the results from Table 9.8, the "endpoints" for the attribute values were set as

$$x_1^w = 320, x_1^b = 185, x_2^w = 95, \text{ and } x_2^b = 60.$$

In assessing the utility function, the project manager wanted to be particularly careful in assessing u_2 by assessing several values for X_2 "close to" 72 days because of the critical nature of this project duration.

Suppose that several points for the project manager's single attribute utility function for X_1, cost in thousands of euros, have been determined, as shown in Table 9.9, with the graph for this function as shown in Figure 9.3. The concave function indicates that the project manager is risk averse with respect to project cost.

TABLE 9.9

Assessed Values for $u_1(x_1)$, Utility Function for Project Cost in Thousands of Euros

X_1, Project Cost in Thousands of Euros	$u_1(x_1)$
320	0.
312	0.125
303	0.25
291	0.375
279	0.5
265	0.625
247	0.75
225	0.875
185	1

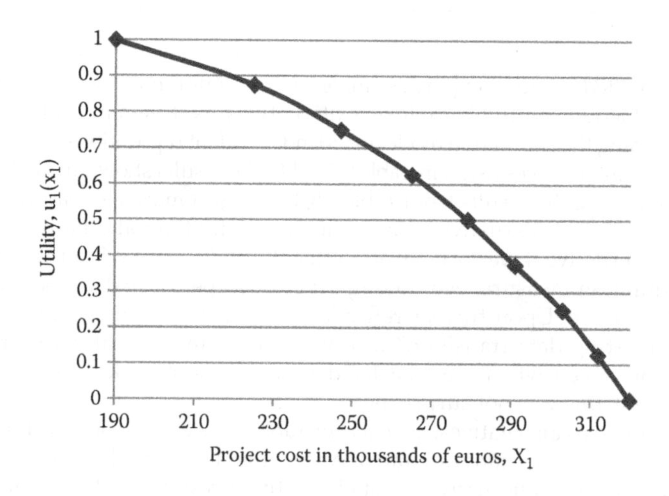

FIGURE 9.3
Graph for $u_1(x_1)$.

Intuition indicates that the utility function for project duration should have a slope that is relatively steep around an x_2 value of 72 days. Because of the importance associated with this duration of 72 days, a response mode involving probability equivalence was employed in determining some of the values on the u_2 utility function curve, as shown in the following dialogue:

Analyst: Suppose that you have two alternatives: A_1, a .5 probability of a project duration of 60 days and a .5 probability of a project duration of 95 days, versus A_2, a certain project duration of 72 days. Which would you prefer?

Project manager: I would prefer A_2, the certain project duration of 72 days.

Analyst: OK, suppose that the probability in A_1 for the 60-day duration was increased to .8 (and therefore, the probability for the 95-day duration was decreased to .2). Which of the two alternatives would you prefer?

Project manager: Then, I would be indifferent between alternatives A_1 and A_2.

At this point, the analyst knows that since the project manager is indifferent between A_1 and A_2, their expected utilities must be equal:

$$Eu(A_1) = Eu(A_2)$$

or

$$.8u_2(60) + .2u_2(95) = u_2(72)$$

or

$$.8 = u_2(72)$$

Continuing with the assessment, the analyst asks the following questions, remembering that utility function values close to a project duration of 72 days are critical for the assessment:

Analyst: OK, suppose that you have the following two alternatives: A_1, a .5 probability of a project duration of 72 days and a .5 probability of a project duration of 95 days, versus A_2, a certain project duration of 78 days. Which would you prefer?

Project manager: I would prefer A_2, the certain project duration of 78 days.

Analyst: OK, suppose that the probability in A_1 for the 72-day duration was increased to .8 (and therefore, the probability for the 95-day duration was decreased to .2). Which of the two alternatives would you prefer?

Project manager: Then, I would be indifferent between alternatives A_1 and A_2.

TABLE 9.10

Assessed Values for $u_2(x_2)$, Utility Function for Project Duration in Days

X_2, Project Duration in Days	$u_2(x_2)$
95	0
88	0.2
78	0.4
74	0.6
72	0.8
70	0.9
65	0.95
60	1

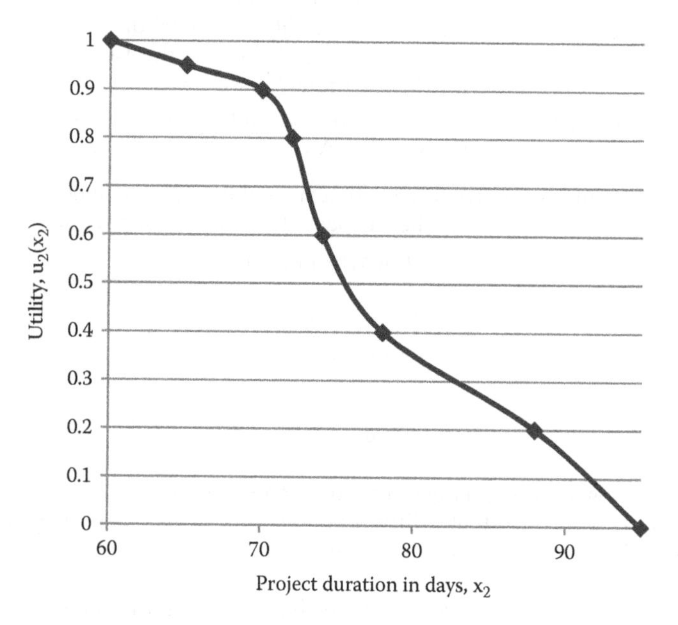

FIGURE 9.4
Graph for $u_2(x_2)$.

So, from these responses, the analyst knows that $u_2(78) = .4$. Note that using the probability response mode in the assessment process allows the analyst to determine the x_2 values that he wants to assess. Continuing, the various values assessed for the u_2 curve are given in Table 9.10, with the corresponding graph given in Figure 9.4. Note that, as expected, the slope of the graph around the value of 72 days is especially steep.

The assessment of the overall utility function, $u(x_1, x_2)$, continues with the analyst showing that X_1 is UI of X_2 and X_2 UI of X_1—that is, (X_1, X_2)

are MUI. Therefore, the analyst knows that the utility function is multiplicative and, upon assessing the values of the scaling constants, arrives at the final form for the utility function:

$$u(x_1, x_2) = .3u_1(x_1) + .5u_2(x_2) + .2u_1(x_1)u_2(x_2)$$

where u_1 and u_2 are as described previously.

A paired-t confidence interval approach was used to compare alternatives 1 and 3 listed in Table 9.7 using expected utility as the criterion. Specifically, the original simulation model was altered slightly to allow for the computation of a utility function value for each project simulation. An estimate of expected utility can then be made by averaging over multiple, independent replications of the model. Note that this estimate for expected utility is obtained not by computing the utility function value of expected cost and expected project duration, but by computing the utility function value of cost and project duration at each replication and then averaging these utility values over all replications. *In other words, we are estimating the expected utility of cost and project duration, not the expected utility of expected cost and expected project duration.*

The utilities for each of the alternatives and for the difference between the alternatives for a few of the 30 replications are shown in Table 9.11. Note that there is a large variation in the expected utility estimates from one replication to another, as a result of the variation in task duration and project cost. Each replication represents only one simulation of the project.

Corresponding to our notation for the paired-t confidence interval, in Table 9.11 the second column is $X_{1,j}$, the third column is $X_{2,j}$ and the fourth column is Y_j. From the Y_j values, we compute values for $\bar{Y}(30) \approx .103$ and, $\widehat{\mathrm{Var}}[\bar{Y}(30)] \approx .000257$. From a table for the t distribution, we find $t_{n-1,\,1-\alpha/2} = t_{29,\,.975} = 2.045$ so that the paired-t 95% confidence interval for

TABLE 9.11

Sample Expected Utilities (and Differences) for Each Alternative for a Subset of the Replications

Replication Number	Expected Utility for Alternative 3	Expected Utility for Alternative 1 (Baseline)	Difference in Expected Utilities (Alternative 3 Minus Alternative 1)
1	0.8358	0.8407	−0.0049
2	0.6675	0.5042	0.1633
3	0.7631	0.5459	0.2172
4	0.3077	0.3690	−0.0613
5	0.5038	0.4513	0.0525
⋮	⋮	⋮	⋮
28	0.6564	0.5005	0.1559
29	0.7426	0.5478	0.1948
30	0.8132	0.8374	−0.0242

TABLE 9.12

Results of Sensitivity Analysis on Weights for Utility Function

Experiment Number	w_1	w_2	95% Paired-t Confidence Interval for $EU_{Alt3} - EU_{Baseline}$	Significance Results
1	.8	.1	(−.28, −.221)	Baseline significantly better than alternative 3
2	.7	.2	(−.201, −.145)	Baseline significantly better than alternative 3
3	.6	.3	(−.123, −.0683)	Baseline significantly better than alternative 3
4	.5	.4	(−.0467, .00994)	No significant difference between the alternatives
5	.4	.5	(.0283, .0895)	Alternative 3 significantly better than baseline
6	.3	.6	(.102, .170)	Alternative 3 significantly better than baseline
7	.2	.7	(.175, .251)	Alternative 3 significantly better than baseline
8	.1	.8	(.248, .333)	Alternative 3 significantly better than baseline

the difference in expected utility values between alternative 3 and the baseline alternative is

$$\bar{Y}(30) \pm t_{30,.975}(\widehat{\text{Var}}\ [\bar{Y}(30)])^{.5}, \text{ or } .103 \pm .0328, \text{ or } (.0702,\ .1360).$$

Since the confidence interval does not contain 0, we say that alternative 3 performs significantly better (at the 95% level) than the baseline alternative with respect to the expected utility criterion.

A sensitivity analysis was performed in which the weights for the individual attribute utility functions were varied. In particular, w_1, the weight for the cost function, was varied from .8 to .1 in increments of .1, while correspondingly, w_2, the weight for project duration, was varied from .1 to .8 in increments of .1; the coefficient for the interaction term in the multiplicative utility function was left constant at $1 - w_1 - w_2$ (= .1). The results are shown in Table 9.12.

The results shown in Table 9.12 are intuitive. That is, as the weight associated with the individual attribute utility function for project duration increases, the attractiveness associated with alternative 3, relative to the baseline, increases. Also, as seen in Table 9.12, the utility function weights for which there is no significant difference between the two alternatives (at the 95% level) is $w_1 = .5$, $w_2 = .4$. However, an increase in the number of replications of the model from the current value of 30 could very well lead to a significant difference between the alternatives.

The results shown in Table 9.12 indicate the importance of obtaining an accurate utility function in order to rank the alternatives correctly.

Example 9.3: Comparing NBA Series Patterns

The National Basketball Association championship involves the playing of a seven-game (at the maximum) series between the two final teams to determine its champion. The first team to win four games is the champion. The series pattern has shifted between a two–three–two series and a two–two–one–one–one series over the years. The two–three–two series involves playing the first two games at the home court of the higher-seeded team, the next three games at the home court of the lower-seeded team (if all three are needed), and the last three games (if needed) at the home court of the higher-seeded team again. The two–two–one–one–one series has games one, two, five (if needed), and seven (if needed) at the home court of the higher-seeded team.

There are obvious advantages and disadvantages associated with each format vis-à-vis the other format. For example, with the two–three–two format, there will be less travel involved; however, some analysts have remarked that the two–three–two format results in an unfair advantage to the lower-seeded team since if they can "steal" one of the first two games, they will be able to close out the series at their home court.

In 2014, the association switched to a two–two–one–one–one series from a two–three–two series. At the time, it was said that this change in series format was more likely to result in a longer series than in a change in outcome (Zillgitt, 2014).

A Monte Carlo simulation model was built to test the difference between the two formats. The model assumes that each team has a constant probability of winning on their home court and a constant probability of winning on the other team's home court. Of course with these assumptions, the differences between the two series could be derived analytically, but let's assume that the NBA commissioner would be more comfortable with a simulation in which he can view the series play out with respect to the number of games won by each team. In addition, with a simulation, one would be able to more easily vary the assumptions—for example, if a team won two games in a row, their probability of winning the next game might decrease because of overconfidence.

The basic input to the models included the probability associated with each team winning on their home court, the number of series played per replication, and the number of replications made. Note that setting up each replication to run multiple series was not difficult since each new series in a replication starts out with an "empty and idle" system. The outputs from the model included estimates of the probability of each team winning and the expected number of games in the series, along with 95% confidence intervals for each of these quantities. For example, if three replications were made, with each replication involving one series, and the outputs for each replication were

1. Higher-seeded team wins in five games in the first replication
2. Lower-seeded team wins in six games in the second replication
3. Higher-seeded team wins in seven games in the third replication

Then the estimates for expected number of games in a series would be six, and the estimate associated with the higher-seeded team winning the series would be 2/3.

If three replications were made, with 100 series played in each replication, and the outputs associated with each replication were

1. Higher-seeded team wins 57 series, with the average number of games per series being 6.24
2. Higher-seeded team wins 63 series, with the average number of games per series being 6.15
3. Higher-seeded team wins 55 series, with the average number of games per series being 6.75

Then the estimates for the expected number of games in a series would be (6.24 + 6.15 + 6.75)/3 = 6.38, and the probability of the higher-seeded team winning the series would be (.57 +.63 +.55)/3 = .5833....

Four separate models (using the Arena simulation software package) were built:

1. A model for the two–three–two series in which each random number in the simulation was generated from stream 10
2. A model for the two–three–two series in which each random number in the simulation was generated from the stream that corresponded to the game number of the series
3. A model for the two–two–one–one–one series in which each random number in the simulation was generated from stream 10
4. A model for the two–two–one–one–one series in which each random number in the simulation was generated from the stream that corresponded to the game number of the series

The idea was that in addition to comparing the series patterns, we also wanted to determine the effect of common random numbers.

The model was set up to run so that the probability of the higher-seeded team winning on their home court was .6 and the probability of the lower-seeded team winning on their home court was .52. (It would be expected that the team with the home court advantage, being the better team, would have a higher probability of winning on their home court than the other team would have of winning on their home court.) One hundred replications, each involving 100 series, were made for each of the four models. The results for the individual models are shown in Table 9.13.

The results associated with the 95% paired-t confidence intervals are shown in Table 9.14.

The first thing to note about the results is that there is very little difference in the two series, either in terms of the probability of the home team winning or in the expected number of games for the series, since each of the paired-t confidence intervals contained 0. The use of common random numbers did have a slight effect on the half-widths of the paired-t confidence intervals, as seen in Table 9.14; specifically, the half-width for the difference in probabilities confidence interval was reduced from .0115 to .00966, and the half-width for the number of games confidence interval was reduced from .0292 to .0237. It should be noted that

TABLE 9.13

Results from NBA Championship Series Simulation with a Probability of Home Court Win by Higher-Seeded Team of .6 and a Probability of Home Court Win by Lower-Seeded Team of .52

Model	95% Confidence Interval for Probability of Series Win by Higher-Seeded Team	95% Confidence Interval for Expected Number of Games in Series
2–2–1–1–1 No CRN	.6071 ± .01	5.8168 ± .02
2–3–2 No CRN	.6039 ± .01	5.8068 ± .02
2–2–1–1–1 CRN	.6056 ± .01	5.7864 ± .02
2–3–2 CRN	.6014 ± .01	5.8003 ± .02

TABLE 9.14

Paired-t 95% Confidence Intervals for the Differences between the 2–2–1–1–1 Series and the 2–3–2 Series

Models	Paired-t 95% CI for Difference in Probability of Series Win by Higher-Seeded Team	Paired-t 95% CI for Difference in Expected Number of Series Games
2–2–1–1–1 No CRN – 2–3–2 No CRN	.0032 ± .0115	.01 ± .0292
2–2–1–1–1 CRN – 2–3–2 CRN	.0042 ± .00966	–.0139 ± .0237

the procedure used for common random numbers in this case (in which the stream number used was the same as the game number) probably would have been more effective had there just been one series per replication; this would have allowed for a better matching of the streams.

One of the more interesting aspects of this experiment would be in considering the benefit of home court advantage in a series. This aspect is addressed in Exercise 9.5.

9.6 Comparing a Few Alternatives

In this section, we discuss methodologies that involve the analysis of a few alternatives using simulation as a modeling tool. By a few alternatives, we mean anywhere between three and about 100. Of course, the specific number of alternatives to consider for any of the methodologies discussed in this section depends upon the computational requirements associated with a replication of the simulation model.

Also, the procedures discussed in this section require the explicit "evaluation" of each alternative through at least one replication of the model representing that alternative—that is, no optimization technique is employed to implicitly evaluate alternatives.

The consideration of multiple objectives in this section can be accomplished through the use of expected utility as the performance measure of interest.

Several categories of techniques are available for this type of analysis, including (1) procedures that involve comparing each alternative to a standard, (2) procedures that involve comparing each alternative to each other alternative in a pairwise fashion, and (3) ranking and selection procedures. The first two categories involve a simple extension of the paired-t confidence interval approach discussed earlier. The ranking and selection techniques represent a different type of approach.

Before discussing these categories of techniques, we will first briefly discuss the Bonferroni inequality, an important result when considering multiple confidence intervals.

9.6.1 Bonferroni Inequality

One important result (for a comparison involving multiple systems with possibly multiple performance measures to consider) is the *Bonferroni inequality* (Banks et al., 2005, p. 449). This result says that if one has n confidence intervals, with each interval at the $(1 - \alpha)$ confidence level, then the probability that all n confidence intervals contain their true performance measure values *simultaneously* will be at least $1 - n\alpha$.

Let's consider the simulation model for the baseline project of Example 9.2. Ten replications (as opposed to the 30 in the example) were made of one simulation each, and the following 95% (or (1 – .05) * 100%) confidence intervals were attained for cost and project duration:

1. For cost, 216.75 ± 8.79 or (207.96, 225.54) thousands of euros
2. For project duration, 81.11 ± 3.54 or (77.57, 84.65) days

From the Bonferroni inequality, we could then say that the probability of the true values for expected cost *and* expected project duration being within these intervals would be at least $((1 - 2(.05)) \times 100)\%$ or 90%.

Note that the Bonferroni inequality only provides a bound on the confidence. Also when using the Bonferroni inequality to make inferences concerning multiple simultaneous confidence intervals, the confidences for the respective individual confidence intervals do not have to be equal. For example, one could form a 2% confidence interval for cost, an 8% confidence interval for project duration, and at least 90% confidence interval (=2% + 8%) in the simultaneous result.

One important advantage associated with the Bonferroni inequality is that it applies in situations in which the common random number variance reduction technique is used for variance reduction.

9.6.2 Comparison to a Standard

In some situations, one has an existing system and also several alternatives which are being considered to modify that system. Such a case is called *comparison to a standard*. The *standard* could be the best system or it could be the existing system. If there are n alternatives to compare to the standard, then if the paired-t confidence interval approach is used, there will be n paired-t confidence intervals formed. So, for example, if there are five alternatives to be compared to a standard, and therefore, five 98% paired-t confidence intervals are formed

Alternative 1 – Standard system
Alternative 2 – Standard system
Alternative 3 – Standard system
Alternative 4 – Standard system
Alternative 5 – Standard system

then the overall confidence will be 100% – 5 * 2% = 90%. That is, the overall probability that each of the five confidence intervals will contain the true differences of the respective mean values of the performance measure simultaneously will be .9.

9.6.3 All Pairwise Comparisons

If one wants to compare each of n alternatives to every other alternative, then using the paired-t confidence interval approach, there will be $(n - 1) + (n - 2) + \cdots + 1 = n(n - 1)/2$ individual confidence intervals to form.

For example, if there are five alternatives, there will be 10 confidence intervals to form if all pairwise comparisons are made. If each of the 10 confidence intervals is at the 2% level, then the overall confidence will be 100% – 10 * 2% = 80%.

Example 9.4: Selecting a Policy for an (s, S) Inventory System Using the All Pairwise Comparisons Approach

This problem is a modified form of a problem described in Kelton et al. (2015, pp. 257–270).

Consider a simple (s, S) inventory system involving one product. The system operates 24 hours per day, 7 days per week. Customers arrive to purchase the product with an interarrival time that is exponentially distributed with a mean value of .05 day. Each customer purchases a number of units that are distributed according to a discrete empirical distribution, as shown in Table 9.15. (The distribution represents two types of customers—"small demand," who demand from 1 to 4 units of the product, and "large demand," who demand 9 or 10 units of the product.)

TABLE 9.15

Units of Demand per Customer

Number of Units of Demand	Probability
1	.1
2	.2
3	.2
4	.1
9	.2
10	.2

An (s, S) policy is followed to replenish the inventory. Specifically, the inventory level is checked at the beginning of each day. If the inventory level is less than s, then an order is placed with an order quantity equal to S minus the current inventory level. If the inventory level is greater than or equal to s, no order is placed. An order placed at the beginning of a day will arrive later in the day, with a lead time that is uniformly distributed between .25 day and 1 day.

Each time an order is placed to replenish inventory, a cost is incurred. This cost has a fixed component of $30 and a variable component of $3. Note that this variable component has nothing to do with the cost of the product itself; it is just the cost associated with ordering the product. So an order for 100 units of the product would have an ordering cost of $30 + $3 * 100 = $330.

There is also an inventory holding cost of $2 per unit of inventory per day. So, for example, if the system had carried 50 units of inventory over a 5-day period, followed by 20 units of inventory over the next 3 days, the holding cost over the entire 8-day period would be 2*50*5 + 2*20*3 = 500 + 120 = $620.

In summary, the parameter values for this problem are shown in Table 9.16.

When a customer arrives with a demand for the product, the entire demand is met if the inventory level is greater than or equal to the demand. If the customer demand is greater than the inventory level, the customer receives a number of units equal to the physical inventory

TABLE 9.16

Parameter Values for the Inventory System

Parameter	Parameter Value
Customer interarrival time	EXPO (.05) day
Customer demand	See Table 9.15
Fixed cost for ordering inventory	$30
Variable cost for ordering inventory	$3 per unit
Lead time from supplier	UNIF (.25, 1) day
Evaluation time for inventory level	Beginning of each day

level but must wait for the remainder of the order until a new delivery of inventory. So, for example, if two units of inventory are on hand, and a customer arrives with a demand for three units of the product, two units of demand will be satisfied, while the other unit will be placed on back order, until additional units arrive from the supplier.

The organization is interested in two performance measures, one having to do with the cost of operating the system and one addressing customer service:

X_1: Average cost per day associated with ordering and holding inventory
X_2: Unit-days of shortage per day

Using two different attributes (or categories of attributes), one to represent the internal system cost and another to represent customer service, is often a desired approach, as has been pointed out previously.

As an example of the second performance measure, suppose that a customer arrives 10 hours into a day and demands 10 units of inventory, but only two units are available; so, after he or she receives the two units available, there is now a shortage of eight units. A second customer arrives at 10.4 hours into the day and demands three units of the product; of course, none of his or her demand can be supplied, so now there is a shortage of 11 units. No other customers arrive between 10.4 and 11 hours into the day. But at 11 hours, an order of 60 units of the product arrives from the supplier, so the two back orders (the first of eight units and the second of three units) are satisfied. During this period of one hour, between 10 and 11 hours into the day, there was a period of .4 hours (between times 10 and 10.4 hour) when there was a shortage of eight units and a period of .6 hour (between times 10.4 and 11 hours) when there was a shortage of 11 units. Therefore, the unit-days of shortage was

.4 hour * (1 day/24 hours) * 8 units of shortage + .6 hour * (1 day/24 hours) * 11 units of shortage = .1333... + .275 = .408 unit-days of shortage.

A simulation model has been developed to study the system in order to determine the best values for s and S. The model has been set up to run for 200 days, with an initial inventory level of 50 units. (Given the parameter values for the situation, no warm-up period for the model is needed in order to allow for a steady-state analysis of the output.) The reader will note that this model is not a Monte Carlo simulation but instead is a "time-dynamic" simulation model, with time-persistent variables such as inventory level.

An "all pairwise comparison" of four policies is to be done:

1. Policy 1: s = 30, S = 60
2. Policy 2: s = 30, S = 120
3. Policy 3: s = 100, S = 250
4. Policy 4: s = 150, S = 250

A scaled multiplicative utility function, representing the inventory manager's preferences, has been developed over X_1 and X_2:

$$u(x_1, x_2) = .18u_1(x_1) + .72u_2(x_2) + .1u_1(x_1)u_2(x_2)$$

where

$$u_1(x_1) = -.0025x_1 + 1.75$$

$$u_2(x_2) = -.00008x_2^2 - .0012x_2 + 1$$

The individual attribute utility functions, u_1 and u_2, have been assessed in such a way that

$$u_1(700) = 0,\ u_1(300) = 1,\ u_2(104.5) = 0, \text{and } u_2(0) = 1.$$

These worst and best values for X_1 (700 and 300, respectively) and for X_2 (104.5 and 0, respectively) were determined by running the simulation model for multiple replications for the various policies and determining the worst and best values from the respective outputs. Note also that the utility function for X_1 is linear, implying that the manager is risk neutral over cost; the function for X_2 is concave, implying that the manager is risk averse over shortages.

Thirty replications were run for each policy. The 99% paired-t confidence intervals for the difference in the expected utilities for each pair of policies are given as follows:

Policy 1–Policy 2: (–.196, –.167)
Policy 1–Policy 3: (–.193, –.169)
Policy 1–Policy 4: (–.162, –.139)
Policy 2–Policy 3: (–.00555, .00658)
Policy 2–Policy 4: (.0257, .0372)
Policy 3–Policy 4: (.0279, .034)

Note that, for example, Policy 1-Policy 2 refers to the 99% confidence interval for the expected utility of Policy 1 minus the expected utility of Policy 2. The overall confidence level is at least (1 – 6 * (.01)) * 100% = 94% from the Bonferroni inequality.

The difference in the policies for any pair of policies, except for Policies 2 and 3, is statistically significant. This can be seen by determining which of the confidence intervals contains 0. In particular, each of Policy 2 and Policy 3 is significantly better than Policy 1 and Policy 4. Policy 4 is significantly better than Policy 1.

Since the difference between Policies 2 and 3 is not significant, one, or more than one, of the following actions can be taken:

1. The number of replications can be increased for the simulations of Policies 2 and 3. (Keep in mind that for a paired-t confidence interval, the number of replications must be the same for each alternative.)
2. The significance levels associated with the individual confidence levels could have been altered.

3. A variance reduction technique such as common random numbers (Banks et al., 2005, pp. 438–446) could be applied to reduce the width of the paired-t confidence interval.

For example, both the 95% confidence interval and the 90% confidence interval for the Policy 2 expected utility minus the Policy 3 expected utility were constructed, yielding the following results:

95% confidence interval (Policy 2–Policy 3): (–.00398, .00502)
90% confidence interval (Policy 2–Policy 3): (–.00322, .00426)

Even though the widths of the confidence intervals decreased in going from 99%, to 95%, to 90% (as they must) each interval still contained 0. Keep in mind that the overall confidence of the simultaneous result decreases, from 94% to 90%, and then to 85%.

In addition, the number of replications associated with the simulations of Policies 2 and 3 was increased, to see if this would have an effect on the result. The paired-t confidence intervals for this experimentation (with 50 replications) were given by

99% confidence interval (Policy 2–Policy 3): (–.00418, .00577)
95% confidence interval (Policy 2–Policy 3): (–.00294, .00452)
90% confidence interval (Policy 2–Policy 3): (–.00232, .0039)

The differences are still not significant. So, at this point, the analyst should spend some time with the manager in looking at the outputs for the individual attributes in more detail. In particular, the estimates (resulting from 30 replications) of the expected values for X_1 for Policies 2 and 3 are $372.04 and $504.08, respectively, while the estimates of the expected values for X_2 for Policies 2 and 3 are 32.49 and 5.702, respectively. This indicates quite a large trade-off between these two policies, and as a result, the difference in preferences between these two policies is highly sensitive to the scaling constants of the utility function.

9.6.4 Ranking and Selection

The methods discussed to this point in Section 9.6 are probably applicable when only a very few alternatives (say 3 to 10) are under consideration. If there are more than 10, but less than, say, 100, alternatives, the analyst may want to use one of the methods of ranking and selection.

The various methods for ranking and selection can be divided into two categories: (1) subset selection (screening) and (2) identifying a best alternative. Both groups of methods often involve a "guarantee" (that the best alternative is contained within an identified subset that contains either multiple alternatives [category 1] or a single alternative [category 2]) according to a probability and possibly a bound on the performance measure.

First, we will provide some notation for the discussion of these procedures. Assume that we have $k \geq 2$ alternatives to consider and that X_{ij} represents the output of interest (i.e., a performance measure value such as utility) from the jth independent replication representing the ith alternative. Suppose that we

TABLE 9.17

Summary of Notation for Ranking and Selection Methodologies

Notation	Definition
k	Number of alternatives under consideration
X_{ij}	Output of interest from the jth independent replication of the simulation representing the ith alternative, for $i = 1,\ldots, k$
$\bar{X}_i(n)$	The sample mean associated with n replications of the ith alternative $\left(= \sum_{j=1}^{n} X_{ij}/n\right)$
$S_i^2(n)$	The sample variance (for the X_{ij}'s) associated with n replications of the ith alternative $= \sum_{j=1}^{n} \left(X_{ij} - \bar{X}_i(n)\right)^2 \Big/ (n-1)$.
θ_i	The true expected value of X_{ij}
$\theta_{[i]}$	The ith largest θ_i

want to maximize this output value. (If the objective involves minimization, the modification is obvious.) Let $\bar{X}_i(n)$ be the sample mean and $S_i^2(n)$ sample variance of the mean for the ith alternative from n replications. In addition, let θ_i be the true expected value and $\theta_{[i]}$ be the ith largest expected value. The notation and associated definitions are shown in Table 9.17.

As an example, suppose that we have four alternatives under consideration and that we want to choose the alternative that yields the largest expected utility. The true expected utilities for alternatives one, two, three, and four are $\theta_1 = .57$, $\theta_2 = .42$, $\theta_3 = .78$, and $\theta_4 = .67$, respectively. This would mean that $\theta_{[1]} = .78$, $\theta_{[2]} = .67$, $\theta_{[3]} = .57$, and $\theta_{[4]} = .42$.

Subset selection, also called screening, means selecting a subset (say 10 out of 100 in total) of the entire set of alternatives such that one is assured with some probability that the best alternative (within a range for the expected value of the performance measure) is contained within the subset. Typically, one wants to employ subset selection when there are "many" alternatives from which a few (up to say about 10) are to be selected for a more intense analysis, for example, involving the second category previously mentioned or all pairwise comparisons.

The concept of *correct selection* is important for both categories of ranking and selection methods. Correct selection as applied in subset selection basically means that the best alternative is contained in the selected subset under some bounds and conditions with at least some specific probability.

For example, Koenig and Law (1985) provide a subset selection procedure, also described in Law (2007, pp. 568–569), which defines correct selection as

> ... selecting a subset of size m from the set of k alternatives such that the subset will contain the alternative with mean $\theta_{[1]}$ with probability $1 - \alpha$ whenever $\theta_{[1]} - \theta_{[2]} \geq \varepsilon$, where $1 \leq m \leq k-1$, $(1 - \alpha) > m/k$, and $\varepsilon > 0$.

The region of the performance measure space, $[\theta_{[2]}, \theta_{[1]}]$, is called the *indifference zone* (IZ). Typically, α might be set at a value of .05 (to give a probability of correct selection of .95), and ε might be set at .01 or .02 or even a slightly larger value if expected (scaled) utility is the performance measure.

One of the difficulties associated with using ranking and selection methods when expected utility is the performance measure has to do with determining the value of ε for the indifference zone. This value of the performance measure should have inherent meaning for the DM. Of course, measures such as expected waiting time, expected number of lost sales, and expected holding cost do have inherent meaning for the DM. Unfortunately, expected utility does not have such inherent meaning. In order to handle this difficulty, Butler et al. (2001) have developed a "utility exchange" method that allows the DM to specify an ε value for one of the attributes of a utility function and then *exchange* that for an ε value for expected utility.

The Koenig and Law procedure assumes that the X_{ij} values are normally distributed and independent; the independence assumption implies that the common random numbers approach to variance reduction is not allowed. The normal distribution assumption is typically not a difficulty. The steps of the procedure are as follows:

1. The DM selects values for α and ε. When the performance measure is expected utility, the choice of a value for ε can be problematic, as discussed earlier.
2. Make $n_0 \geq 2$ replications of each alternative simulation and compute values for the first-stage sample means, $\bar{X}_i^{(1)}(n_0)$ and sample variances, $S_i^2(n_0)$ from the n_0 replications for $i = 1,\ldots, k$.
3. Compute the total number of replications, N_i, for the ith alternative for $i = 1,\ldots, k$, from the following formula:
$$N_i = \max\left\{n_0 + 1, \left\lceil h^2 S_i^2(n_0)/\varepsilon \right\rceil\right\}$$
where $\lceil x \rceil$ refers to the smallest integer greater than or equal to x and h is a constant that can be obtained from Table 10.12 in Law (2007) or from Koenig and Law (1985). Note that the total number of replications will depend upon the alternative.
4. Make $(N_i - n_0)$ additional replications for $i = 1,\ldots, k$, in order to obtain the second-stage sample means: $\bar{X}_i^{(2)}(N_i - n_0)$.
5. Compute a weighted sum of the two sample means for each alternative, where the weights are as given by a formula from Law (2007, p. 564).
6. Select the m alternatives associated with the best m weighted sums as computed in Step 5.

A Bonferroni approach to subset selection, which guarantees with a probability of $1 - \alpha$ that the system with a performance measure value of $\theta_{[1]}$ is in the subset and also allows for the use of common random numbers, is described in Banks et al. (2005, p. 457). This procedure is due to Nelson et al. (2001).

In addition to allowing the use of common random numbers, the approach has the advantages of not requiring the use of (1) a table of constant values (e.g., h in the Koenig and Law approach) or (2) the weights required for the weighted sum of the two sample means for each alternative. As with the Koenig and Law approach, however, the data from the replications are required to be normally distributed. Finally, the procedure does not result in a specific number of alternatives in the selected subset. The steps of this Bonferroni approach to screening are as follows:

1. Select values for α, typically .05 and n_0, the initial number of replications for each alternative. Typically, n_0 might be set to 5 or 10.
2. Make n_0 independent replications for each alternative.
3. Compute the sample mean for alternative i, $\bar{X}_i(n_0)$, for i = 1,…, k, and for all i ≠ j, calculate the sample variance of the difference:

$$S_{ij}^2 = \frac{1}{n_0 - 1}\sum_{r=1}^{n_0}\left(X_{ir} - X_{jr} - \left(\bar{X}_i(n_0) - \bar{X}_j(n_0)\right)\right)^2.$$

4. Retain system i in the selected subset if $\bar{X}_i(n_0) \geq \bar{X}_j(n_0) - t\left(S_{ij}/\sqrt{n_0}\right)$ for all j ≠ i, where $t = t_{\frac{\alpha}{k-1}, n_0-1}$.

Note that this procedure, unlike that of Koenig and Law, does not have an indifference zone associated with it—in effect, $\varepsilon = 0$.

The second category of ranking and selection methods, involving the identification of a best alternative, is typically used to address situations with only a "few" alternatives (say, less than or equal to 20). Chau et al. (2014) note that there are three categories of these methods: indifference zone methods, value of information procedures, and optimal computing budget allocation methods. In this section, we will focus on the most popular category, the indifference zone (IZ) methods.

Two of the earliest popular procedures for identifying a best alternative were developed by Dudewicz and Dalal (D & D, 1975) and Rinott (1978), respectively. Both the D & D and Rinott methods assume that the X_{ij} values are normally distributed, with (perhaps) unknown and unequal (for different alternatives) variances. As mentioned earlier, the normal distribution assumption is generally not a problem for simulation model output, especially since typically the performance measure value associated with a single replication is obtained over many observations (e.g., average customer waiting time). The allowance for unequal variances for the different respective alternatives also fits the usual situation for simulation model output.

In addition, both of the D & D and Rinott methods assume that the X_{ij} values are generated independently, which precludes the use of common random numbers as a variance reduction technique for increasing the accuracy of the estimates, thereby reducing the number of replications required to differentiate between the alternatives.

Law (2007, p. 565) notes that the D & D procedure typically requires fewer replications than the Rinott procedure, but this is at the expense of being more complicated from a computational standpoint. Specifically, the D & D procedure requires a weighted sum of the sample means computed from the two sets of replications where the weights arise from a relatively complicated formula. The D & D procedure is much like the Koenig and Law approach for subset selection described earlier, except the identified subset is of size one. Nelson et al. (2001) note that the indifference zone procedures like Rinott's are based on a "worst-case" analysis in which all the alternatives except the best are assumed to be tied for second, exactly ε away from the best; this can result in much (unneeded) computational effort.

The two-stage Rinott procedure is described as follows:

1. Identify a practically significant difference, ε, and a probability of correct selection, $1-\alpha$.
2. Make n_0 replications for each alternative, in order to obtain X_{ij} for $i = 1,\ldots, k$ and $j = 1,\ldots, n_0$.
3. Compute the first-stage sample means and sample variances:

$$\bar{X}_i^{(1)}(n_0) = \sum_{j=1}^{n_0} X_{ij}/n_0$$

and

$$S_i^2(n_0) = \sum_{j=1}^{n_0} \left(X_{ij} - \bar{X}_i^{(1)}(n_0)\right)^2/(n-1), \quad \text{for } i = 1,\ldots,k.$$

4. Compute the total sample size needed for alternative i as $N_i = \max\left\{n_0, \left\lceil (gS_i(n_0)/\varepsilon)^2 \right\rceil\right\}$ where $\lceil x \rceil$ is the smallest integer greater than or equal to x, and g is a constant that solves the "Rinott integral."
5. Make $(N_i - n_0)$ additional replications for alternative i for $i = 1,\ldots, k$, and select the alternative with the largest overall sample mean.

Nelson and Matejcik (1995) developed a ranking and selection procedure that allows for the use of common random numbers as a variance reduction technique as long as the covariance matrix associated with the random variables $X_{1j}, X_{2j},\ldots, X_{kj}$ satisfies a condition called sphericity (see Law, 2007, p. 565). This two-stage Bonferroni procedure allows for comparisons of all alternatives with the best and is described in Banks et al. (2005, pp. 454–458):

1. Identify a practically significant difference, ε, and a probability of correct selection, $1-\alpha$.
2. Make n_0 replications for each alternative, in order to obtain $X_{i,j}$ for $i = 1,\ldots, k$ and $j = 1,\ldots, n_0$.

3. Calculate the sample mean for alternative i, $\bar{X}_i(n_0)$, for i = 1,…, k, and for all i ≠ j, calculate the sample variance of the difference:

$$S_{ij}^2 = \frac{1}{n_0 - 1}\sum_{r=1}^{n_0}(X_{ir} - X_{jr} - (\bar{X}_i(n_0) - \bar{X}_j(n_0)))^2,$$

   And let $\hat{S}^2$ = the largest of the S_{ij}^2's.
4. Calculate the number of replications to make for each alternative for the second stage of the procedure:

   $N = \max\left\{n_0, t^2\hat{S}^2/\varepsilon^2\right\}$, where $\lceil x \rceil$ refers to the largest integer less than x.
5. Make $N - n_0$ additional replications for each alternative and calculate the overall sample mean for each alternative: $\bar{X}_i(N)$ for i = 1,…, k.
6. Select the alternative with the largest sample mean as the best.

If i* denotes the best alternative, then if $X_i(N) - X_{i^*}(N) + \varepsilon \leq 0$ for any i ≠ i*, then alternative i is inferior to the best alternative. But if $\bar{X}_i(N) - \bar{X}_{i^*}(N) + \varepsilon > 0$ for any i ≠ i*, then alternative i is statistically indistinguishable from i*, and in fact, alternative i might be the best alternative.

There is much research and many different perspectives on ranking and selection that have not been covered in this section of the book. For example, Sullivan and Wilson (1989) developed a two-stage restricted subset selection procedure that requires independent replications and results in a random-size subset, but with at most m alternatives. Chen and Kelton (2005) developed a sequential approach to ranking and selection that employed their enhanced two-stage selection procedure (Chen and Kelton, 2000), which accounts for the differences in the sample means (as well as the variances of the samples) in determining the sample sizes.

Finally, Mattila and Virtanen (2015) addressed the use of expected utility in ranking and selection in which incomplete preference information was available from the DM; the incomplete preference information arose from linear constraints placed on the weights of the utility function.

Example 9.5: Selecting a Subset of Policies for an (s, S) Inventory System Using the Bonferroni Approach to Screening

Consider Example 9.4, but with the four additional policies to examine. Specifically, we are now considering the following policies:

1. Policy 1: s = 30, S = 60
2. Policy 2: s = 30, S = 120
3. Policy 3: s = 100, S = 250
4. Policy 4: s = 150, S = 250
5. Policy 5: s = 50, S = 100
6. Policy 6: s = 50, S = 150
7. Policy 7: s = 120, S = 200
8. Policy 8: s = 120, S = 300

Now, in order to use the Bonferroni approach to screening, the X_{ij} outputs should be normally distributed. This was tested with the output for the first policy and was shown to be approximately true for 30 data points obtained from 30 independent replications.

We set $\alpha = .05$ and n_0, the number of replications for each of the eight alternative systems, to 5. The results for the five replications made over each of the eight alternatives, along with the sample expected utility values, are shown in Table 9.18. (Because of the closeness of several of the numerical values in this problem, the numbers shown in the various tables for this example are carried out to several decimal places.)

The sample variance values associated with the differences from the policies are shown in Table 9.19. Note, for example, that S_{12}^2, the sample variance associated with the difference between Policy 1 and Policy 2, is given by

$$(1/(5-1)) * (((.6745-.8304)-(-.1917))^2 + ((.6681-.8688)-(-.1917))^2$$

$$+\cdots+((.6588-.8545)-(-.1917))^2 = .000565.$$

The "–.1917" figure in the this calculation is the difference in the sample means between Policies 1 and 2 (.6611 – .8528). Note also that, by definition, $S_{ij}^2 = S_{ji}^2$; hence, only the upper right-hand portion of the matrix needs to be shown.

TABLE 9.18

Utility Values from Five Independent Replications of Eight Inventory Policies

Replication Number	Policy 1 Output	Policy 2 Output	Policy 3 Output	Policy 4 Output	Policy 5 Output	Policy 6 Output	Policy 7 Output	Policy 8 Output
1	.6745	.8304	.8524	.8157	.8650	.8739	.8564	.7979
2	.6681	.8688	.8556	.8243	.8747	.8777	.8674	.7979
3	.6690	.8547	.8488	.8158	.8673	.8758	.8567	.8032
4	.6349	.8557	.8478	.8139	.8563	.8707	.8581	.8116
5	.6588	.8545	.8466	.8029	.8574	.8870	.8687	.8019
Sample expected utility	.6611	.8528	.8502	.8145	.8642	.8770	.8615	.8025

TABLE 9.19

Sample Variances, S_{ij}^2, Associated with Differences in Policies

i/j	1	2	3	4	5	6	7	8
1		0.000565	0.000196	0.000236	0.0001373	0.000235508	0.00027871	0.000439843
2			0.000196	0.000196	0.0001897	0.00019243	0.000129028	0.000196592
3				2.37E–05	2.2165E–05	0.000062597	0.000045677	7.2973E–05
4					2.0047E–05	0.00015165	0.000113343	0.000114112
5						0.000103627	8.3657E–05	0.000147588
6							1.5058E–05	9.5867E–05
7								8.9458E–05
8								

TABLE 9.20

Values for $\bar{X}_j(n_0) - t\frac{S_{ij}}{\sqrt{n_0}}$ for $i \neq j$

i/j	1	2	3	4	5	6	7	8
1		0.80868	0.824274	0.786031	0.84238805	0.848531436	0.830468399	0.763567116
2	0.61692		0.824241	0.788516	0.83857193	0.851268406	0.840373247	0.776471409
3	0.635094	0.826821		0.805478	0.8554002	0.862332615	0.848913675	0.786641985
4	0.632571	0.826816	0.841198		0.85582825	0.854159331	0.841696444	0.782669513
5	0.639308	0.827252	0.8415	0.806208		0.858122507	0.844480735	0.779947575
6	0.632571	0.827068	0.835553	0.791659	0.84524251		0.85425637	0.784323833
7	0.630068	0.831733	0.837694	0.794756	0.84716074	0.86981637		0.784941908
8	0.622127	0.826791	0.834382	0.79469	0.84158757	0.858843833	0.843901908	

The value for t (= $t_{.05/7,4} = t_{.007142854,4}$) for this analysis is 4.151. The values for $\bar{X}_j(n_0) - t(S_{ij}/\sqrt{n_0})$ for i, j = 1,…, 8 and i ≠ j are shown in Table 9.20.

As an example of an entry in Table 9.20, $\bar{X}_5(n_0) - t\left(S_{25}/\sqrt{n_0}\right)$ is the entry given in the second row, fifth column of the matrix: .83857193. The computation (with some round off) for this value is given by $.8642 - 4.151 * \sqrt{0.0001897}/\sqrt{5} = .8642 - .0256 = .8386$.

In order to determine whether or not alternative i is in the selected subset, we need to see if $\bar{X}_i(n_0) \geq \bar{X}_j(n_0) - t\left(S_{ij}/\sqrt{n_0}\right)$ for j = 1,…, 8 and i ≠ j. For example, to determine whether alternative 1 is in the selected subset, we would see if .6611 (= $\bar{X}_i(n_0)$) is greater than or equal to each of the numbers in the first row of the matrix given in Table 9.20: 0.80868, 0.824274, 0.786031, 0.84238805, 0.848531436, 0.822527116, and 0.763567116, which clearly it is not. Proceeding with this comparison for each of the sample means with each respective row of the matrix, we determine that alternatives 2, 5, and 6, that is, the (s, S) policies of (30, 120), (50,100), and (50, 150), are in the selected subset.

What is perhaps surprising about this conclusion is that Policy 2 (with a sample expected utility of .8528) is in the selected subset, while Policy 7 (with a larger sample expected utility of .8615) is not. This can be at least partially explained by looking at the values for the sample variances of the differences between Policy 2 and Policy 6 (= 0.00019243) and between Policy 7 and Policy 6 (= 1.5058E−05 = .000015058). The former is about 13 times the size of the latter, making the value for $\bar{X}_6(n_0) - t\left(S_{76}/\sqrt{n_0}\right)$ larger than $\bar{X}_6(n_0) - t\left(S_{26}/\sqrt{n_0}\right)$; this results in Policy 2 being included in the set, while Policy 7 is not. In effect, the sample variance values indicate that Policy 2 is closer to Policy 6 in performance than Policy 7, and Policy 6 has by far (relatively speaking) the largest sample expected utility.

As with Example 9.4, it is clear the differences in expected utilities for the various policies are fairly small, indicating that the ranking of policies will be highly sensitive to the values for the scaling constants of the utility function. In addition, this example points out that it may be desirable, when employing the ranking methodologies for a situation involving expected utility as a performance measure, to scale the function from 0 to 100, rather than from 0 to 1.

Example 9.6: Using the Two-Stage Bonferroni Approach to Select the Best Policy for an (s, S) Inventory System

Let's apply the two-stage Bonferroni approach to see if we can find a "best" alternative for our inventory system. Let ε be .01, and let's start with all eight alternative policies given for Example 9.5, so k = 8; we will also employ our initial 5 replications, so $n_0 = 5$. In viewing Table 9.19, the largest S_{ij}^2 is $S_{12}^2 = .000565$, so $\hat{S}^2 = .000565$. Hence, the number of replications to make for the second stage is given by

$$R = \max\left\{5, \left\lceil \frac{4.151^2(.000565)}{.01^2} \right\rceil\right\} = \max\{5, 98\} = 98.$$

By making 93 additional replications for each policy (or just making 98 replications total), we obtain the following sample means for the eight alternatives:

$$\bar{X}_1(98) = .67, \bar{X}_2(98) = .8523, \bar{X}_3(98) = .8500, \bar{X}_4(98) = .8197, \bar{X}_5(98) = .8627,$$

$$\bar{X}_6(98) = .8808, \bar{X}_7(98) = .8619, \bar{X}_8(98) = .8066.$$

The largest sample mean is given by Policy 6 with $\bar{X}_6(98) = .8808$. Since none of the other policies have a sample mean within .01 (= ε) of $\bar{X}_6(98)$, we can declare that each of the other policies is inferior to Policy 6, with values of 50 and 150 for s and S, respectively.

Example 9.7: Scheduling and Resource Allocation for a Medical Clinic

(This example is derived from the PhD dissertation of Sun, 2015.)

An ambulatory internal medical clinic, located in the downtown of a metropolitan area, serves as a teaching clinic for a local medical school. As a teaching clinic, it serves low-income patients with major medical issues at a much reduced rate from what would normally be paid for medical services. Mainly because of the long wait times experienced by the clinic's patients, the management was interested in analyzing alternative policies for resource assignment and scheduling.

The clinic addresses different medical issues at different time periods of the week (e.g., cardiovascular issues on Tuesday mornings, pulmonary on Wednesday afternoons). This study addressed the period of Tuesday morning, one of the busiest periods for the clinic. The clinic's personnel during this period included 2 receptionists, 5 nurses, 12 resident doctors (which included 4 first-year residents, 5 second-year residents, and 3 third-year residents), and four attending physicians. First-year residents, as compared to second-year and third-year residents, required greater supervision from an attending physician and longer amounts of time to treat patients.

In addition, the clinic had 15 exam rooms, 5 triage areas, and a waiting room.

The process of treating the patients generally followed the sequence of activities: check-in at receptionist, nurse triage examination, wait, examination by a resident, consultation between resident and attending physician, possible examination by attending physician, and wait in waiting room for reports and lab results.

The management of the clinic was interested in investigating new policies with respect to (1) the interarrival times of patients, (2) the more flexible use of examination rooms, and (3) the number of patients assigned to first-, second-, and third-year residents. More specifically, the clinic's management wanted to investigate:

1. Allowing mean interarrival times of 4, 5, 6, and 7 minutes
2. Using the current approach in which third-year residents were assigned two exam rooms each and other residents were allowed one exam room each versus allowing any resident to use any exam room available, that is, first come, first served (denoted as exam rooms preassigned and exam rooms not preassigned, respectively)
3. Having the current approach of each resident being assigned the same number of patients, versus allowing the third-year residents to have more patients than the second-year residents, and the second-year residents being assigned more patients than the first-year residents (denoted as original resident allocation and modified resident allocation, respectively)

Varying the interarrival times would allow the clinic to see more/fewer patients. Allowing flexible use of the exam rooms would allow for increased efficiency, at the expense of the current privilege allowed third-year residents. Initial data collection indicated that third-year residents were able to complete tasks faster than second-year residents and that second-year residents were able to complete tasks faster than first-year residents, hence, the investigation of the third pair of subpolicies.

There were four objectives of interest to the clinic's management, as measured by the following four attributes:

X_1: Expected waiting time of patients
X_2: Staff utilization
X_3: Exam room utilization
X_4: Amount of overtime required to see all patients

The fourth attribute was used because typically the clinic continued to operate past the Tuesday morning time frame until all patients were seen. Overtime was a common occurrence at the clinic for its Tuesday morning schedule. An additive utility function was employed to represent preferences over these four attributes:

$$u(x_1,x_2,x_3,x_4) = \sum_{i=1}^{4} w_i u_i(x_i),$$

where

$$w_1 = .4, w_2 = .3, w_3 = .2, \text{ and } w_4 = .1.$$

Such an additive function while probably not an exact representation of the preference structure of the clinic's management was thought to be good enough to rank the alternatives, listed later.

TABLE 9.21

Best and Worst Values for Attributes

Attribute	Best Value	Worst Value
X_1, expected waiting time in minutes	0	70
X_2, staff utilization	1	0
X_3, exam room utilization	1	0
X_4, overtime in minutes	0	120

The individual attribute utility functions were given by

$$u_i(x_i) = A_i - B_i \exp(C_i x_i)$$

where A_i, B_i, and C_i were constant set so that the functions would have the appropriate minimum (0) and maximum (1) values and risk nature to represent the situation.

The best and worst values for the attributes are shown in Table 9.21.

The simulation model developed for the clinic was a complex one due to the required interaction among the resources (e.g., among residents, attending physicians, and examination rooms). The utility exchange method mentioned previously (Butler et al., 2001) was used to estimate a meaningful indifference zone for expected utility. Specifically, an indifference between 30 and 37 minutes in waiting time resulted in an ε value of .09 for the expected utility measure. This indifference zone was employed with the Dudewicz and Dalal (1975) ranking and selection scheme.

The various alternatives considered involved combinations of the three types of subpolicies mentioned previously, as shown in Table 9.22. The parameters used for the D & D method, in addition to $\varepsilon = .09$, were $\alpha = .05$, $k = 12$ (since there were 12 alternatives to consider), and $n_0 = 10$

TABLE 9.22

Examined Alternatives in the Ambulatory Internal Medical Clinic Example

Alternative	Mean Interarrival Time in Minutes	Exam Room Assignment	Resident Allocation
1	4	Preassigned	Original
2	4	Not preassigned	Original
3	4	Preassigned	Modified
4	4	Not preassigned	Modified
5	5	Preassigned	Original
6	5	Not preassigned	Original
7	5	Preassigned	Modified
8	5	Not preassigned	Modified
9	6	Preassigned	Original
10	6	Not preassigned	Original
11	6	Preassigned	Modified
12	6	Not preassigned	Modified

TABLE 9.23

Quantities for the Dudewicz and Dalal Ranking and Selection Procedure Used for the Ambulatory Internal Medicine Clinic Example

Alternative Number	Initial Expected Utility Estimates	$N_i - n_0$, Number of Additional Replications	Second Set of Expected Utility Estimates	Weights for Initial Expected Utility Estimates	Weights for Second Set of Utility Estimates	Weighted Sum of Expected Utilities
1	0.53	271	0.59	0.04	0.96	0.59
2	0.55	189	0.60	0.05	0.95	0.59
3	0.78	17	0.79	0.37	0.63	0.79
4	0.81	17	0.81	0.37	0.63	0.81
5	0.75	10	0.76	0.51	0.49	0.76
6	0.76	10	0.72	0.49	0.51	0.74
7	0.86	38	0.85	0.21	0.79	0.86
8	0.86	38	0.83	0.21	0.79	0.84
9	0.68	26	0.69	0.28	0.72	0.69
10	0.68	25	0.69	0.29	0.71	0.68
11	0.77	11	0.79	0.48	0.52	0.78
12	0.77	11	0.76	0.48	0.52	0.77

(since initially 10 replications were made for each alternative). This gave an initial set of estimates for the expected utility values as shown in Table 9.23.

For each alternative, the expected utility estimates for the initial set of 10 replications, the number of additional replications required, the expected utility estimates for the second set of replications, the weights for each set of replications, and the weighted sum of the expected utilities are given in Table 9.23. The reader will note that the weights and expected utility estimates are carried out to two decimal places only. Also, the N_i values are dependent on the sample variances for the initial set of replications, among other quantities, which are not shown in the table.

Alternative 7, with the subpolicies of 5-minute interarrival times, preassigned exam rooms, and a modified resident allocation, gives the highest weighted expected utility and is therefore denoted as the best policy.

9.7 Comparing Many Alternatives

In Sections 9.5 and 9.6, which discuss the comparison of two alternatives or a few alternatives, every alternative in the feasible set could be simulated at least twice (i.e., at least two replications). There are many situations in which there are, in effect, a very large number, or even an infinite number of alternatives. These situations can occur when one has either a combination of a number of integer/discrete control variables and/or continuous control variables to consider. Examples of these situations include the following:

1. Determining the number of staff personnel to assign to various shifts and staff categories of a service system such as an emergency department of a hospital, a call center, or a fast-food restaurant. Objectives in this case could be divided into two categories: those related to the cost of operating the system (e.g., staff person-hours) and those relating to customer service (e.g., average wait time for customers or fraction of customers who wait longer than a threshold value).
2. Determining the values for s and S for an (s, S) inventory policy. For example, if s can range from 20 to 50 and S can range from 70 to 150, there would be 31 * 81 = 2511 different policies to evaluate. Again, objectives could be divided into two categories: those related to internal operating cost (holding costs plus ordering costs) and those related to customer service (back orders or lost sales). If the (s, S) policies related to an entire distribution system with multiple branches, the number of control variables would be multiplied by the number of branch offices. If there were 10 branches, each with 2511 different policies to evaluate, the number of alternatives would not be 10 * 2511, but 2511^{10}, or approximately $9.965 * 10^{33}$.
3. Determining the number of resources of various types to have in a manufacturing system. Such resources could include assembly machines, drills, lathes, and AGVs. If there were N types of resources and n_i number of units of resource possible for i = 1,…, N, then the number of different policies/alternatives to examine would be $\prod_{i=1}^{N} n_i$.
4. Determining the routes for a set of delivery trucks. Each route would consist of a number of stops; in addition to determining the number of trucks needed, the stops would have to be sequenced into routes, and the routes assigned to the trucks. The sequencing of the stops to form the routes would affect the delivery/pickup times. Objectives would relate to service (e.g., tardiness/earliness of delivery) as well as cost for operating the system (e.g., as measured by the number of trucks needed and distance traveled).

There are numerous additional examples that could be mentioned, some of a very specific nature and some more general in nature. But the basic idea is the same: design and/or operation of complex systems requires the determination of values for numerous control variables, some continuous in nature and some discrete. Each combination of values results in a policy/alternative, and all combinations result in a number too large to investigate explicitly with a simulation, when a simulation model is desired as a representation of the system.

The type of problem addressed in this section is difficult for several reasons:

1. A closed-form relationship (an explicit function) does not exist relating the values for the control variables to the attribute values. (This is one reason why a simulation model is being used.)
2. The outputs (i.e., the attribute values) are random in nature. One can never be sure of the exact values for the attributes.
3. Multiple objectives mean that trade-offs must be made between the attribute values.

The typical optimization approach used for these problems is a metaheuristic or a combination of metaheuristics. As with heuristics in general, metaheuristics do not guarantee an optimal solution but typically will give a very good solution to a problem. These metaheuristics are designed so that a closed-form representation of the relationship between the control variable values and the attribute values is not required. In addition, many of them are designed to avoid getting "trapped" at a local optimal solution.

Examples of metaheuristics include simulated annealing (Kirkpatrick et al., 1983), tabu search (Glover, 1986), variable neighborhood search (Mladenović and Hansen, 1997), scatter search (Marti et al., 2006), genetic algorithms (Holland, 1975), and particle swarm optimization (Kennedy and Eberhart, 2001), among others. See Gendreau and Potvin (2010) for an overview of various metaheuristics.

These metaheuristic algorithms are embedded in software packages that are interfaced with various simulation software packages. Examples of these optimization packages include OptQuest, Witness Optimizer™ (Lanner Group, Inc., 2005), SimRunner2™, AutoStat™, and Extend Optimizer™. Another package, which might be considered more of a tool for sensitivity analysis, is Palisade software's TopRank™, which is meant to be used with their @Risk™ software.

OptQuest is a product of OptTek Systems Incorporated (http://opttek.com/OptQuest) and was designed by Fred Glover. It has been interfaced with many different simulation software packages including Arena, Simio™, Flexsim™, and ProModel™, among others. OptQuest employs a combination of scatter search, tabu search, and artificial neural networks to seek an optimal solution for a criterion model. Its use requires the specification of control variables, responses, constraints on functions of the responses (if needed), and an objective that is specified as either of the minimization or of the maximization type. Each control variable requires a specification of a minimum and a maximum value, as well as being defined as continuous, discrete, or integer in nature.

Eskandari et al. (2011) did a comparative analysis of Witness Optimizer and OptQuest and found their performances to be similar. In most cases, the choice of a particular optimizer will be dictated by the simulation software being used.

The user is given the option of stopping the search process of OptQuest (1) by specifying a fixed number of alternative simulations or (2) by specifying that the process stop after a particular number of alternatives have been simulated with no improvement. The user can also employ a combination of these methods.

Many researchers (e.g., see Nelson et al., 2001) have suggested that the user of one of these optimization packages apply a ranking and selection procedure to the top few alternatives (say top 20) found through the optimization process. Such an approach, which involves a more in-depth analysis of the remaining individual alternatives than the optimization process, could very well result in an alternative, other than the top-ranked optimization alternative, being ranked first.

Example 9.8: Selecting a Best Construction Contract Bid

Let's consider a modified version of Example 7.1. Suppose that additional information has led A1 to revise its estimate of the probability for a competing bid of .6. If there is at least one competing bid, A1 estimates that the lowest competing bid will have an uncertain value, corresponding to a random variate with a uniform distribution, with a minimum value of \$11 and a maximum value of \$14 million. The cost of construction has also been revised to be estimated as a random variable, distributed according to a triangular distribution with parameters as \$10 million, \$11 million, and \$15 million. A1 wants to determine what their optimal bid should be, within the range of \$11–\$14 million. Input data for the problem are summarized in Table 9.24.

A1 wants to consider two attributes in their decision concerning the amount to bid, if a bid is made:

X_1: Probability of winning the contract

X_2: Net profit

One reason why A1 wants to consider the probability of winning the contract as a separate attribute is that there is some prestige associated with the outcome of winning the contract—as a result, this increased prestige may very well put A1 in line for additional business.

As an example of these attributes, let's suppose that A1 bids \$12 million for the contract. If they win the contract, they will make a profit of

TABLE 9.24

Parameters for a Bid Decision Problem

Parameters	Value
Probability of a competing bid	.6
Cost of construction	TRIA (10, 11, 15) million dollars
Lowest competing bid, if competing bids occurs	UNIF (11, 14) million dollars
Cost of preparing bid	\$50,000

\$12,000,000–\$50,000 – cost of construction, which can be anywhere between \$10 and \$15 million. If they do not win the contract, they will have a loss of \$50,000, the cost of preparing the proposal.

A multiplicative utility function representing A1's preferences over X_1 and X_2 has been developed:

$$u(x_1, x_2) = .1u_1(x_1) + .8u_2(x_2) + .1u_1(x_1)u_2(x_2),$$

where

$$u_1(x_1) = x_1,$$

and

$$u_2(x_2) = .0011x_2^3 + .0064x_2^2 + .0602x_2 + .3741.$$

Both of the individual attribute utility functions are scaled from 0 to 1 and increasing in the attribute, as they should. The function for X_1, the probability of winning the contract, is linear, so it is risk neutral; the function for X_2, the profit, is concave for negative profit and convex for positive profit.

Now, the value for X_1 *could* be estimated by the simulation model. However, this is not needed since the probability of A1 winning the contract is a straightforward deterministic calculation. Specifically, if B is the bid by A1 in millions of dollars (remember B is between \$11 million and \$14 million), and $X_1(B)$ denotes the probability of A1 winning the contract at a bid of B, then

$$\begin{aligned} X_1(B) &= P\,(\text{A1 wins contract} \mid \text{at least one competing bid}) \\ &\quad \times P\left(\text{at least one competing bid}\right) \\ &\quad + P(\text{A1 wins contract} \mid \text{no competing bids}) \\ &\quad \times P\left(\text{no competing bids}\right) \\ &= ((14-B)/(14-11)) * .6 + (1) * (1-.6) \end{aligned}$$

So, if A1 bids \$12 million, their probability of winning the contract is (2/3) * .6 + 1 * .4 = .8.

Now, A1 may or may not actually win the contract, but the probability of them winning is just a deterministic function of their bid.

This calculation for X_1 was included in the Monte Carlo simulation, which also calculated values for X_2 and utility for each replication of the model. The OptQuest optimization software was interfaced with the model in order to determine the bid that would maximize expected utility. The optimization was set up to simulate 100 different design points, with 30 replications each. The best bid value found was \$12,613,850, which gave a probability for a successful bid of .6772 and an estimate for expected profit of \$248,800. The estimate for expected utility with this bid was .4125. It should be noted that the \$248,800 is only the *estimate* for *expected* profit.

TABLE 9.25

Estimates of Optimal Bids with Various Constraints on the Probability of a Successful Bid

Constraint on the Probability of a Successful Bid	Estimates of an Optimal Bid and of Expected Profit
≥.5	\$13.5 million, \$440,800
≥.7	\$12.5 million, \$172,900
≥.8	\$12 million, −\$161,737
≥.9	\$11.5 million, −\$589,522
≥1	\$11 million, −\$1,107,500

With a probability of .3228, A1 will not get the contract and will therefore be out of the \$50,000 for preparing the proposal. Even if A1 receives the contract, they may very well have a construction cost that is more than their bid of \$12,613,850. This points out the importance of understanding the outputs associated with a simulation of a decision situation.

In some cases with two attributes, a DM may prefer to constrain one attribute, while optimizing the expected utility of the other. This approach was tried here; specifically, the expected utility for profit (X_2) was optimized, while the value for X_1 was constrained. Normally, constraining the value of an attribute, which is output from a simulation model, is problematic, since typically the attribute value is not deterministic. In this case though, as shown earlier, the value for X_1 is a deterministic function of the bid.

Specifically, the following optimization models were solved for the problem:

> Find the value of the bid, which maximizes the expected utility of expected profit, subject to the probability of receiving the contract being greater than or equal to each of the following values: .5, .7, .8, .9, and 1.

In effect, we solved five different optimization problems. The results are shown in Table 9.25.

As one would expect, as the constraint on the probability of a successful bid becomes tighter, the expected profit for the optimal bid value decreases. Also, a simulation is actually not needed to project these results (see Exercise 9.11).

Example 9.9: Optimization of s and S in an (s, S) Inventory Policy Using Expected Utility as a Criterion

Consider Example 9.6 involving the choice of an (s, S) inventory policy from a small set of policies, using expected utility as the criterion. Now suppose that we want to find the best values for s and S within some range of integer values. In particular, using the same simulation model as was used in Example 9.6 and the OptQuest optimization software, we specify the control variables as s and S, with both being integer valued.

TABLE 9.26

95% Confidence Intervals (with 10 Replications) for Two Inventory Policies

Policy	95% CI for Average Unit-Days of Shortage	95% CI for Average Ordering plus Holding Cost	95% CI for Expected Utility
s = 50, S = 150	22.4831 ± .70	395.43 ± 2.21	.8793 ± .00
s = 90, S = 136	12.7161 ± .47	406. ± 1.39	.9033 ± .00

The ranges for these control variables were set at 10–200 for s and 60–500 for S. In addition, a constraint, $S - s \geq 30$, was added to the model in order to prevent the optimization from searching solutions, which did not make sense, such as those with $s > S$. The objective was specified as "maximize utility." (By default, the actual objective is to maximize expected utility.)

For the optimization run, the number of simulations was set at 100, with 10 replications per simulation. Because of the low value for variance in the output, this number of replications seemed sufficient.

The optimal solution found was s = 90 and S = 136, with an estimate of expected utility of .901906. This compares to an estimate of expected utility of .8808 (after 98 replications) for the best solution (s = 50, S = 150) found in Example 9.6, where we were selecting from only a few policies. A more detailed comparison of these two policies is shown in Table 9.26.

The reader will note that the half-widths for the expected utility confidence intervals in Table 9.26 are only carried out to two decimal places—that is, these half-widths are not actually 0, but they are greater than .00.

The results in Table 9.26 show that the policy found by OptQuest with s = 90 and S = 136 gives a markedly better value for average unit-days of shortage, but a slightly worse value for average ordering plus holding cost.

Example 9.10: Staffing Optimization at an Emergency Call Center

Let's return to the emergency (911) call center referred to in Example 8.3. For this system, the problem was to determine the number of call takers to assign to weekly shifts at an emergency call center of a metropolitan area. Data relating to the number of calls by type (e.g., requesting assistance from police, fire, etc.) and time of week (day and hour of the day) were collected. Data were also collected on the activities of the call takers and dispatchers of the call center. A simulation model representing a typical week's operation was constructed. Among other quantities, inputs to the system included weekly shift schedules and the number of call takers assigned to these shifts. One of the shift schedules examined was the following:

1. Shift 1: Monday through Wednesday, 12 midnight to 12 noon; Thursday, 12 midnight to 4 a.m.
2. Shift 2: Monday through Wednesday, 12 noon to 12 midnight; Thursday, 4 p.m. to 8 p.m.
3. Shift 3: Friday through Sunday, 12 midnight to 12 noon; Thursday, 4 a.m. to 8 a.m.

4. Shift 4: Friday through Sunday, 12 noon to 12 midnight; Thursday, 8 p.m. to midnight
5. Shift 5 (overtime shift): Thursday, 8 a.m. to 4 p.m.

Note that every hour of the week was covered by these various shifts, as was required for the system. In addition, note that for any particular shift, the time periods are distributed in a way that allows for adequate recuperation between work periods. The problem then was to determine the number of call takers to assign to each of these shifts.

Two basic objectives were of interest to the center's management. The first one had to do with answering as many of the calls as possible within 12 seconds of the call being placed. Since there were on average about 25,000 calls per week to the call center, and the variability of calls arriving during one period of the week as compared to another was high, it was impossible to answer every call through the week within 12 seconds.

The second objective had to do with the internal cost of operating the system. This was represented as the number of person-hours assigned during the week. So, for example, if 10 persons were assigned to each of the five shifts listed earlier, this would represent $10*(40+40+40+40+8)=1680$ person-hours assigned.

Instead of combining these two objectives into a utility/value function, the system's management was interested in optimizing the first objective, subject to a constraint on the second objective. This was accomplished through the use of the following optimization model:

Minimize fraction of incoming calls exceeding a 12-second ring time, subject to number of call taker weekly hours ≤1200, 1400, or 1600 hours.

The control variables employed for this optimization were the numbers (restricted to integer values) of call takers assigned to each of the five shifts listed earlier.

The OptQuest software package, interfaced with the simulation model, was employed to solve the optimization problem, with the first-ranked solutions in each case as shown in Table 9.27.

The results allowed the management to choose a staffing policy with an implicit trade-off between the two objectives related to service and internal system cost.

TABLE 9.27

Results of an Optimization of Call Taker Assignments for Emergency Call Center

Constraint on Call Taker Hours	Number of Call Taker Hours in Solution	Fraction of Calls Exceeding 12 Second Ring Time	No. of Call Takers Assigned to Shift 1	No. of Call Takers Assigned to Shift 2	No. of Call Takers Assigned to Shift 3	No. of Call Takers Assigned to Shift 4	No. of Call Takers Assigned to Shift 5
≤1200	1176	0.238	7	10	7	4	7
≤1400	1344	0.152	8	8	8	8	8
≤1600	1584	0.048	9	10	8	11	8

Example 9.11: Optimization for the Lead Poison Testing Problem

Let's reconsider Example 6.4, involving testing children for lead poisoning. In that example, there were two tests: an imperfect urine test and a "perfect" blood test. In this example, let's consider the strategy where we apply the urine test to a set of randomly selected children from the population, followed by the blood test for those children who test positive on the urine test. The data for the problem are reproduced in Table 9.28.

The basic problem here is to determine the number of urine tests to give, followed by the requisite blood tests for those who test positive on the urine test. Following Example 6.4, we want to consider the four attributes of

X_1 = The cost for administering the tests for lead poisoning in thousands of dollars
X_2 = The number of children definitely identified as having lead poisoning
X_3 = The number of children incorrectly identified as having lead poisoning through the urine test (but later correctly identified as not having lead poisoning through the blood test)
X_4 = The number of children who are incorrectly identified as *not* having lead poisoning through the urine test (but actually do have lead poisoning)

Given the strategy of a specified number of urine tests followed by the requisite blood tests, the values for these attributes will be uncertain.

There are at least two approaches for modeling this problem. One approach would be to develop an influence diagram, with the decision node of "Number of Urine Tests to Administer." The influence diagram would also contain various outcome/consequence nodes and chance event nodes corresponding to the number of urine-tested children with lead poisoning, number of urine-tested children without lead poisoning, number of children with lead poisoning who test positive on the urine test, total cost for testing, and so on. The values for many of the nodes would be determined through the use of random variate generation. For example, the number of urine-tested children with lead poisoning would be the value generated from either a hypergeometric random variable or a binomial random variable if an approximation is used.

TABLE 9.28

Input Data for Lead Poison Testing Problem

Problem Parameter	Value
Sensitivity of urine test	0.9
Specificity of urine test	0.85
Cost of urine test	$10
Sensitivity of blood test	1
Specificity of blood test	1
Cost of blood test	$100
Prior probability of lead poisoning	0.1

A second approach would be to employ a Monte Carlo simulation in which the entities would represent children going through the testing process. This approach would represent the appropriate hypergeometric/binomial distributions only implicitly through the sampling process. That is, the number of entities generated would be the number of children to be urine tested. These would be probabilistically routed (e.g., through the use of a Decide module if the Arena simulation software package is being employed), to group those children being tested as having or not having lead poisoning. These two groups would each be further segmented into those testing positive and those testing negative on the urine test, according to the sensitivity and specificity of the urine test. Appropriate variable values would be collected by the model. This was the approach used in modeling the problem for this example. With this approach, we were implicitly modeling the binomial distribution with the simulation model.

A more accurate, but also more complex, model could have been built by implicitly representing the hypergeometric distribution. This could have been accomplished by tracking the values of the number of children in the population with lead poisoning, and without lead poisoning, as the urine testing was progressing; then the probability associated with the next child tested having lead poisoning could have been changed after each test. Given the relatively small size of the sample however, as compared to the size of the population, modeling the process as a binomial distribution was appropriate.

The OptQuest optimization software was used to optimize on the number of urine tests to give for the problem. In particular, a constrained optimization was performed:

$$\text{Maximize } X_2, \text{ subject to } X_1 \leq 4, X_3 \leq 8, X_4 \leq 2.$$

In words, we want to maximize the number of children correctly identified with lead poisoning, subject to spending at most $4000 on testing and having at most eight children incorrectly identified (temporarily until they receive the blood test) as having lead poisoning when they really do not and at most two children incorrectly identified as not having lead poisoning when they do.

The optimization was set up so that the number of urine tests to give was constrained to be between 50 and 150, inclusive. Since the optimization was also set up to perform 100 simulations (i.e., design points) with 30 replications at each point, in effect, we were performing a complete enumeration of all of the design points.

Another thing to note about this problem is that we are actually using *estimates of the expected values* of our attributes in our optimization model. However, the number of replications resulted in sufficiently small confidence intervals for our output.

Upon solving this problem, we obtained the following solution:

Give 58 urine tests followed by the requisite blood tests, resulting in estimates of $X_2 = 5.733$, $X_1 = 1.94$ (thousands of dollars), $X_3 = 7.866$, and $X_4 = .9$.

TABLE 9.29

Results for Three Optimization Problems

Model #	X_1 Constraint and X_1 Value	X_3 Constraint and X_3 Value	X_4 Constraint and X_4 Value	No. of Urine Tests	X_2^*
1	≤4, 1.94	≤8, 7.87	≤2, .9	58	5.733
2	≤4, 2.74	≤11, 10.86	≤2, 1.03	82	8.33
3	≤4, 3.45	≤14, 14	≤2, 1.03	103	10.233

In other words, giving 58 urine tests, we could expect to identify 5.733 children with lead poisoning at a cost of $1940; we would also expect to temporarily misidentify 7.866 children as having lead poisoning when they do not and also misidentify .9 child as not having lead poisoning when he or she does. Again, remember that these are estimates of expected values.

Looking at the results, it appears that our only "close to binding" constraint is the one on X_3. By relaxing this constraint and leaving the others as they are, we could expect to obtain improved results for our objective. Hence, two other optimization problems, each involving a modification of the constraint on X_3, were solved. The results are shown in Table 9.29.

As an example, in interpreting Table 9.29, the third optimization model, with the constraint on X_3 of $X_3 \leq 14$, resulted in 103 urine tests to give, with expected values of 10.233 children correctly identified with lead poisoning, at a cost of $3450, 14 children temporarily misidentified as having lead poisoning when they do not, and 1.03 child misidentified as not having lead poisoning when they do.

This approach of optimizing one attribute while varying the constraints on the other attributes is similar to using the STEP method (STEM) with a simulation model (Mollaghasemi and Evans, 1994).

9.8 Final Thoughts

Much of the methodology discussed in this chapter has been covered in other sources, many of which have been cited here. The reader is referred to those sources, and others, such as Fu (2014) and Hong and Nelson (2006).

There are also several other areas not discussed in this chapter that can be useful for optimization via simulation. One of these is the area of *metamodeling*, in which the relationship between the input variables and the output of a simulation is replaced by a simpler, closed-form, model than the simulation. Examples of these approaches are regression analysis, kriging, and stochastic kriging. Zakerifar et al. (2011) provide an example of kriging in multiple objective simulation optimization.

Material Review Questions

9.1 What is a simulation?

9.2 What is the difference between a discrete event simulation and a continuous simulation?

9.3 Simulation models used to evaluate influence diagrams and decision trees are typically "static" in nature (*true* or *false*).

9.4 What is a key aspect of a Monte Carlo simulation?

9.5 Methods to generate random variates are typically programmed into simulation software packages (*true* or *false*).

9.6 What is the most popular method for generating random variates?

9.7 Different probability distributions might employ different methods for the generation of random variates associated with that distribution (*true* or *false*).

9.8 The choice of a method to generate random variates for a particular distribution function is dependent upon which three performance measures?

9.9 Outputs from simulaion models are typically *estimates* of something (*true* or *false*).

9.10 Give two examples of time-persistent variables from a simulation model.

9.11 Replications of a simulation model representing a particular alternative are typically independent in nature (*true* or *false*).

9.12 The sample mean of an output, obtained from averaging over n replications of a simulation model, is itself a random variable (*true* or *false*).

9.13 If X_i is an output from the ith independent replication of a simulation model, and $S^2(n)$ is the estimate of the variance of the X_i's as given by (9.6), then the expected value of $S^2(n)$ will remain constant as n increases (*true* or *false*).

9.14 The value of $\widehat{\text{Var}}\left[\bar{X}(n)\right]$ will usually increase or decrease (choose one) as n increases.

9.15 If one computes 1000 confidence intervals at the 95% confidence interval from experimentation with a simulation model, about how many of these confidence intervals will contain the variable of interest?

9.16 For the same set of output data generated by independent replications of a simulation model, the 90% confidence interval will be smaller or larger (choose one) than a 95% confidence interval.

9.17 For values of $n \geq 30$, the value for $t_{n-1,\,1-\alpha/2}$ approximates the value of $z_{1-\alpha/2}$ (*true* or *false*).

9.18 What is the main benefit of a variance reduction technique?

9.19 How would you implement the variance reduction technique of common random numbers in a simulation model?

9.20 If the value of 0 is contained in the paired-t confidence interval for the difference of two alternatives, what can you say about the difference between these alternatives?

9.21 For what reason would one use a two-sample t-test?

9.22 Shortening the duration of one of the activities of a project will always shorten the project duration (*true* or *false*).

9.23 The Bonferroni inequality provides a bound on the overall confidence associated with a group of confidence intervals (*true* or *false*).

9.24 In performing an all-pairwise comparisons approach for several alternatives, what three actions can be taken if the difference between two alternatives is not statistically significant?

9.25 In what general situation would one want to use subset selection (or screening) as opposed to a method for finding the best alternative in a group of alternatives?

9.26 What is the *indifference zone* in a ranking and selection method?

9.27 What is the difficulty associated with using ranking and selection methods when the performance measure used is expected utility?

9.28 Which of the following approaches for ranking and selection allow for the use of common random numbers as a variance reduction technique: Koenig and Law subset selection procedure, Bonferroni approach to screening due to Nelson et al., Dudewicz and Dalal approach for identifying a best alternative, Rinott approach for identifying a best alternative, and Nelson and Matejcik procedure for identifying a best alternative.

9.29 The Dudewicz and Dalal procedure for finding a best alternative typically requires more replications of the alternative simulations than the Rinott procedure (*true* or *false*).

9.30 Procedures like Rinott's for ranking and selection are based on a "worst-case analysis" in terms of the alternatives under consideration (*true* or *false*).

9.31 Describe an example of a situation in which a simulation model is used for a design problem for which an optimization procedure is also required.

Exercises

9.1 A simulation model for a project has been constructed. The model allows the representation of various policies for reducing the durations of various respective tasks of the project. The main output of interest is the project duration. Twenty independent replications for a particular policy have been run, with the following project durations (in weeks) for each replication:

62.5, 64.3, 60.8, 52.3, 56.7, 56.1, 83.7, 69.1, 78.5, 78.2, 56.8, 77.2, 84.9, 55.9, 67.1, 68.9, 62.7, 59.2, 58.9, 87.4.

a. From these data, calculate a 95% confidence interval for the probability that the project, under the simulated policy, will have a duration of longer than 65 weeks.
b. Calculate 95% confidence intervals for the respective probabilities that the project will have a duration of longer than 70, 75, and 80 weeks.
c. Calculate a 95% confidence interval for the expected project duration under the simulated policy. Use (9.10) for this calculation.
d. Calculate a 90% confidence interval for the expected project duration under the simulated policy. Again, use (9.10) for this calculation.

9.2 Consider two alternatives for execution of a project. The alternatives have to do with the choice of a major subcontractor for the project. One of the subcontractors is less reliable than the other, but this subcontractor might actually provide better performance, at least with respect to project duration.

A simulation model for the project has been constructed, which allows for the representation of each of the alternatives. Five replications have been made of each alternative, with the following results for project duration (in weeks):

Alternative 1: 67.8, 85.5, 59.2, 78.5, 61.2
Alternative 2: 75.5, 73.4, 81.5, 74.2, 73.2

The project manager's utility function over project duration has been constructed. Utility function values for the replication values are shown in the following:

Project duration	59.2	61.2	67.8	73.2	73.4	74.2	75.5	78.5	81.5	85.5
Utility	.9	.88	.8	.75	.7	.65	.6	.05	.02	.01

The utility function reflects the fact that the company has a deadline corresponding to a project duration of 76 weeks.

Compute a point estimate and a 90% confidence intervals for each alternative for project duration and expected utility based upon these output data. What conclusions can you reach?

9.3 Consider the Thief of Baghdad problem in Example 9.1. Compute the probability, without a simulation, of the thief requiring more than 6 hours to achieve freedom. Now construct a simulation of the process. Estimate this probability with 100 replications of your model. Estimate the probability with 1,000 replications and then with 10,000 replications of your model.

9.4 A simulation model has been developed to study various scheduling policies in a job shop. The model has been run for five independent replications, with the following output for average job lateness in hours: 3.4, 2.9, 2.8, 3.1, 3.8.

Using (9.10), compute a 95% confidence interval for expected lateness based on these data. Using the formula given in (9.16), estimate the total number of replications needed to reduce the half-width for the original confidence interval by 50%. Make the same estimate using the formula in (9.17).

9.5 Consider Example 9.3: Comparing NBA Series Patterns. Develop a Monte Carlo simulation for a 2–2–1–1–1 finals series. Consider the team with the home court advantage. Suppose that the probability that each team wins on their home court is equal to p. Hence, the probability that the visiting team wins is 1 – p. Using trial and error, determine an estimate for the value for p such that the team with the home court advantage has the same chance of winning the series (50%) as the other team. Set the number of replications for each experiment at 1000. Note that this value for p must be less than .5.

9.6 Consider various values for the probability that the team with the home court advantage in a 2–2–1–1–1 series wins the series as a function of the value for p, as defined in Exercise 9.5. More specifically, determine an estimate for the expected probability (and the 95% confidence intervals) that the home court advantage team wins the series for p equal to .4, .55, and .7. Determine estimates for the number of games in the series for the same probabilities (and the 95% confidence intervals). Use 1000 replications to determine your estimates and confidence intervals in each case.

9.7 Suppose that you are performing an analysis involving simulation for an inventory system. There are two performance measures of interest:

X_1: Average cost per day associated with ordering and holding inventory
X_2: Unit-days of shortage per day

You are analyzing a particular policy with your model, and with one set of replications, you have determined two 95% confidence intervals for the two performance measures:

X_1: (\$257, \$290)
X_2: (9.2, 16.7)

You are satisfied with the simultaneous confidence of your intervals (what is it?) and also (more than) satisfied with the half-width of the confidence interval for X_1. However, you are not satisfied with the half-width of the confidence interval for X_2 and would like to make it smaller. Without running additional replications of your model, what can you do?

9.8 Develop the simulation model described for Example 9.4, involving the (s, S) inventory policy. Suppose that the utility function is given by

$$u(x_1, x_2) = .7u_1(x_1) + .2u_2(x_2) + .1u_1(x_1)u_2(x_2),$$

with the same individual attribute utility functions.

Perform the same all-pairwise comparisons analysis as was done in Example 9.4, except with this new utility function.

9.9 Suppose you had an inventory with a single item. The system operates 24 hours per day, 7 days per week. The inventory level was 25 items, at 10 a.m. At 10 a.m., one demand order was placed for 35 items, and at 11 a.m., another demand order was placed for 50 items. At 12 noon, a shipment of 500 items arrived to replenish the inventory. What was the unit-days of shortage over this period of 10 a.m. to 12 noon?

9.10 Consider Example 9.6. Perform the same analysis with an ε value of .02 instead of .01. How many total replications are needed to identify the best policy for this new value of ε? Does the best policy identified change?

9.11 Consider Example 9.8, involving the optimal bid to make for a contract. Determine, analytically (i.e., without a simulation), the maximum bid that A1 should make if they want the probability of their bid being successful to be equal to 1, given that the probability of a competing bid is .5. Determine the maximum bid that A1 should make for various values of X and Y, where X is the minimum probability of their bid being successful and Y is the probability of a competing bid, for X = .5, .8, and 1. and Y = .5, .8, and .9.

9.12 Consider Example 9.9. Using the simulation you developed for Exercise 9.8 and an appropriate optimization package, determine optimal values for s and S using the following utility function:

$$u(x_1, x_2) = .72u_1(x_1) + .18u_2(x_2) + .1u_1(x_1)u_2(x_2).$$

Discuss how the values found for s and S relate to the values found in Example 9.9, within the context of trading off between the attributes X_1 and X_2.

9.13 Consider Example 9.10. Suppose that the manager of the emergency call center had a multiattribute value function over the two attributes:

X_1 = Fraction of incoming calls exceeding a 12-second ring time
X_2 = Number of call taker weekly hours

This multiattribute value function is given by

$$v(x_1, x_2) = .7v_1(x_1) + .3v_2(x_2),$$

with

$$v_1(x_1) = -12.298x_1^2 - 1.0645x_1 + 1.0602$$

and

$$v_2(x_2) = -.0017x_2 + 2.7.$$

Treating the relationship between the solutions (number of call takers assigned to the various shifts) found and the values for X_1 and X_2 as deterministic, rank the three solutions given in the example by computing their respective values using this value function.

9.14 Build the simulation model corresponding to Example 9.11, the lead poison testing problem. Use the approach described in the example in which entities are used to represent the tested children. Find the optimal solutions for the three criterion models given in the problem for three varying values of the sensitivity and specificity of the urine test:

a. Sensitivity = .85, specificity = .8
b. Sensitivity = .95, specificity = .9
c. Sensitivity = .98, specificity = .95

Discuss the intuitive sense of the results.

References

Ackermann, F., C. Eden, and T. Williams. 1997. Modeling for litigation: Mixing qualitative and quantitative approaches. *Interfaces* 27(2):48–65.

Ackermann, F. 2012. Problem structuring methods "In the Dock": Arguing the case for soft OR. *European Journal of Operational Research* 219:652–658.

Ackoff, R. L. 1979. The future of operational research is past. *Journal of the Operational Research Society* 30:93–104.

Ackoff, R. L. 1981. The art and science of mess management. *Interfaces* 11:20–26.

Allais, M. 1953. Behavior of the rational man before risk-criticism of American school postulates and axioms. *Econometrica* 21:503–546.

Al-Shemmeri, T., B. Al-Kloub, and A. Pearman. 1997. Model choice in multicriteria decision aid. *European Journal of Operational Research* 97:550–560.

Alonso, J. A. and M. T. Lamata. 2006. Consistency in the analytical hierarchy process: A new approach. *International Journal of Uncertainty, Fuzziness and Knowledge-Based Systems* 14(4):445–459.

Aouni, B., C. Colapinto, and D. La Torre. 2014. Financial portfolio management through the goal programming model: Current state of the art. *European Journal of Operational Research* 234(2):536–545.

Banks, J., J. S. Carson II, B. L. Nelson, and D. M. Nicol. 2005. *Discrete Event System Simulation*, 4th ed. Upper Saddle River, NJ: Prentice Hall.

Barichard, V., M. Ehrgott, X. Gandibleux, and V. T'kindt. 2009. *Multiobjective Programming and Goal Programming: Theoretical Results and Practical Applications.* Berlin, Germany: Springer Verlag.

Bart, C. K. 1997. Sex, lies, and mission statements. *Business Horizons* 40(6):9–18.

Barzilai, J. and B. Golany. 1994. AHP rank reversal, normalization, and aggregation rules. *INFOR* 32:57–64.

Barzilai, J. and F. A. Lootsma. 1997. Power relations and group aggregation in the multiplicative AHP and SMART. *Journal of Multi-Criteria Decision Analysis* 6:155–165.

Basadur, M., S. J. Ellspermann, and G. W. Evans. 1994. A new methodology for formulating Ill-structured problems. *Omega* 22:627–645.

Behzadian, M., R. B. Kazemzadeh, A. Albadvi, and M. Aghdasi. 2010. PROMETHEE: A comprehensive literature review on methodologies and applications. *European Journal of Operational Research* 200:198–215.

Behzadian, M., S. Khanmohammadi Otaghsara, M. Yazdani, and J. Ignatius. 2012. A state of the art survey of TOPSIS applications. *Expert Systems with Applications* 39:13051–13069.

Bell, D. E. and P. H. Farquhar. 1986. Perspectives on utility theory. *Operation Research* 34(1):179–183.

Belton, V. 1986. A comparison of the analytic hierarchy process and a simple multi-attribute value function. *European Journal of Operational Research* 26(1):7–21.

Belton, V. and T. Gear. 1983. On a shortcoming of Saaty's method of analytic hierarchies. *Omega* 11:228–230.

Benayoun, R., J. de Montgolfier, J. Tergny, and O. Laritchev. 1971. Linear programming and multiple objective functions. *Mathematical Programming* 1(3):366–375.
Bennett, J. and D. Worthington. 1998. An example of a good but partially successful OR engagement: Improving outpatient clinic operations. *Interfaces* 28(5):56–69.
Bennett, L. M. and M.A. Kerr. 1996. A systems approach to the implementation of total quality management. *Total Quality Management* 7(6):631–665.
Bennett, P. 1994. Designing a Parliamentary briefing system—An OR look at the Commons. *Journal of the Operational Research Society* 45(11):1221–1232.
Billman, B. and J. F. Courtney. 1993. Automated discovery in managerial problem formulation—Formation of causal hypotheses for cognitive mapping. *Decision Sciences* 24(1):23–41.
Boardman, J. T. and A. J. Cole. 1996. Integrated process improvement in design and manufacture using a systems approach. *IEEE Proceedings—Control Theory and Applications* 143(2):171–185.
Bodily, S. E. 1980. Analysis of risks to life and limb. *Operations Research* 28(1):156–175.
Bodily, S. E. and M. S. Allen. 1999. A dialogue process for choosing value creating strategies. *Interfaces* 29(6):16–28.
Bolton, R. and J. Gold. 1994. Career management—Matching the needs of individuals with the needs of organizations. *Personnel Review* 23:6–24.
Bond, S., K. Carlson, and R. Keeney. 2008. Generating objectives: Can decision makers articulate what they want? *Management Science* 54(1):56–70.
Bond, S., K. Carlson, and R. Keeney. 2010. Improving the generation of decision objectives. *Decision Analysis* 7(3):238–255.
Boran, F. E., S. Genc, and D. Akay. 2011. Personnel selection based on intuitionistic fuzzy sets. *Human Factors and Ergonomics in Manufacturing and Service Industries* 21:493–503.
Borison, A. 1995. Oglethorpe power corporation decides about investing in a major transmission system. *Interfaces* 25(2):25–36.
Bossert, J. 1991. *Quality Function Deployment—A Practitioner's Approach*. Milwaukee, WI: ASQC Quality Press.
Bouyssou, D., T. Marchant, M. Pirlot, P. Perny, A. Tsoukias, and P. Vincke. 2000. *Evaluation and Decision Models: A Critical Perspective*. Boston, MA: Kluwer Academic Publishers.
Bouyssou, D., T. Marchant, M. Pirlot, A. Tsoukias, and P. Vincke. 2010. *Evaluation and Decision Models with Multiple Criteria: Stepping Stones for the Analyst*. New York: Springer.
Bramel, J. and D. Simchi-Levi. 1997. *The Logic of Logistics: Theory, Algorithms, and Applications for Logistics Management*. New York: Springer-Verlag.
Brans, J. P. and P. H. Vincke. 1985. Note-A preference ranking organization method (The Promethee Method for Multiple Criteria Decision Making). *Management Science* 31:647–656.
Brans, J. P., P. Vincke, and B. Mareschal. 1986. How to select and how to rank projects—The Promethee method. *European Journal of Operational Research* 24:228–238.
Brightman, H. J. 1988. Group problem solving: An improved managerial approach. Atlanta, GA: Business Publishing Division, College of Business Administration, Georgia State University.
Brito, A. J., A. T. de Almeida, and C. M. M. Mota. 2010. A multicriteria model for risk sorting of natural gas pipelines based on ELECTRE TRI integrating utility theory. *European Journal of Operational Research* 200:812–821.

Brocklesby, J. 1995. Using soft systems methodology to identify competence requirements in HRM. *International Journal of Manpower* 16:70–84.
Brocklesby, J. and S. Cummings. 1996. Designing a viable organization structure. *Long Range Planning* 29(1):49–57.
Brown, J. R. and N.D. Macleod. 1996. Integrating ecology into natural resource management policy. *Environmental Management* 20(3):289–296.
Brown, S. M. 1992. Cognitive mapping and repertory grids for qualitative survey research—Some comparative observations. *Journal of Management Studies* 29(3):287–307.
Buede, D. M. 1986. Structuring value attributes. *Interfaces* 16:52–62.
Buede, D. M. and D. T. Maxwell. 1995. Rank disagreement: A comparison of multi-criteria methodologies. *Journal of Multi-Criteria Decision Analysis* 4:1–21.
Bunn, D. W. 1984. *Applied Decision Analysis.* New York: McGraw-Hill.
Burr, I. W. 1953. *Engineering Statistics and Quality Control.* New York: McGraw-Hill.
Butler, J., A. N. Chebeskov, J. S. Dyer, T. A. Edmunds, J. Jia, and V. I. Oussanov. 2005. The United States and Russia evaluate plutonium disposition options with multiattribute utility theory. *Interfaces* 35(1):88–101.
Butler, J., J. S. Dyer, and J. Jia. 2006. Using attributes to predict objectives in preference models. *Decision Analysis* 3:100–116.
Butler, J., D. J. Morrice, and P. W. Mullarkey. 2001. A multiple attribute utility theory approach to ranking and selection. *Management Science* 47:800–816.
Buzan, T. 1991. *Use Both Sides of Your Brain*, 3rd ed. New York: E.P. Dutton.
Caballero, R., T. Gomez, and F. Ruiz. 2009. Goal programming: Realistic targets for the near future. *Journal of Multi-Criteria Decision Analysis* 16:79–110.
Caballero, R., M. Luque, J. Molina, and F. Ruiz. 2005. MOPEN: A computational package for linear multiobjective and goal programming problems. *Decision Support Systems* 41:160–175.
Caballero, R., F. Ruiz, M. V. Rodriguez Uria, and C. Romero. 2006. Interactive meta-goal programming. *European Journal of Operational Research* 175(1):135–154.
Callahan, S. 1972. Dr. Land's magic camera. *Life Magazine*, October 27, 1972: 42–50.
Calori, R., G. Johnson, P. Sarnin. 1994. CEOs, cognitive maps and the scope of the organization. *Strategic Management Journal* 15(6):437–457.
Cambron, K. and G. W. Evans. 1991. Use of the analytic hierarchy process for multi-objective facility layout. *Computers and Industrial Engineering* 20:211–229.
Charnes, A. and W. W. Cooper. 1961. *Management Models and Industrial Applications of Linear Programming.* New York: Wiley.
Charnes, A., W. W. Cooper, and R. Ferguson. 1955. Optimal estimation of executive compensation by linear programming. *Management Science* 1:138–151.
Chau, M., M. C. Fu, H. Qu, and I. O. Ryzhov. 2014. Simulation optimization: A tutorial overview and recent developments in gradient-based methods. *Proceedings of the 2014 Winter Simulation Conference* (A. Tolk, S. Y. Diallo, I. O. Ryzhov, L. Yilmaz, S. Buckley, and J. A. Miller, eds.), pp. 21–35, Piscataway, NJ: IEEE.
Checkland, P. 2001. *Soft Systems Methodology in Rational Analysis for a Problematic World Revisited*, 2nd ed. Chichester, UK: John Wiley & Sons Ltd.
Checkland, P. and J. Scholes. 1990. *Soft Systems Methodology in Action.* Chichester, UK: Wiley.
Chen, C. T., C. T. Lin, and S. F. Huang. 2006. A fuzzy approach for supplier evaluation and selection in supply chain management. *International Journal of Production Economics* 102:289–301.

Chen, E. J. and W. D. Kelton. 2000. An enhanced two-stage selection procedure. *Proceedings of the 2000 Winter Simulation Conference* (J. A. Joines, R. Barton, P. Fishwick, and K. Kang, eds.), pp. 727–735. Piscataway, NJ: IEEE.

Chen, E. J. and W. D. Kelton. 2005. Sequential selection procedures: Using sample means to improve efficiency. *European Journal of Operational Research* 166(1):133–153.

Cherif, M. S., H. Chabchoub, and B. Aouni. 2008. Quality control system through the goal programming model and the satisfaction functions. *European Journal of Operational Research* 186(3):1084–1098.

Churchman, C. W., R. L. Ackoff, and E. L. Arnoff. 1957. *Introduction to Operations Research*. New York: John Wiley & Sons.

Clemen, R. T. and T. Reilly. 2001. *Making Hard Decisions with Decision Tools*. Mason, OH: South-Western Cengage Learning.

Clemen, R. T. and T. Reilly. 2013. *Making Hard Decisions with Decision Tools*. 3rd ed. Mason, OH: South-Western Cengage Learning.

Coyle, R. G., and M. D. W. Alexander. 1997. Two approaches to qualitative modeling of a nation's drugs trade. *Systems Dynamics Review* 13(3): 205–222.

De Leeneer, I. and H. Pastijn. 2002. Selecting land mine detection strategies by means of outranking MCDM techniques. *European Journal of Operational Research* 139:327–338.

Deason, J. 1984. A multi-objective decision support system for water project portfolio selection, PhD dissertation, University of Virginia, Charlottesville, VA.

Diakoulaki, D., C. H. Antunes, and A. G. Martins. 2005. MCDA and energy planning: Chapter 21. *Multiple Criteria Decision Analysis: State of the Art Surveys* (J. Figuerira, J. Greco, and M. Ehrgott, eds.). New York: Springer Science and Business Media.

Dickover, N. 1994. Reflection in action—Modeling a specific organization through the viable systems model. *Systems Practice* 7(1):43–62.

DuBois, P., J. P. Brans, F. Cantraine, and B. Mareschal. 1989. MEDICIS: An expert system for computer aided diagnosis using the PROMETHEE method. *European Journal of Operational Research* 39:284–292.

Dudewicz, E. J. and S. R. Dalal. 1975. Allocation of observations in ranking and selection with unequal variances. *Sankhya* B27:28–78.

Dunning, D. J., S. Lockfort, Q. E. Ross, P. C. Beccue, and J. S. Stonebraker. 2001. New York power authority uses decision analysis to schedule refueling of its Indian point 3 nuclear power plant. *Interfaces* 31(5):121–135.

Dyer, J. S. 1990a. A clarification of "Remarks on the Analytic Hierarchy Process." *Management Science* 36(3):274–275.

Dyer, J. S. 1990b. Remarks on the analytic hierarchy process. *Management Science* 36(3):249–258.

Dyer, J. S., T. Edmunds, J. C. Butler, and J. Jia. 1998. A multiattribute utility analysis of alternatives for the disposition of surplus weapons-grade plutonium. *Operations Research* 46(6):749–762.

Dyer, J. S. and R. K. Sarin. 1979. Measurable multiattribute value functions. *Operations Research* 27:810–822.

EasyFit. 2013. Mathwave: Data Analysis and Simulation Software. http://www.mathwave.com/products/easyfit.html. Accessed August 12, 2015.

Eden, C. and F. Ackermann. 2001. SODA—The principles. *Rational Analysis for a Problematic World Revisited*, 2nd ed. (J. Rosenhead and J. Mingers, eds.), pp. 21–42. Chichester, UK: John Wiley & Sons.

Eden, C. and F. Ackermann. 2004. Cognitive mapping expert views for policy analysis in the public sector. *European Journal of Operational Research* 152:615–630.

Edwards, W. 1977. How to use multiattribute utility measurement for social decision making. *IEEE Transactions on Systems, Man, Cybernetics* SMC-7:326–340.

Edwards, W. and F. H. Barron. 1994. SMARTS and SMARTER: Improved simple methods for multiattribute utility measurement. *Organizational Behavior Human Decision Processes* 60(3):306–325.

Ehrgott, M. 2000. *Multicriteria Optimization*. Berlin, Germany: Springer-Verlag.

Ellspermann, S. J., G. W. Evans, and M. Basadur. 2007. The impact of training on the formulation of ill structured problems. *Omega* 35(2):221–236.

Elton, E. J. and M. J. Gruber. 1997. Modern portfolio theory, 1950 to date. *Journal of Banking and Finance* 21:1743–1759.

Eskandari, H., E. Mahmoodi, H. Fallah, and C. D. Geiger. 2011. Performance analysis of commercial simulation-based optimization packages: OptQuest and witness optimizer. *Proceedings of the 2011 Winter Simulation Conference* (S. Jain, R. R. Creasey, J. Himmelspach, K. P. White, and M. Fu, eds.), pp. 2363–2373. Piscataway, NJ: IEEE.

Evans, G. W. 1984. An overview of techniques for solving multiobjective mathematical programs. *Management Science* 30(11):1268–1282.

Evans, G. W. and S. M. Alexander. 1987. Multiobjective decision analysis for acceptance sampling plans. *Institute of Industrial Engineers Transactions* 19:308–316.

Evans, G. W. and R. Fairbairn. 1989. Selection and scheduling of advanced missions for NASA using 0–1 integer linear programming. *Journal of the Operational Research Society* 40:971–982.

Evans, J. R. 1991. *Creative Thinking in the Decision and Management Sciences*. Cincinnati, OH: South-Western Publishing.

Fielden, D. and J. K. Jacques.1998. Systemic approach to energy rationalisation in island communities. *International Journal of Energy Research* 22(2):107–129.

Figuerira, J., V. Mousseau, and B. Roy. 2005. ELECTRE methods: Chapter 4. *Multiple Criteria Decision Analysis: State of the Art Surveys* (J. Figuerira, J. Greco, and M. Ehrgott, eds.)., pp. 133–162. New York: Springer Science and Business Media.

Fine, L. G. 2009. *The Swot Analysis: Using Your Strength to Overcome Weaknesses, Using Opportunities to Overcome Threats*. CreateSpace Publishing.

Fishburn, P. C. 1970. *Utility Theory for Decision Making*. New York: Wiley.

Flavell, R. B. 1976. A new goal programming formulation. *Omega* 4:731–732.

Forman, E. H. and S. I. Gass. 2001. The analytic hierarchy process—An exposition. *Operations Research* 49(4):469–486.

Forman, E. H. and K. Peniwati. 1998. Aggregating individual judgments and priorities with the analytic hierarchy process. *European Journal of Operational Research* 108:165–169.

Fu, M. C. (ed.). 2014. *Handbook of Simulation Optimization*. Springer.

Gass, S. I. 2005. Model world: The great debate—MAUT versus AHP. *Interfaces* 35(4):308–312.

Gass, S. I. and A. A. Assad. 2005. *An Annotated Timeline of Operations Research: An Informal History*. Boston, MA: Kluwer Academic Publishers.

Gendreau, M. and J.-Y. Potvin. 2010. *Handbook of Metaheuristics*. New York: Springer.

Geoffrion, A., J. Dyer, and A. Feinberg. 1972. An interactive approach for multicriterion optimization, with an application to the operation of an academic department. *Management Science* 19:357–368.

Gershon, M. 1981. Model choice in multi-objective decision making in natural resource systems, PhD dissertation, University of Arizona, Tucson, AZ.

Glover, F. 1986. Future paths for integer programming and links to artificial intelligence. *Computers and Operations Research* 13(5):533–549.

Goicoechea, A., D. R. Hansen, and L. Duckstein. 1982. *Multiobjective Decision Analysis with Engineering and Business Applications.* New York: John Wiley & Sons.

Goodwin, P. and G. Wright. 2009. *Decision Analysis for Management Judgment*, 4th ed. Chichester, UK: John Wiley & Sons.

Gough, J. D. and J. C. Ward. 1996. Environmental decision making and lake management. *Journal of Environmental Management* 48(1):1–15.

Gregory, A. J. and M. C. Jackson. 1992. Evaluating organizations—A systems and contingency approach. *Systems Practice* 5:37–60.

Grubbs, F. E. 1969. Procedures for detecting outlying observations in samples. *Technometrics* 11(1):1–21.

Grushka-Cockayne, Y. and B. De Reyck. 2009. Towards a single European sky. *Interfaces* 39(5):400–414.

Gupta, A. and G. W. Evans. 2009. A goal programming model for the operation of closed loop supply chains. *Engineering Optimization* 41(8):713–735.

Harker, P. T. and L. G. Vargas. 1987. The theory of ratio scale estimation: Saaty's analytic hierarchy process. *Management Science* 33(11):1383–1403.

Harker, P. T. and L. G. Vargas. 1990. Reply to the "Remarks on the Analytic Hierarchy Process by J. S. Dyer." *Management Science* 36(3):269–273.

Haynes, M. G., A. G. McGregor, and N. D. Stewart. 1997. The business team standard: A means of improving the effectiveness of individual businesses in a multibusiness corporation. *Systems Practice* 10(3):219–239.

Heizer, J. and B. Render. 2006. *Operations Management*, 8th ed. Upper Saddle River, NJ: Pearson–Prentice Hall.

Hesse, R. and G. Woolsey. 1980. *Applied Management Science: A Quick and Dirty Approach.* Chicago, IL: Science Research Associates.

Hillier, F. S. and G. J. Lieberman. 2010. *Introduction to Operations Research*, 9th ed. New York: McGraw-Hill.

Hindle, T., P. Checkland, M. Mumford, and D. Worthington. 1995. Development of a methodology for multidisciplinary action research—A case-study. *Journal of the Operational Research Society* 46(4):453–464.

Hokkanen, J. and P. Salminen. 1997. Choosing a solid waste management system using multicriteria decision analysis. *European Journal of Operational Research* 98:19–36.

Holland, J. H. 1975. *Adaptation in Natural and Artificial Systems.* Ann Arbor, MI: University of Michigan Press.

Hong, L. J. and B. L. Nelson. 2006. Discrete optimization via simulation using COMPASS. *Operations Research* 54(1):115–129.

Horner, P. 2004. The science of better synergy (an interview with Richard Larson). *OR/MS Today* 31(6):36–43.

Huge, E. C. 1990. *Total Quality: An Executive Guide for the 1990s.* Homewood, IL: Business One.

Hunink, M., M. C. Weinstein, E. Wittenberg, M. F. Drummond, J. S. Pliskin, J. B. Wong, and P. F. Glasziou. 2001. *Decision Making in Health and Medicine: Integrating Evidence and Values*. New York: Cambridge University Press.

Hwang, C. L., Y. Lai, and T. Y. Liu. 1993. A new approach for multiple objective decision making. *Computers and Operational Research* 20:889–899.

Hwang, C. L. and K. Yoon. 1981. *Multiple Attribute Decision Making: Methods and Applications*. New York: Springer-Verlag.

Ignizio, J. P. 1976. *Goal Programming and Extensions*. Lexington, MA: Lexington Books.

Ignizio, J. P. 1982. *Linear Programming in Single and Multiple Objective Systems*. Upper Saddle River, NJ: Prentice Hall.

Ignizio, J. P. 1985. An algorithm for solving the linear goal programming problem by solving its dual. *Journal of the Operational Research Society* 36:507–515.

Ignizio, J. P. and T. Cavalier. 1994. *Linear Programming*. Upper Saddle River, NJ: Prentice Hall.

Ishikawa, K. 1990. *Introduction to Quality Control*. Tokyo, Japan: 3A Corporation.

Joiner, C. 1980. Academic planning through the goal programming model. *Interfaces* 10(4):86–92.

Joldersma, C. and E. Roelofs. 2004. The impact of soft OR-methods on problem structuring. *European Journal of Operational Research* 152:696–708.

Jones, D. F. and M. Tamiz. 2002. Goal programming in the period 1990–2000. *Multi-Criteria Optimization: State of the Art Annotated Bibliographic Surveys* (M. Ehrgott and X. Gandibleux, eds.), pp. 129–170. Dordrecht, the Netherlands: Kluwer.

Jones, D. F. and M. Tamiz. 2010. *Practical Goal Programming*. New York: Springer.

Kahneman, D. and A. A. Tversky 1972. Subjective probability: A judgment of representativeness. *Cognitive Psychology* 3:430–454.

Kangas, A., J. Kangas, and J. Pykäläinen. 2001a. Outranking methods as tools in strategic natural resources planning. *Silva Fennica* 35(2):215–227.

Kangas, J., A. Kangas, P. Leskinen, and J. Pykäläinen. 2001b. MCDM methods in strategic planning of forestry on state-owned lands in Finland: Applications and experiences. *Journal of Multi-Criteria Decision Analysis* 10:257–271.

Kartowisastro, H. and K. Kijima. 1994. An enriched soft systems methodology (SSM) and its application to cultural-conflict under a paternalistic value system. *Systems Practice* 7(3):241–253.

Keefer, D. L., Jr. and S. E. Bodily. 1983. Three point approximations for continuous random variables. *Management Science* 29(5):595–609.

Keeney, R. L. 1992. *Value Focused Thinking: A Path to Creative Decision Making*. Cambridge, MA: Harvard University Press.

Keeney, R. L. 1999. Developing a foundation for strategy at seagate software. *Interfaces* 29(6):4–15.

Keeney, R. L. 2002. Common mistakes in making value trade-offs. *Operations Research* 50(6):935–945.

Keeney, R. L. 2008. Personal decisions are the leading cause of death. *Operations Research* 56(6):1335–1347.

Keeney, R. L. 2012. Value-focused brainstorming. *Decision Analysis* 9(4):303–313.

Keeney, R. L. and R. S. Gregory. 2005. Selecting attributes to measure the achievement of objectives. *Operations Research* 53(1):1–11.

Keeney, R. L., J. F. Lathrop, and A. Sicherman. 1986. An analysis of Baltimore gas and electric company's technology choice. *Operations Research* 34(1):18–39.

Keeney, R. L., T. L. McDaniels, and C. Swoveland. 1995. Evaluating improvements in electric utility reliability at British Columbia hydro. *Operations Research* 43(6):933–947.

Keeney, R. L. and K. Nair. 1977. Evaluating potential nuclear power plant sites in the Pacific northwest using decision analysis: Chapter 14. *Conflicting Objectives in Decisions.* (D. E. Bell, R. L. Keeney, and H. Raiffa, eds.). London, UK: Wiley.

Keeney, R. L. and H. Raiffa. 1993. *Decisions with Multiple Objectives: Preferences and Value Tradeoffs*, 2nd ed. Cambridge, UK: Cambridge University Press.

Keisler, J. M., W. A. Buehring, P. D. McLaughlin, M. A. Robershotte, and R. G. Whitfield. 2004. Allocating vendor risks in the Hanford waste cleanup. *Interfaces* 34(3):180–190.

Keller, L. R., J. Simon, and Y. Wang. 2009. Multiple-objective decision analysis involving multiple stakeholders. Chapter 14 (pp. 139–155) in *2009 Tutorials in Operations Research: Decision Technologies and Applications.* (M. R. Oskoorouchi, ed.). Baltimore, MA: INFORMS.

Kelly, G. A. 1955. *The Psychology of Personal Constructs.* New York: Norton.

Kelton, W. D., R. P. Sadowski, and N. B. Zupick. 2015. *Simulation with Arena*, 6th ed. New York: McGraw-Hill Education.

Kennedy, B. 1996. Soft systems methodology in applying psychology. *Australian Psychologist* 31(1):52–59.

Kennedy, J. and R. C. Eberhart. 2001. *Swarm Intelligence.* San Francisco, CA: Morgan Kaufman.

Kepner, C. H. and B. B. Tregoe. 1981. *The New Rational Manager.* Princeton, NJ: Princeton Research Press.

Khisty, C. J. 1995. Soft systems methodology as learning and management tool. *Journal of Urban Planning and Development* 121(3):91–107.

Kirkpatrick, S., C. D. Gelatt, Jr., and M. P. Vecchi. 1983. Optimization by simulated annealing. *Science* 220:671–680.

Kleijnen, J. P. C. and J. Wan. 2007. Optimization of simulated systems: OptQuest and alternatives. *Simulation Modelling Practice and Theory* 15:354–362.

Knowles, J. 1993. A soft systems analysis of a CD-Rom network for a multisite polytechnic. *Journal of Librarianship and Information Science* 25(1):15–21.

Koenig, L. W. and A. M. Law. 1985. A procedure for selecting a subset of size m containing the l best of k independent normal populations. *Communications in Statistics—Simulation and Computation* 14:719–734.

Korhonen, P. 1997. Comments on Barzilia and Lootsma. *Journal of Multi-Criteria Decision Analysis* 6:167–168.

Korhonen, P. and J. Laakso. 1986. A visual interactive method for soling the multiple criteria problem. *European Journal of Operational Research* 24:277–287.

Korhonen, P., H. Moskowitz, and J. Wallenius. 1986. A progressive algorithm for modeling and solving multiple criteria decision problems. *Operations Research* 34:726–731.

Krohling, R. A. and V. C. Campanharo. 2011. Fuzzy TOPSIS for group decision making: A case study for accidents with oil spill in the sea. *Expert Systems with Applications* 38:4190–4197.

Lanner Group, Inc. 2005. Witness Optimizer Module, 4.3. Houston, TX: Lanner Group, Inc.

Larichev, O. I. 1997. Comments on Barzilia and Lootsma. *Journal of Multi-Criteria Decision Analysis* 6:166.

Lartindrake, J. M. and C. R. Curran, 1996. All together now—The circular organization in a university hospital. *Systems Practice* 9(5):391–401.

Law, A. M. 2007. *Simulation Modeling & Analysis*, 4th ed. New York: McGraw-Hill.

Ledington, P. 1992. Intervention and the management process—An action-based research study. *Systems Practice* 5(1):17–36.

Ledington, P. and J. Donaldson. 1997. Soft OR and management practice: A study of the adoption and use of soft systems methodology. *Journal of the Operational Research Society* 48(3):229–240.

Lee, S. M. 1972. *Goal Programming for Decision Analysis*. Philadelphia: Auerbach.

Lee, S. M. and M. Schniederjans. 1983. A multicriteria assignment problem: A goal programming approach. *Interfaces* 13(4):75–81.

Lee, S., J. F. Courtney, R. M. O'Keefe. 1992. A system for organizational learning using cognitive maps. *Omega* 20(1):23–36.

Lehaney, B. and R. J. Paul. 1994. Using soft systems methodology to develop a simulation of outpatient services. *Journal of the Royal Society of Health* 114(5):248–251.

Lehaney, B. and R. J. Paul. 1996. The use of soft systems methodology in the development of a simulation of outpatient services at Watford General Hospital. *Journal of the Operational Research Society* 47(7):864–870.

Lehaney, B. and V. Hlupic. 1995. Simulation modeling for resource allocation and planning in the health sector. *Journal of the Royal Society of Health* 115(6):382–385.

Leyva-López, J. C. and E. Fernández-González. 2003. A new method for group decision support based on ELECTRE III methodology. *European Journal of Operational Research* 148:14–27.

Liberatore, M. J. and R. L. Nydick. 2008. The analytic hierarchy process in medical and health care decision making: A literature review. *European Journal of Operational Research* 189:194–207.

Lootsma, F. A. 1993. Scale sensitivity in a multiplicative variant of the AHP and SMART. *Journal of Multi-Criteria Decision Analysis* 2:87–110.

Lootsma, F. A. 1996. A model for the relative importance of the criteria in the multiplicative AHP and SMART. *European Journal of Operational Research* 94:467–476.

Lootsma, F. A. and J. Barzilai. 1997. Response to the comments by Larichev, Korhonen and Vargas on "Power Relations and Group Aggregation in the Multiplicative AHP and SMART." *Journal of Multi-Criteria Decision Analysis* 6:171–174.

Lootsma, F. A. and H. Schuijt. 1997. The multiplicative AHP, SMART and ELECTRE in a common context. *Journal of Multi-Criteria Decision Analysis* 6:185–196.

Luce, R. D. and H. Raiffa. 1957. *Games and Decisions*. New York: Wiley.

Macadam, R., R. Vanasch, B. Hedley, and E. Pitt. 1995. A case study in development planning using a systems learning approach—Generating a master plan for the livestock sector in Nepal. *Agricultural Systems* 49(3):299–323.

Maciaschapula, C. A. 1995. Development of a soft systems model to identify information values, impact and barriers in a health-care information system. *Journal of Information Science* 21(4):283–288.

MacCrimmon, K. R. 1969. Improving the system design and evaluation process by the use of tradeoff Information: An application to northeast corridor transportation planning, RM-5877-DOT. Santa Monica, CA: The RAND Corporation.

Macharis, C., J. Springael, K. De Brucker, and A. Verbeke. 2004. PROMETHEE and AHP: The design of operational synergies in multicriteria analysis. Strengthening PROMETHEE with Ideas of AHP. *European Journal of Operational Research* 153:307–317.

Machol, R. E. 1996. Flying scared: How much spending on safety makes sense. *OR/MS Today* 23(5), October 1996.

Maciaschapula, C. A. 1995. Development of a soft systems model to identify information values, impact and barriers in a health-care information system. *Journal of Information Science* 21(4):283–288.

Magee, J. F. 1964. Decision trees for decision making. *Harvard Business Review* 42(4):126–138.

Magidson, J. 1992. Systems practice in several communities in Philadelphia. *Systems Practice* 5(5):493–508.

Mareschal, B. 1986. Stochastic PROMETHEE multicriteria decision making under uncertainty. *European Journal of Operational Research* 26:58–64.

Mareschal, B. and J. P. Brans. 1988. Geometric representations for MCDM (GAIA). *European Journal of Operational Research* 34:69–77.

Mareschal, B. and J. P. Brans. 1991. BANKADVISER: An industrial evaluation system. *European Journal of Operational Research* 54:318–324.

Marti, R., M. Laguna, and F. Glover. 2006. Principles of scatter search. *European Journal of Operational Research* 169:351–372.

Matheson, D. and J. E. Matheson. 1999. Outside-in strategic modeling. *Interfaces* 29(6):29–41.

Mattila, V. and K. Virtanen. 2015. Ranking and selection for multiple performance measures using incomplete preference information. *European Journal of Operational Research* 242(2):568–579.

McDaniels, T. L. 1995. Using judgement in resource management: A multiple objective analysis of a fisheries management decision. *Operations Research* 43(3):415–426.

McHale, I. G., P. A. Scarf, and D. E. Folker. 2012. On the development of a soccer player performance rating system for the English premier league. *Interfaces* 42(4):339–351.

Mehrota, V. 1997. Ringing up big business. *OR/MS Today* 24(4):18–25.

Meredith, J. R. and S. J. Mantel Jr. 2012. *Project Management: A Managerial Approach,* 8th ed. New York: John Wiley & Sons.

Metty, T., R. Harlan, Q. Samelson, T. Moore, T. Morris, R. Sorensen, A. Schneur et al. 2005. Reinventing the supplier negotiation process at motorola. *Interfaces* 35:7–23.

Midgley, G. and A. Milne. 1995. Creating employment opportunities for people with mental health problems—A feasibility study for new initiatives. *Journal of the Operational Research Society* 46(1):35–42.

Miettinen, K. and P. Salminen. 1999. Decision-aid for discrete multiple criteria decision making problems with imprecise data. *European Journal of Operational Research* 119:50–60.

Miller, A. C., M. W. Merkhofer, R. A. Howard, J. E. Matheson, and T. R. Rice. 1976. *Development of Automated Aids for Decision Analysis.* Menlo Park, CA: Stanford Research Institute.

Millspacko, P. A., K. Wilson, and P. Rotar. 1991. Highlights from the use of the soft systems methodology to improve agrotechnology transfer in Kona, Hawaii. *Agricultural Systems* 36(4):409–425.

Mingers, J. and J. Rosenhead. 2004. Problem structuring methods in action. *European Journal of Operational Research* 152:530–554.

Mingers, J. and L. White. 2010. A review of the recent contribution of systems thinking to operational research and management science. *European Journal of Operational Research* 207:1147–1161.

Minguez, M. I., C. Romero, and J. Domingo. 1988. Determining optimum fertilizer combinations through goal programming with penalty functions: An application to sugar beet production in Spain. *Journal of the Operational Research Society* 39(1):61–70.

Miser, H. J. and E. S. Quade. 1985. *Handbook of Systems Analysis.* New York: John Wiley & Sons.

Mladenović, N. and P. Hansen. 1997. Variable neighborhood search. *Computers and Operations Research* 24:1097–1100.

Mlodinow, L. 2008. *The Drunkard's Walk: How Randomness Rules Our Lives.* New York: Vintage Books.

Moder, J. J., C. R. Phillips, and E. W. Davis. 1995. *Project Management with CPM, PERT and Precedence Diagramming,* 3rd ed. Middleton, WI: Blitz Publishing.

Mollaghasemi, M. and G. W. Evans. 1994. Multicriteria design of manufacturing systems through simulation optimization. *IEEE Transactions on Systems, Man, and Cybernetics* 24:1407–1411.

Moskowitz, H., G. W. Evans, and I. Jimenez-Lerma. 1978. Development of a multiattribute value function for long range electrical generation expansion. *IEEE Transactions on Engineering Management* 25:78–87.

Myung, I. J. 2003. Tutorial on maximum likelihood estimation. *Journal of Mathematical Psychology* 47:90–100.

Nadler, G. and S. Hibino. 1990. *Breakthrough Thinking.* Rocklin, CA: Prima Publishing.

Nakano, M., H. Nobutomo, M. Okada, and A. Mizushima. 1997. Research on how to develop R&D strategies using the soft systems approach. *International Journal of Technology Management* 14:822–833.

Nelson, B. L. and F. J. Matejcik. 1995. Using common random numbers for indifference-zone selection and multiple comparisons in simulation. *Management Science* 41:1935–1945.

Nelson, B. L., J. Swann, D. Goldsman, and W. Song. 2001. Simple procedures for selecting the best system when the number of alternatives is large. *Operations Research* 49:950–963.

Nutt, P. C. 2001. A taxonomy of strategic decisions and tactics for uncovering alternatives. *European Journal of Operational Research* 132:505–527.

O'Connor, A. D. 1992. Soft systems methodology—A case study of its use within an Australian organization. *Australian Computer Journal* 24(4):130–138.

Olson, D. L. 2001. Comparison of three multicriteria methods to predict known outcomes. *European Journal of Operational Research* 130:576–587.

Opricovic, S. and G. H. Tzeng. 2004. The compromise solution by MCDM methods: A comparative analysis of VIKOR and TOPSIS. *European Journal of Operational Research* 156:445–455.

Opricovic, S. and G. H. Tzeng. 2007. Extended VIKOR method in comparison with outranking methods. *European Journal of Operational Research* 178:514–529.

Ormerod, R. 1996a. Information systems strategy development at Sainsbury's supermarkets using soft OR. *Interfaces* 26(1):102–130.

Ormerod, R. 1996b. Putting soft OR methods to work—Information systems strategy development at Richards Bay. *Journal of the Operational Research Society* 47(9):1083–1097.

Ormerod, R. 1998. Putting soft OR methods to work: Information systems strategy development at Palabora. *Omega* 26(1):75–98.

Ozernoy, V. M., D. R. Smith, and A. Sicherman. 1981. Evaluating computerized geographic information systems using decision analysis. *Interfaces* 11(5):92–100.

Palisade StatTools: Advance Statistical Analysis for Excel. Palisade StatTools 2013. Available from http://www.palisade.com/stattools/.

Palmer, B. 1999. Fortune 500 companies are discovering new software that better manages the process of making complex decisions. *Fortune* 139:153–156.

Parnell, G. S. 2001. Work-package-ranking system for the department of energy's office of science and technology, in D. L. Keefer, Practice Abstracts. *Interfaces* 31(4):109–111.

Pauley, G. and R. Omerod. 1998. The evolution of a performance measurement project at RTZ. *Interfaces* 28:94–118.

Peerenboom, J. P., W. A. Buehring, and T. W. Joseph. 1989. Selecting a portfolio of environmental programs for a synthetic fuels facility. *Operations Research* 37(5):689–699.

Pidd, M. 1989. From problem-structuring to implementation. *Journal of the Operational Research Society* 39(2):115–121.

Pirlot, M. 1995. A characterization of "min" as a procedure for exploiting valued preference relations and related results. *Journal of Multi-Criteria Decision Analysis* 4:37–56.

Pitz, G. F., N. J. Sachs, and J. Heerboth. 1980. Procedures for eliciting choices in the analysis of individual decisions. *Organizational Behavior and Human Performance* 26:396–408.

Platt, D. G. 1996. Building process models for design management. *Journal of Computing in Civil Engineering* 10(3):194–203.

Pliskin, J. S., D. S. Shepard, and M. C. Weinstein. 1980. Utility functions for life years and health status. *Operations Research* 28(1):206–224.

Popper, R., K. Andino, M. Bustamante, B. Hernandez, and L. Rodas.1996. Knowledge and beliefs regarding agricultural pesticides in rural Guatemala. *Environmental Management* 20(2):241–248.

Pritsker, A. A. B. 1986. *Introduction to Simulation and SLAM II*, 3rd ed. New York: John Wiley & Sons.

Raiffa, H. 1968. *Decision Analysis*. Reading, MA: Addison-Wesley.

Rasegard, S. 1991. A comparative-study of Beer and Miller systems designs as tools when analyzing the structure of a municipal organization. *Behavioral Science* 36(2):83–99.

Ravindran, A., W. S. Shen, J. L. Arthur, and H. Moskowitz. 1986. Nonlinear integer goal programming models for acceptance sampling. *Computers and Operations Research* 13:611–622.

Rinott, Y. 1978. On two-stage selection procedures and related probability inequalities. *Communications in Statistics* A7:799–811.

Rittel, H. W. J. and M. M. Webber. 1973. Dilemmas in a general theory of planning. *Policy Science* 4(2):155–169.

Robbins, L. 1994. Using interactive planning in the entrepreneurial class—A live fieldwork-based case and simulation exercise. *Simulation & Gaming* 2(3):353–367.

Rodriguez Uria, M. V., R. Caballero, F. Ruiz, and C. Romero. 2002. Meta-goal programming. *European Journal of Operational Research* 136:422–429.

Rogers, M. and M. Bruen. 1998. Choosing realistic values of indifference, preference, and veto thresholds for use with environmental criteria within ELECTRE. *European Journal of Operational Research* 107:542–551.

Romero, C. 1991. *A Handbook of Critical Issues in Goal Programming*. Oxford, UK: Permagon Press.

Ronen, B. and J. S. Pliskin. 1981. Decision analysis in microelectronic reliability: Optimal design and packaging of a diode array. *Operations Research* 29(2): 229–242.

Rosenhead, J. 1996. What's the problem? An introduction to problem structuring methods. *Interfaces* 26(6):117–131.

Rosenhead, J. 2006. The past, present, and future of problem structuring methods. *Journal of the Operational Research Society* 57:759–765.

Ross, S. 1998. *A First Course in Probability*. 5th Edition. Upper Saddle River, NJ: Prentice Hall.

Roy, B. 1968. Classement et Choix en Présence de Points de Vue Multiples (la Méthode ELECTRE). *Revue Francaise d'Informatique de la Recherche Opérationnelle* 8:57–75.

Roy, B. 1991. The outranking approach and the foundations of the ELECTRE methods. *Theory and Decisions* 31:49–73.

Roy, B. and J. Hugonnard. 1982. Ranking of suburban line extension projects for the Paris metro system by a multicriteria method. *Transportation Research* 16(A):301–312.

Saaty, T. L. 1980. *The Analytic Hierarchy Process*. New York: McGraw-Hill.

Saaty, T. L. 1986. Axiomatic foundation of the analytic hierarchy process. *Management Science* 32(7):841–855.

Saaty, T. L. 1990a. *Decision Making for Leaders: The Analytic Hierarchy Process for Decisions in a Complex World*. Pittsburg, PA: RWS Publications.

Saaty, T. L. 1990b. An exposition of the AHP in reply to the paper "Remarks on the Analytic Hierarchy Process." *Management Science* 36(3):259–268.

Saaty, T. L. and L. G. Vargas. 1998. Diagnosis with dependent symptoms: Bayes theorem and the analytic hierarchy process. *Operations Research* 46(4):491–502.

Saaty, T. L. and L. G. Vargas. 2006. *Decision Making with the Analytic Network Process*. New York: Springer.

Samson, D. 1988. *Managerial Decision Analysis*. Homewood, IL: Richard Irwin Inc.

Sandholm, T., D. Levine, M. Concordia, P. Martyn, R. Hughes, J. Jacobs, and D. Begg. 2006. Changing the game in strategic sourcing at Procter & Gamble: Expressive competition enabled by optimization. *Interfaces* 36(1):55–68.

Schniederjans, M. J., N. K. Kwak, and M. C. Helmer. 1982. An application of goal programming to resolve a site location problem. *Interfaces* 12(2):65–72.

Schon, D. A. 1987. *Educating the Reflective Practitioner: Toward a New Design for Teaching and Learning in the Professions*. San Francisco, CA: Jossey-Bass.

Schoner, B., E. U. Choo, and W. C. Wedley. 1997. A comment on "Rank Disagreement: A Comparison of Multi-Criteria Methodologies." *Journal of Multi-Criteria Decision Analysis* 6:197–200.

Schoner, B. and W. C. Wedley. 1989. Ambiguous criteria weights in AHP: Consequences and solutions. *Decision Sciences* 20:462–475.

Schoner, B., W. C. Wedley, and E. U. Choo. 1993. A unified approach to AHP with linking pins. *European Journal of Operational Research* 64:384–392.

Schuhman, W. 1990. Strategy for information systems in the film division of Hoechst AG. *Systems Practice* 3(3):265–287.

Schuwirth, N., P. Reichert, and J. Lienert. 2012. Methodological aspects of multi-criteria decision analysis for policy support: A case study on pharmaceutical removal from hospital wastewater. *European Journal of Operational Research* 220(2):472–483.

Schwenk, C. and H. Thomas. 1983. Formulating the mess: The role of decisions aids in problem formulation. *Omega* 11(1):15–26.

Shachter, R. D. 1986. Evaluating influence diagrams. *Operations Research* 34(6):871–882.

Shapiro, J. 2007. *Modeling the Supply Chain*, 2nd ed. Belmont, CA: Thomson Higher Education.

Shaw, D., M. Westcombe, J. Hodgkin, and G. Montibeller. 2004. Problem structuring methods for large group interventions. *Journal of the Operational Research Society* 55(5):453–463.

Siemens, N. and C. Gooding. 1975. Reducing project duration at minimum cost: A time/cost trade-off algorithm. *Omega* 3:569–581.

Simon, H. A. 1960. *The New Science of Management*. New York: Harper & Row.

Skaf, M. A. 1999. Portfolio management in an upstream oil and gas organization. *Interfaces* 29(6):84–104.

Smith, G. F. 1988. Towards a heuristic theory of problem structuring. *Management Science* 34(12):1489–1506.

STATA. 2013. STATA: Data Analysis and Statistical Software. http://www.stata.com/. Accessed August 12, 2015.

Stephens, M. A. 1974. EDF statistics for goodness of fit and some comparisons. *Journal of the American Statistical Association* 69:730–737.

Steuer, R. E. 1986. *Multiple Criteria Optimization: Theory, Computation, and Application*. New York: John Wiley & Sons.

Stevens, S. S. 1946. On the theory of scales of measurement. *Science* 103:677–680.

Stonebraker, J. S. 2002. How Bayer makes decisions to develop new drugs. *Interfaces* 32(6):77–90.

Strumpfer, J. P. 1997. Planning as a means of social change: The Durban functional region forum case. *Systems Practice* 10(5):549–566.

Sullivan, D. W. and J. R. Wilson. 1989. Restricted subset selection procedures for simulation. *Operations Research* 37:52–71.

Sun, B. 2015. Operational decision making for medical clinics through the use of simulation and multi-attribute utility theory, PhD dissertation, Department of Industrial Engineering, University of Louisville, Louisville, KY.

Tamiz, M. and D. F. Jones. 1997. Interactive frameworks for investigation of goal programming models: Theory and practice. *Journal of Multi-Criteria Decision Analysis* 6:52–60.

Tamiz, M., D. F. Jones, and E. El-Darzi. 1995. A review of goal programming and its applications. *Annals of Operations Research* 58:39–53.

Thoren, M. 1996. Systems approach to clothing for disabled users—Why is it difficult for disabled users to find suitable clothing? *Applied Ergonomics* 27(6): 389–396.

Thunhurst, C. and C. Ritchie. 1992. Housing in the Dearne Valley: Doing community OR with the Thurnscoe tenants housing co-operative. Part 2—An evaluation. *Journal of the Operational Research Society* 43:677–690.

Toland, R. J., J. M. Kloeber, and J. A. Jackson. 1998. A comparative analysis of hazardous waste remediation alternatives. *Interfaces* 28(5):70–85.

Trainor, T. E., G. S. Parnell, B. Kwinn, J. Brence, E. Tollefson, and P. Downes. 2007. The US army uses decision analysis in designing its us installation regions. *Interfaces* 37(3):253–264.

Triantaphyllou, E. 2001. Two new cases of rank reversals when the AHP and some of its additive variants are used that do not occur with the multiplicative AHP. *Journal of Multicriteria Decision Analysis* 10:11–25.

Tversky, A. A. and D. Kahneman. 1974. Judgment under uncertainty: Heuristics and biases. *Science* 185:1124–1131.

Ulengin, F. and I Topcu. 1997. Cognitive map: KBDSS, integration in transportation planning. *Journal of the Operational Research Society* 48(11):1065–1075.

Vaida, O. S. and S. Kumar. 2006. Analytic hierarchy process: An overview of applications. *European Journal of Operational Research* 169:1–29.

VanGundy, A. B. 1988. *Techniques of Structured Problem Solving,* 2nd ed. New York: Van Nostrand Reinhold.

Vargas, L. G. 1997. Comments on Barzilia and Lootsma why the multiplicative AHP is invalid: A practical counterexample. *Journal of Multi-Criteria Decision Analysis* 6:169–170.

Von Neumann, J. and O. Morgenstern. 1947. *Theory of Games and Economic Behavior,* 2nd ed. Princeton, NJ: Princeton University Press.

Von Winterfeldt, D. and W. Edwards. 1986. *Decision Analysis and Behavioral Research.* London, UK: Cambridge University Press.

Von Winterfeldt, D. and B. Fasolo. 2009. Structuring decision problems: A case study and reflections for practitioners. *European Journal of Operational Research* 199:857–866.

Von Winterfeldt, D. and T. M. O'Sullivan. 2006. Should we protect commercial airplanes against surface to air missile attacks by terrorists? *Decision Analysis* 3(2):63–75.

Vos, B. and H. Akkermans. 1996. Capturing the dynamics of facility allocation. *International Journal of Operations & Production Management* 16(11):57.

Walker, J. 1990. Diagnosis and implementation: How a large cooperative employed a series of proposals for restructuring based on the viable systems model. *Systems Practice* 3(5):441–451.

Wallenius, J., J. S. Dyer, P. C. Fishburn, R. E. Steuer, S. Zionts, and K. Deb. 2008. Multiple criteria decision making, multiattribute utility theory: Recent accomplishments and what lies ahead. *Management Science* 54(7):1336–1349.

Walpole, R. E. and R. H. Myers. 1993. *Probability and Statistics for Engineers and Scientists,* 5th ed. New York: MacMillan.

Welch, J. 2005. *Winning.* New York: HarperCollins.

Wells, J. S. G. 1995. Discontent without focus—An analysis of nurse management and activity on a psychiatric inpatient facility using a soft systems approach. *Journal of Advanced Nursing* 21(2):214–221.

Wenstöp, F. E. and A. J. Carlsen. 1988. Ranking hydroelectric power projects with multicriteria decision analysis. *Interfaces* 18(4):36–48.

White, L. 2009. Understanding problem structuring methods interventions. *European Journal of Operational Research* 199:823–833.

Winch, G. W. 1993. Consensus building in the planning process—Benefits from a hard modeling approach. *System Dynamics Review* 9(3):287–300.

Yarnold, J. K. 1970. The minimum expectation in χ^2 goodness-of-fit tests and the accuracy of approximation for the null distribution. *Journal of American Statistical Association* 65:864–886.

Yoon, K. 1987. A reconciliation among discrete compromise solutions. *Journal of the Operational Research Society* 38:277–286.

Zahedi, F. 1986. The analytic hierarchy process—A survey of the method and its applications. *Interfaces* 16:96–108.

Zakerifar, M., W. E. Biles, and G. W. Evans. 2011. Kriging metamodeling in multiple objective simulation optimization. *Simulation: Transactions for the Society for Modeling and Simulation International* 87(10):843–856.

Zillgitt, J. 2014. NBA basketball format is good for business. *USA Today*, June 6, 2014. http://www.usatoday.com/story/sports/nba/playoffs/2014/06/06/nba-finals-format-heat-spurs/10033681/ (accessed October 16, 2015).

Zionts, S. and J. Wallenius. 1983. An interactive multiple objective linear programming method for a class of underlying nonlinear utility functions. *Management Science* 29:519–529.

Index

S

T

U

V

W